Earth

Edited by
Russell O. Utgard
and Garry D. McKenzie
The Ohio State University

BURGESS PUBLISHING COMPANY • MINNEAPOLIS, MINNESOTA

This book was typeset in Press Roman
by Burgess-Beckwith, printed by the Colwell Press.
The design and cover were developed by Dennis Tasa.
Editors were Robert E. Lakemacher and Marcia Bottoms.
Sondra L. Decker researched the photos and permissions,
and Morris Lundin supervised production.

Typesetting commenced on October 4, 1973, and bound books
were available on February 15, 1974.

Copyright © 1974 by Burgess Publishing Company
Printed in the United States of America
Library of Congress Card Number 73-90807
SBN 8087-2104-6

All rights reserved.
No part of this book may be reproduced
in any form whatsoever, by photograph or mimeograph
or by any other means, by broadcast or transmission,
by translation into any kind of language,
nor by recording electronically or otherwise,
without permission in writing from the publisher,
except by a reviewer, who may quote brief
passages in critical articles and reviews.

0 9 8 7 6 5 4 3 2 1

*For today's little ones
Keith, Louise, Tom, Jane
What will they see in 2050?*

Preface

This collection of readings was prepared in response to the need for current materials in the rapidly expanding field of earth science. Recent discoveries in environmental geology, sea-floor spreading, and astrogeology have necessitated organization of new material for students, who in many cases are aware of the latest developments in the above fields as presented in magazines and newspapers and as discussed on television. Generally the various textbooks, although well written and fact filled, have limitations on the number and depth of topics in the above categories. These readings are intended to partly fill this gap and to provide supplementary readings with the emphasis on environmental science. In this field of environmental earth science we include topics such as geologic hazards, mineral resources, waste disposal, water supply, and the geological aspects of environmental health, land reclamation, and regional planning. This orientation and format should be useful to a wide variety of students, including those in geology, geography, and the earth sciences in general.

Sources of the articles are largely scientific journals, although some reports were taken from government publications and news magazines. Several articles from our earlier book, *Man and His Physical Environment: Readings in Environmental Geology*, have been included. In preparation of the present set of readings, many good articles had to be omitted in the interest of space and the need for broad coverage of earth science topics. In many cases, where personal or school libraries permit, this limitation can be alleviated by assigning additional readings from current journals.

The book could not have become a reality without the gracious cooperation of the authors and their publishers whose works are included within. Our sincere thanks to them and to our colleagues and students who have funnelled information into our sieves and who have helped to evalute these materials.

January 1974 *ROU*
GDM

Contents

Part One   Atmospheric Science   1

1. The Atmospheric Circulation *Abraham H. Oort*   7
2. Dust, Goats, and Deserts *Reid A. Bryson*   20
3. When Will the Present Interglacial End? *G. J. Kukla and R. K. Matthews*   28
4. Climatic Effects of Man's Activities *Report of the Study of Critical Environmental Problems (SCEP)*   34
5. Air Pollution in the Inner City *Council on Environmental Quality*   44

Part Two   Oceanography   51

6. Chemical Oceanography *Francis A. Richards*   56
7. Marine Sediments *Rhodes W. Fairbridge*   68
8. The Red and the Black Seas *David A. Ross*   77

Part Three   Plate Tectonics   89

9. Sea-Floor Spreading—New Evidence *F. J. Vine*   93
10. Plate Tectonics, Sea-Floor Spreading, and Continental Drift *Robert S. Dietz*   112

Part Four   Astrogeology   125

11. Apollo 15: Scientific Journey to Hadley-Apennine *Joseph P. Allen*   130

CONTENTS

12 Puzzling Facts from Lunar Exploration *Thornton Page*   150

13 The New Mars: Volcanism, Water, and a Debate over Its History *Allen L. Hammond*   160

14 Mariner 9 Television Reconnaissance of Mars and Its Satellites: Preliminary Results *Harold Masursky et al*   167

Part Five   Environmental Geology   173

15 A Geologist Views the Environment *John C. Frye*   178

16 Hydrology for Urban Land Planning *Luna B. Leopold*   189

17 Earthquake "Briefs" *U. S. Geological Survey*   193

18 Land Subsidence *Frank Forrester*   199

19 Supplemental Areas for Storage of Radioactive Wastes in Kansas *Charles K. Bayne (compiler) and John C. Halepaska and Ernest E. Angino*   203

20 Medical Geology *Harry V. Warren and Robert Delavault*   209

21 A Progress Report on Mercury *John M. Wood*   213

22 Good Coffee Water Needs Body *Wayne A. Pettyjohn*   227

Part Six   Earth Resources   237

23 Mineral Resource Estimates and Public Policy *V. E. McKelvey*   246

24 Energy Conservation *Philip H. Abelson*   266

25 Geothermal Energy *L. J. P. Muffler and D. E. White*   268

26 Solar Energy—Prospects for Its Large-Scale Use *Peter E. Glaser*   278

27 Soil *Paul B. Sears*   287

28 Inventorying Soil Resources *F. L. Himes*   296

Part Seven   Man and the Finite Earth   307

29 Population *Council on Environmental Quality*   312

30 After the Population Explosion *Harrison Brown*   314

31 Declaration on the Human Environment *United Nations Conference on the Human Environment*   324

Appendices

    I Geologic Time Chart   333

   II Periodic Table and Nuclide Data   334

  III Units and Conversions   339

  IV Surface Water Criteria for Public Water Supplies   343

Glossary   345

Credits and Acknowledgments   365

Man's Finite Earth

Part One

Atmospheric Science

Although the direct effect of man's activities on the global climate is the subject of controversy, his effect on local climate is established, and his ability to affect global change is at least probable.

James D. Hays, 1973

For hundreds of years man has been modifying climate through his elimination of forests, development of irrigation systems and reservoirs, and building of cities; however, these changes have mainly been local. Such changes have also been inadvertent.

This part of the book is concerned with climate and focuses on changes in climate. The first article provides an introduction to the heating and cooling of the earth and the circulation of the atmosphere. It also briefly describes the importance of oceans in the heat balance of the earth and in the creation of different climates on earth.

At first glance, the title of the second article, "Dust, Goats, and Deserts," appears strange, but it makes more sense after a few paragraphs when it is realized that man, through his management of the land, can cause fairly substantial modifications in climate. Whenever changes in the climate of a region have been determined, either by historical, archaeological, or geological analysis, there is always a question of how much of this change has been caused by man. Climates are continually changing. They have changed for millions of years as any glacial geologist will tell you, and they have changed for about a hundred years as any climatologist will testify. In the latter case, the climatic change is indicated by changes in average annual surface temperatures for the earth. From 1880 to 1940, the earth's average temperature increased by about $0.4°C$, and from 1940 to the present, the temperature decreased by about $0.2°C$. Although these changes may seem very small indeed, they are important when the effects of these changes on daily life and the calculated temperature change needed for glaciation are considered.

Small changes in temperature have been associated with changes in atmospheric circulation resulting in changes in frost boundaries, dust bowl conditions, or increased precipitation. Observations that such "small" climatic changes are taking place are not restricted to the thermometer. Recent increases in North Atlantic sea ice, expansion of permanent snow cover in the Arctic, and migration of armadillos southward also signal changes in climate. Although such changes may not seem small by the standards of the fishermen and farmers who are directly affected by them, they are of minor proportions compared to a possible glaciation.

The possibility of glaciation is raised in the paper by Kukla and Matthews, which is a report on a conference dealing with the possible end of the present interglacial. During the Quaternary there have been several interglacials, or times when the average temperature of the earth was as warm or warmer than it is today. Punctuating these warm periods have been glacials or cold periods when the glaciers of the world advanced far beyond their present positions. This periodic advance and retreat of glaciers has been ascribed to many causes, and the fluctuations in climate now experienced are similar to those recorded in the stratigraphic record. According to the participants of the conference, within the next few millenia or even centuries we can expect global cooling and related rapid changes in environment which exceed fluctuations experienced by man in historical times. One might speculate on the numerous problems that would develop in a world with an expanding population and a decreasing temperature.

At this point it should be mentioned that not everyone agrees that we are headed for another ice age and that, in addition to variations in climate and temperature that are of natural origin, there are possible large-scale man-made changes in climate. The paper on Climatic Effects of Man's Activities by the Study of Critical Environmental Problems (SCEP) reiterates the delicacy of the global energy balance and the ways in which man may alter this balance at certain leverage points. The most commonly discussed problem of man's altering the global climate is the introduction into the atmosphere of carbon dioxide (CO_2) from the burning of buried sunlight, i.e., fossil fuels. With increased use of energy, more fuels have been burned. About half of this output has gone into the biosphere and the oceans; the rest has gone into the atmosphere at the rate of 0.2 percent per year between 1958 and 1969. Carbon dioxide is important in regulating the temperature of the earth because of its effect on some of the outgoing or long-wave radiation from the earth's surface. Although CO_2 absorbs little of the incoming solar energy, it is opaque to some of the energy reradiated by the earth. Thus an increase in the amount of CO_2 would decrease the loss of heat by radiation with a resulting increase in temperature of the earth. This is known as the "greenhouse" effect. The increase in CO_2 concentration in the atmosphere is not directly related to the increase in temperature *because there are other factors, such as cloudiness, water vapor absorption, and ozone, that tend to reduce the effect of the CO_2 increase to about half of what it would otherwise be.*

A second major factor considered in the discussions of man's alteration of climate is particle pollution, produced by industrial or agricultural activity. The particles decrease the transmissivity of the atmosphere to incoming solar radiation but also affect outgoing radiation. The radiation balance depends upon the size, altitude, and distribution of the particles, but estimates indicate that the general effect is one of cooling. Geologists are interested in this aspect of climate modification and draw upon data regarding volcanic eruptions—natural events—that have changed stratospheric temperatures in the past. Some scientists have indicated that dustiness has also affected glacier fluctuations.

The conclusions of SCEP with regard to environmental impact of supersonic transports (SSTs) were of general concern; however, one of the conclusions of this report, that of the impact of a reduction in ozone (O_3), seems to have been understated. In the text of the full report, the calculated changes on ozone were within the limits of day-to-day and geographical variability; however, it was emphasized that calculations were provisional. A report released by the National Academy of Sciences in 1973 indicated that the importance of ozone changes, as a result of SST activity, was underrated. A fleet of SSTs might reduce the ozone in the stratosphere by introduction of water and nitrogen oxides which would react with the ozone to reduce its concentration. What are the possible effects of such a reduction? Probably the main effect would be the increase in ultraviolet radiation, that portion of the sunlight spectrum between the longest X rays and the shortest visible rays. Ozone serves to block this radiation and a 5 percent decrease in ozone over the central United States could cause a 26 percent increase in the amount of radiation reaching the surface. This might be equivalent to 8000 extra cases of skin cancer per year in the United States and destruction of DNA resulting in cell death or mutations. It is uncertain at this time whether consideration is still being given to development of SSTs.

Some changes in the atmosphere are deliberately induced by man. Weather modification is now a way of life in some parts of the world where techniques are used to reduce hail and fog or to increase precipitation. The legal aspects of hazards resulting from weather changes have not yet been solved. If one area benefits, does another area always lose? And what are the possibilities of tampering with the weather to gain a military advantage?

Finally, there are climatic modifications that might develop through diversion of large rivers or changes in interocean circulation. Proposals to modify the climate of the Soviet Arctic by damming the Bering Straits have been made, and the possible climatic changes caused by changes in rivers that flow into the Soviet portion of the Arctic Ocean have been considered. Calculations on the effect of using coal dust or other covering on arctic sea ice to increase melting show some interesting climatic changes in the north. But such changes would also affect atmospheric circulation and climate on a worldwide basis. All of these possible man-induced changes, whether deliberate or inadvertent, provide a basis for contemplation of the future and data for computer simulation of world weather and climate. The fact that airsheds and atmospheric effects are on international and intercontinental scales suggests the need for cooperation not only in the reduction of hazards but also in the monitoring of the changes that man is making.

1 The Atmospheric Circulation

Abraham H. Oort

Abraham H. Oort is a research scientist for the Geophysical Fluid Dynamics Laboratory of the National Oceanic and Atmospheric Administration in Princeton, New Jersey. He is also a visiting lecturer in the Department of Geological and Geophysical Sciences at Princeton University.

Climate can be discussed from different points of view. We shall limit the discussion to some large-scale features of climate from the point of view of a dynamic climatologist.

It is clear that the meteorological conditions experienced by an observer at ground level give a good description of the climate at that location. For example, the annual temperature range with its extremes and similar statistics for wind speed, wind direction, humidity, and precipitation are basic features of the climate at the earth's surface. However, if one wishes to get a deeper understanding of how the climate near the ground is maintained, one has to measure the meteorological parameters throughout the entire atmosphere.

Both observational and theoretical studies after World War II have shown that one part of the atmosphere cannot be studied very well without taking into account its interaction with surrounding atmosphere. This is particularly evident in the field of numerical weather prediction. For example, for a midlatitude forecast of the weather a day or two in advance, one needs data from an area roughly the size of the United States. However, if one wants to extend the forecast to several days or a week, one needs data from most of the Northern Hemisphere. Beyond a week one should probably include interactions between the Northern and Southern Hemispheres, and for longer periods there is a need for global atmospheric data. Moreover, for forecasts of a week or more, data in the surface layer of the ocean are necessary.

The realization that the climate at one place on the globe is to a certain extent influenced by the climate at all other places has made it necessary to modify the older, more static concepts in climatology. The modern climatologist is presently faced with the challenging task of having to incorporate the effects of atmospheric and oceanic circulations as one of the basic factors in the creation of climate. The interaction between the different regions takes place through the operation of the general circulation in the atmosphere (and to some extent also by the general circulation in the oceans).

In the next section of this article we shall outline how the atmospheric and oceanic heat engines are driven by the incoming solar energy. It will be shown that powerful large-scale circulation systems operate to accomplish the "status quo" in the present climate. Any change in climate would have to occur through a change in the character of this large-scale circulation. In a subsequent section we shall discuss some large-scale properties of the circulation and climate with the aid of north-south cross sections and hemispheric maps of the east-west wind component, the temperature, and the humidity. The final section consists of a summary and some concluding remarks.

The Energy Cycle

Incoming solar energy is the primary energy source feeding the atmospheric and oceanic heat engines. On the average a unit area of the earth receives 0.5 calories of solar radiation per cm^2 per minute. The left-hand side of Figure 1 shows schematically how this energy is used. If one assigns a value of 100 units to the incoming flux of solar energy, about 30 units of this are directly reflected by clouds and dust in the atmosphere or by the earth's surface. The remaining 70 units are available to heat the atmosphere (20) and the earth (50). For brevity, we shall refer here to the solid earth and the oceans as "the earth."

In a steady state the temperature of the atmosphere and the earth do not increase, because heating due to the absorption of (short-wave) solar radiation is offset by an equally strong cooling

ATMOSPHERIC SCIENCE 9

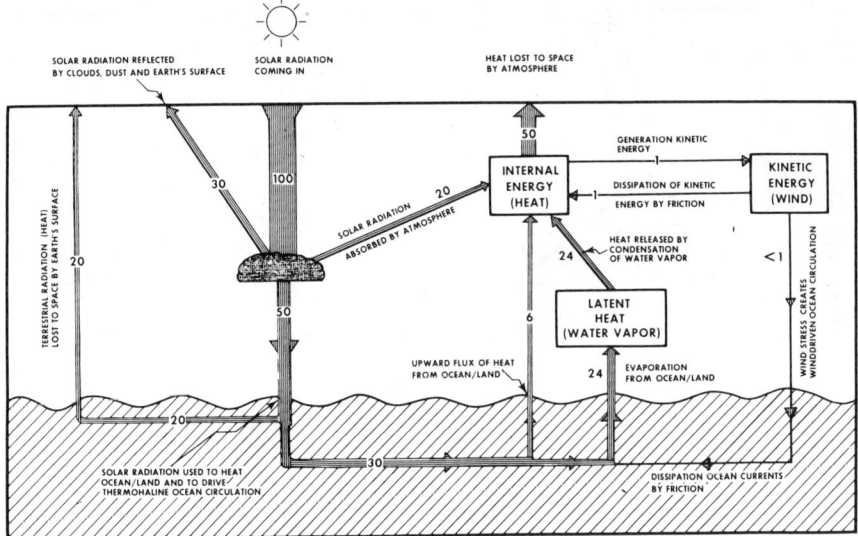

Figure 1. The "weather machine" of the earth, a schematic drawing of the flow of energy in the atmosphere-earth system. A value of 100 units is assigned to the incoming flux of solar energy (0.5 cal cm^{-2} min^{-1}). All values represent annual averages for the entire atmosphere. For simplicity, no energy boxes have been drawn for the ocean or land.

due to the emission of (long-wave) terrestrial radiation. Thus, the system of the atmosphere plus the earth is practically always in so-called radiative equilibrium, gaining as much energy as it loses.

Considering the atmosphere and earth separately, however, there is no balance, as is illustrated in Figure 2. In this figure the observed vertical temperature distribution averaged over a year and over the Northern Hemisphere is compared with the calculated temperature distribution assuming radiative equilibrium conditions. Near the earth's surface the atmosphere is—fortunately for life—about 45°C colder than the radiative equilibrium temperature, and in the middle and upper troposphere (2 to 12 km), much warmer. The actual temperature distribution comes about through the action of convective mixing by dry thermals and by clouds, which carry the heat upward away from the earth's surface.

Returning to Figure 1 we see that from the 50 units of solar radiation absorbed by the earth's surface, about 20 units are lost

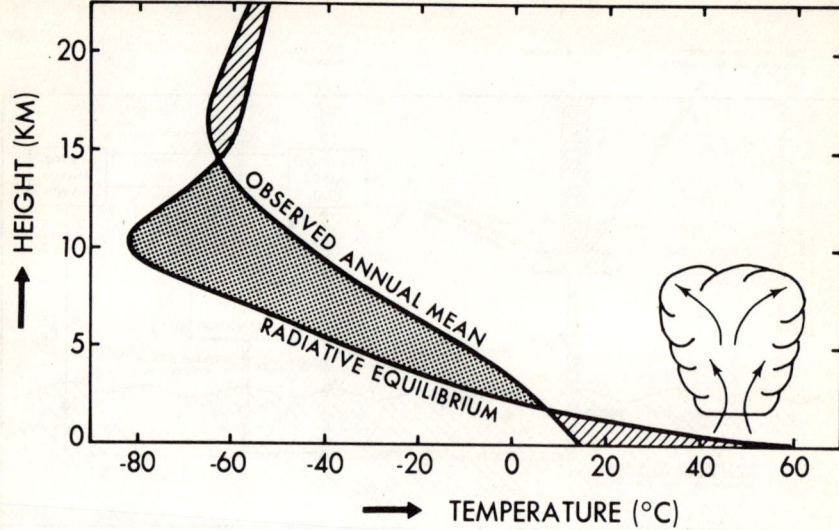

Figure 2. Imbalance in the vertical direction. The diagram shows a comparison of the observed vertical temperature distribution averaged for a year and over the entire Northern Hemisphere with the calculated temperature distribution assuming radiative equilibrium conditions. The radiative equilibrium computations were carried out by S. Manabe and R. F. Strickler for annual mean sunshine conditions, assuming no clouds and zero heat capacity for the ground. The computed vertical profile is unstable, and the inclusion of convective processes (clouds) which transport heat upward away from the earth's surface is needed to make the temperature profile realistic.

to space as long-wave radiation from the surface (long-wave radiation emitted by the surface minus the back-radiation from the atmosphere). The remaining 30 units are transferred to the atmosphere in the form of latent (24) and sensible heat (6). Through the condensation process the 24 units are made available to the basic atmospheric energy cycle as illustrated on the right-hand side of Figure 1.

Three boxes labeled internal energy, latent heat, and kinetic energy indicate the basic energy forms.[1] The major action in-

---

[1] Note that for simplicity the fourth basic energy form, namely potential energy, is left out in the diagram. The conversions from potential to kinetic energy and *vice versa* become very important when we consider portions of the atmosphere, but these conversions can be left out if we discuss the entire atmosphere as is done in Figure 1.

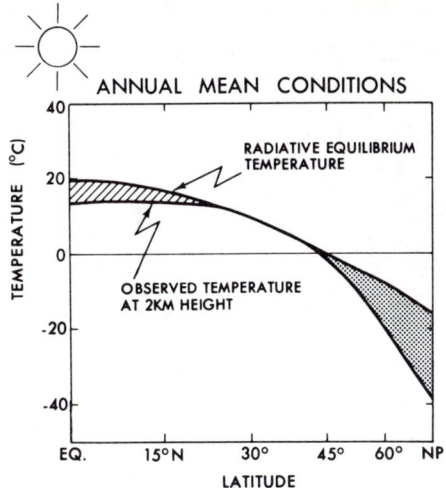

Figure 3. Imbalance in the north-south direction. The diagram shows as a function of latitude a comparison of the observed temperature at 2 km height with the temperature calculated assuming radiative equilibrium. Just as was the case in Figure 2, the computed profile is dynamically unstable. In this case, the large-scale atmospheric and oceanic circulations perform the necessary poleward transport of energy.

volves, on the one hand, heating due to convection from the earth's surface (24 + 6 units) and to direct absorption of solar radiation (20 units), and, on the other hand, cooling due to long-wave emission (50 units).

Surprisingly enough, only 1 percent of the original solar input is used to drive the atmospheric wind systems. From this low level of "efficiency" the general circulation is a feature of secondary importance in determining the climate on earth. This becomes evident if one again computes the radiative equilibrium temperature of the earth-atmosphere system, but in this case as a function of latitude. Figure 3 shows the north-south distributions of the radiative equilibrium temperature and of the observed temperature at 2 km height,[2] both for the annual mean conditions. It is clear

---

[2] The temperature at about 2 km height was chosen for comparison with the radiative equilibrium temperature because the hemispheric mean temperature at 2 km is approximately equal to the hemispheric mean radiative equilibrium temperature (assuming a cloudless earth).

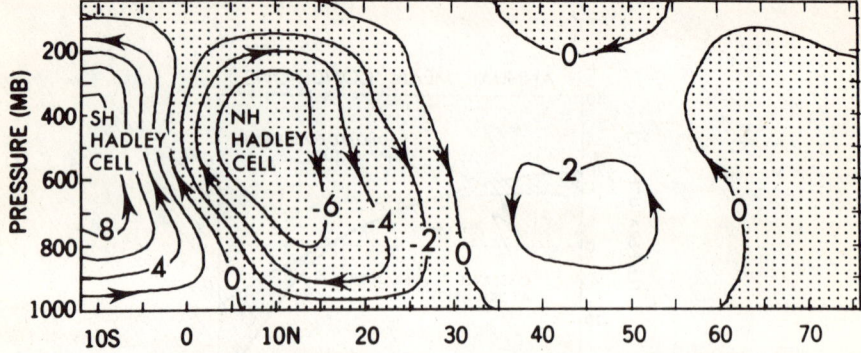

Figure 4. Observed annual mean streamlines in a north-south section through the Northern Hemisphere. In each hemisphere one observes one strong cell, the tropical Hadley cell, and two weak cells at middle and high latitudes. In the cells the air moves at all longitudes in the same manner.

that radiation alone would tend to heat the atmosphere and oceans at low latitudes and to cool them at high latitudes. A steady state is maintained because the large-scale circulation systems transport sufficient amounts of energy poleward to compensate for the radiational effects.

The mechanism by which the necessary poleward transport of energy is accomplished is quite different at tropical and extratropical latitudes. We shall now describe in a simplified and schematic fashion how the two basic transport mechanisms operate. Between equatorial latitudes and the subtropics the energy transport takes place mainly by one large overturning, the so-called Hadley cell (Figure 4). In this cell the air rises at all longitudes near the equator and sinks in the subtropics (20-30° latitude). The net effect of this circulation is a transport of energy toward the subtropics.[3]

At the middle and high latitudes overturnings on the scale of the low-latitude Hadley cell are weak and rather ineffective

---

[3] The situation is fairly complicated. The lower branch of the Hadley cell transports large amounts of energy, in the form of latent and sensible heat, toward the intertropical convergence zone near the equator. (In this zone, heat is converted into potential energy.) The upper return branch is ineffective in transporting energy in these forms because the upper air is dry and cold. However, at these high levels the potential energy is very large. In fact, the return branch at the upper levels transfers so much potential energy poleward that it overcompensates for the equatorward transport of sensible and latent heat in the lower branch. Thus, the net effect is a transport of total energy poleward.

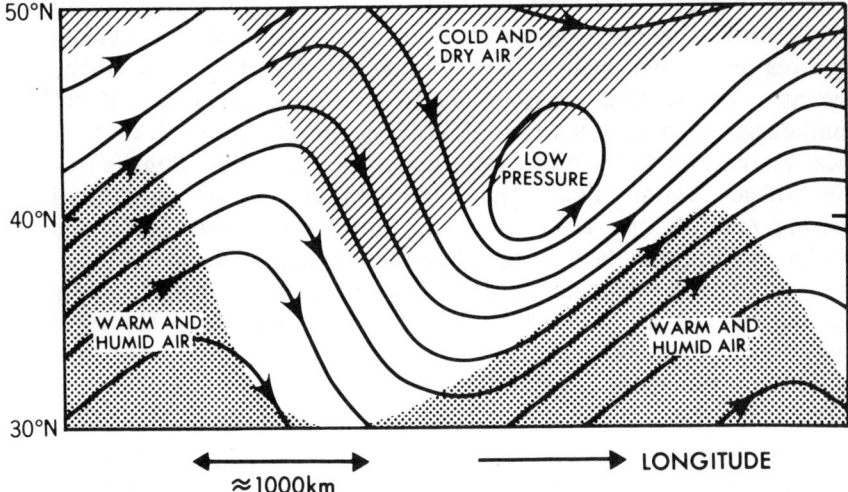

Figure 5. A schematic picture of a typical configuration of horizontal streamlines at several km's height. Waves of the type shown are responsible for the poleward transport of energy at middle and high latitudes, where the mean cells (Figure 4) are quite weak. The transport occurs both in the form of sensible and latent heat.

(Figure 4). At these latitudes smaller-scale waves or eddies take over the role of transporting energy poleward. The waves have a wavelength of the order of 3000 to 6000 km. They are the upper-level extensions of the familiar low and high pressure systems. These moving weather systems form an efficient mechanism to transfer heat poleward (both in sensible and latent form). Figure 5 illustrates schematically that the north-south exchange occurs through the northward movement of relatively warm and humid air and the simultaneous southward movement of cold and dry air across the same latitude circle. It is of interest to note that, in contrast with the situation in the tropical Hadley cell, almost no net mass transport takes place across a latitude circle. For more information concerning the basic reasons for the occurrence of two, distinctly different transport mechanisms in the atmosphere, the reader is referred to an earlier article of the author.[4]

Before discussing some of the consequences of the general circulation on the climate, let us return once more to the energy

---

[4] "The Energy Cycle of the Earth." *Scientific American* 223:54-63; September 1970.

diagram in Figure 1. Regarding the oceans, one can distinguish two mechanisms for creating ocean currents. The first mechanism involves radiational heating and cooling in the surface layer of the oceans. The heating is used largely to evaporate water. If no compensation occurs by rainfall, this evaporation will lead to a local increase in the salinity (and consequently in the density) of the surface water. Thus, the combined effects of radiational heating and cooling tend to create horizontal density gradients which, in turn, may lead to so-called thermo-haline circulations.

A second mechanism to drive ocean currents is by the stress exerted on the water by atmospheric winds. Although the total energy used to drive ocean currents is probably of the order of 1 percent of the incoming solar radiation or less, the effect of the currents on the climate can be profound. For example, it is often claimed that the northward flow of warm water in the Gulf Stream system is partly responsible for the mild climate in western Europe. Therefore, one cannot escape the conclusion that a complete understanding of the climate can only be brought by studying the ocean and atmosphere as one integrated system. A courageous, first attempt in this direction was made by S. Manabe and K. Bryan, Jr., of the National Oceanic and Atmospheric Administration, in a numerical simulation experiment of a combined ocean-atmosphere model.[5]

Large-Scale Climate

Let us continue the line of thinking started in the preceding section with regard to the importance of the general circulation in establishing the earth's climate.

First, we will discuss the influence of the mean cells (in the yearly average one finds three cells in each hemisphere) as depicted in Figure 4. The powerful low-latitude Hadley cell can be easily recognized in the climatic zones it imposes at the earth's surface. Thus, one finds in the rising branch of the cell the intertropical convergence zone with an abundance in rainfall. Going northward one encounters a latitude belt with steady (at least over the oceans) northeasterly trade winds which form the extension of the lower branch of the Hadley cell. Finally in the sinking branch of the cell between roughly 20 and 30°N in the subtropics one encounters very dry, often desert-like conditions.

[5] "Climate Calculations with a Combined Ocean-Atmosphere Model." *The Journal of the Atmospheric Sciences* 26:786-789; July 1969.

The midlatitude so-called Ferrel cell and the polar cell are quite weak in comparison with the Hadley cell and, as we have argued earlier, they play only a secondary role in the poleward flow of energy needed for an overall balance. One of the few noteworthy features is a slow mean rising in the latitude belt between about 50 and 60°N. Here one finds a secondary maximum in the rainfall, but this precipitation maximum is mainly due to upward motion in the cyclone waves. Further, little precipitation occurs over the polar regions.

It should be mentioned that the picture as given above is much simplified. For example, by averaging over the year, the annual cycle was filtered out; thereby masking some significant variations in the local climate. In this context, one has only to think of the dramatic switch in circulation regime between winter and summer over the Asian monsoon region.

Poleward of the climatic regimes dictated by the Hadley cell, traveling weather systems determine the climate (Figure 5). The more or less regular succession of low and high pressure systems with their warm and cold fronts gives the element of variability to the weather pattern at our latitudes. This type of climate is, of course, basically different from the steady and dependable climate in the tropics.

As an example, the distributions of some of the important climatological parameters over the Northern Hemisphere are shown in Figures 6 and 7. Figures 6a, b, and c show north-south sections through the atmosphere of the west-to-east wind component, the temperature, and the specific humidity, respectively. For the same parameters, Figures 7a, b, and c depict their horizontal distribution averaged in the vertical direction. We shall not go into an elaborate discussion of the diagrams; they are more or less self-explanatory.

At our latitudes the annual mean conditions, shown in Figures 6 and 7, depict a situation which is probably never encountered on any particular day of the year. These figures give only the *steady* component of the climate. To illustrate the magnitude of the unsteady component, let us consider Figures 6a and 7a in somewhat more detail. These figures show, respectively, the 5-year mean west-to-east flow averaged along a latitude circle (Figure 6a) and averaged in the vertical direction between the earth's surface and 25 km in height (Figure 7a). Evident are the easterlies at low latitudes and the strong westerlies at midlatitudes. At 30°N one finds the subtropical jet stream at about 12 km height with major

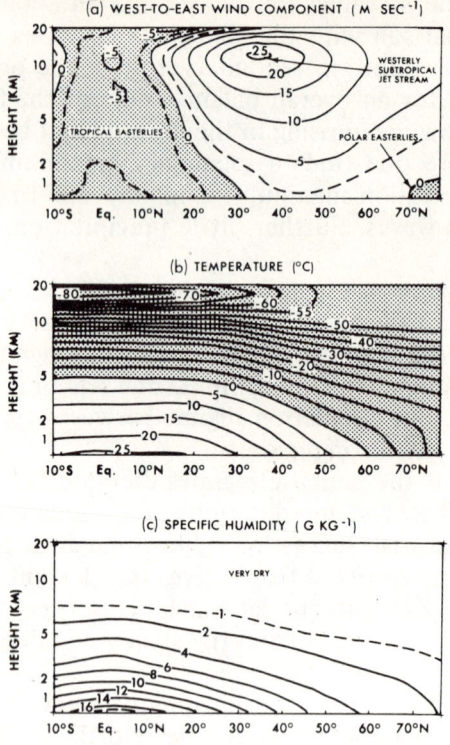

Figure 6. Observed north-south sections through the atmosphere of three important meteorological parameters. All values are averaged along a latitude circle and over a period of five years (May 1958 to April 1963). The parameters are (a) the west-to-east wind component in m sec^{-1}, (b) the temperature in °C, and (c) the specific humidity in g of water vapor per kg air.

maxima east of the Asian and American continents. If one calculates the standard deviation in the wind speed maximum east of Japan (18 m sec^{-1}), one obtains a value of 10 m sec^{-1}. In other words, during more than 100 days of the year the vertical-mean wind speed east of Japan is either smaller than 8 or larger than 28 m sec^{-1}.

Although the climatic elements shown in Figures 7a, b, and c are primarily a function of latitude, there are also significant differences along a latitude circle. These differences are connected with the ocean-continent distributions. They are called standing or quasi-stationary waves and are quite important in the sense that

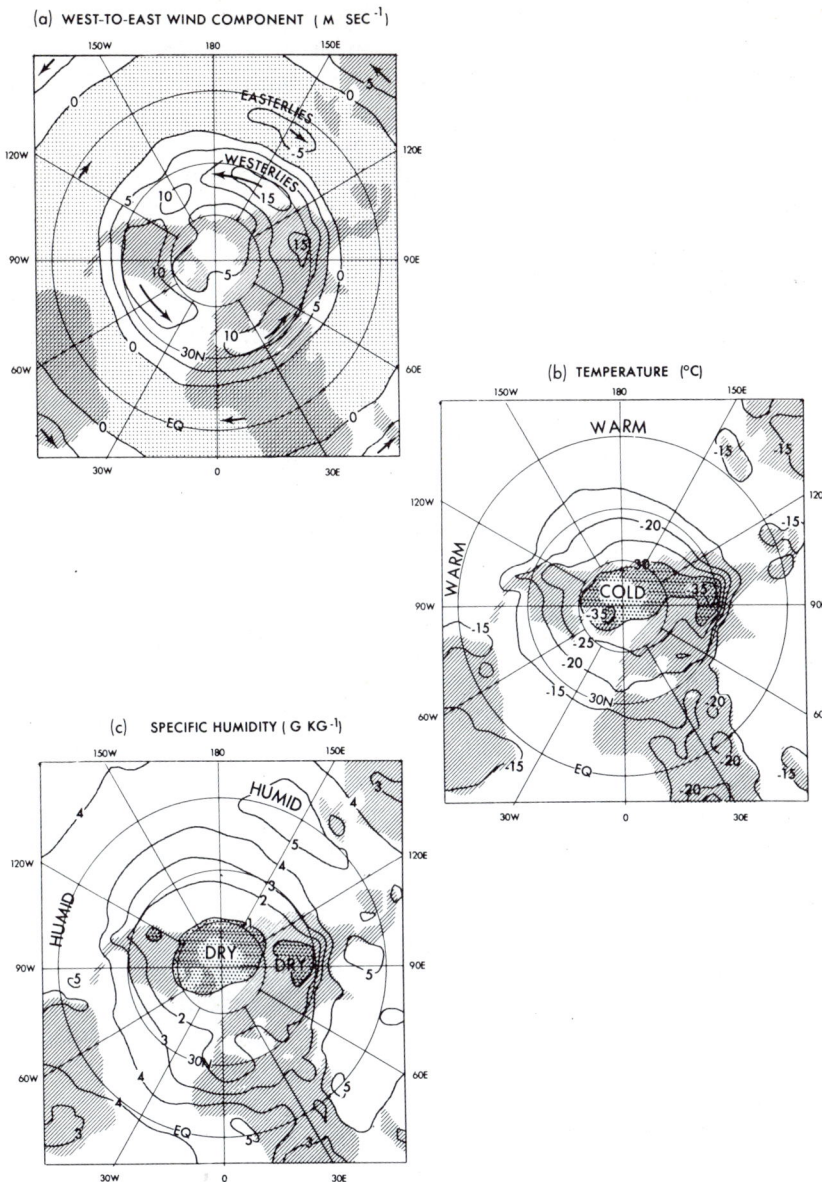

Figure 7. Observed horizontal maps of the vertical mean values of (a) the west-to-east wind component in m sec^{-1}, (b) the temperature in °C, and (c) the specific humidity in g of water vapor per kg air. (The maps are smoothed in order to take out some of the fine structure in mountainous terrain, which is irrelevant in this discussion.)

they—as do the transient waves—contribute to the poleward transfer of energy. The exact position of the standing waves may further be affected by changes in the conditions of the earth's surface, such as the temperature of the ocean surface, the extent of sea ice, the area of land covered by snow, etc. In this respect, the standing waves play a crucial role in the theories of climatic change.

Concluding Remarks

I have tried to emphasize the importance of the general circulation in the atmosphere and the oceans in creating the different climates on earth. A slight change in the character of the general circulation could eventually lead to important changes in climate. Through the general circulation the climate at one place on the globe is to some extent affected by the climate at all other places. Ideally, ocean and atmosphere have to be studied as an integrated system since the circulation in either of the two media is much dependent on the other.

The atmospheric and ocean circulations are, of course, driven ultimately by the incoming sunlight. However, the energy cycle proves to be a rather complex machine, as is illustrated in Figure 1. Only about 1 percent of the original influx of solar energy is used to drive the powerful atmospheric and oceanic currents. Without any motions, the earth would be everywhere in radiative equilibrium; in other words, at each point in the atmosphere as much radiation would be absorbed as emitted. Because of the more extreme temperature conditions (see Figures 2 and 3), most of the earth would be quite inhospitable to life as we know it now.

The temperature gradients in both the vertical and north-south directions resulting from radiative equilibrium are unstable. Convection in, for example, cumulus clouds brings about the observed distribution in the vertical. The necessary poleward transport of energy occurs by two mechanisms. At low latitudes a gigantic overturning involving all longitudes, the so-called Hadley cell, accomplishes the transport from equatorial latitudes to the subtropics (Figure 4).[6] In middle and high latitudes asymmetric waves with wavelengths of the order of 3000 to 6000 km take over the role of the Hadley cell. These waves form the upper air extensions of the familiar weather systems (Figure 5).

---

[6]This is an idealized picture. In reality, there are often large asymmetries around the latitude circle, e.g., in the region of the Asian monsoon.

The character of the weather and climate in the different latitude belts can be understood to a certain degree by studying these two modes of energy transfer. The abundance of rainfall near the equator, the dryness of the subtropics, and the steady conditions in the trade winds are, of course, related to, respectively, the upward, downward, and lower branches of the Hadley cell. On the other hand, the variable weather with cold and warm fronts passing over at rather regular intervals is a typical feature of the unstable waves at middle and high latitudes. In the last section of the paper, three of the important meteorological elements, namely the west-to-east wind speed, the temperature, and the specific humidity, are discussed in some detail (Figures 6, 7).

The fields of meteorology and oceanography can be understood using very simple physical and hydrodynamical principles. Although the principles are simple, the solutions to specific problems turn out to be far from easy. The complications are due to the nonlinear character of the equations prescribing the motions. Fascinating problems arise as to the interaction between different scales of motion, for example, between the cumulus-scale and the cyclone-scale waves. Many of these problems are still unresolved.

2

Dust, Goats, and Deserts

Reid A. Bryson

Reid A. Bryson is a professor in the Departments of Meteorology and Geography at the University of Wisconsin-Madison, where he is also director of the Institute for Environmental Studies.

The Dusty Regions

Polynesian sea voyagers used to sail towards islands over the horizon by watching for the characteristic features of land clouds—features that differentiate them from sea clouds. Similarly, air pilots may see a distant city indicated by its "dust dome"— usually brown by day and glowing with the internal city lights at night. But the pilot with an educated eye may see far more than local pollution. He may see city dust plumes merging into regional palls that extend from hundreds to thousands of kilometers downwind, perhaps rising along a sloping internal atmospheric surface and flattening into a dust layer at a height of 10 km or more—a layer that the ground-bound observer would not see or understand, for dust in the atmosphere is not routinely measured. The pilot knows that the upper surface of the hazy or dusty layer near the ground marks an inversion and smooth flying. He knows that the dusty layers produce a glare of scattered sunlight and that a dusty region is a bright region.

Traveling about the world, an air pilot will see tremendous variations in the turbidity of the atmosphere, ranging from the crystal clarity of arctic skies with 300-kilometer visibility to the brown air and bronze-blue zenith of West Pakistan and northwest India, where the ground may not be visible at all from a height of only 3 km. He will wonder about the source of the dust and its significance. From the ground he might watch the red twilight glow not on the western horizon, but as a weird orange-red oval some 15° above the black, dust-obscured horizon. If he is also a climatologist, the pilot will ask himself what effect this dense dust might have on the climate.

Without instruments, the attenuation of the sunlight is obvious. Looking at the sun is not painful in these dusty regions, but the glare of scattered sunlight is great. Prolonged exposure of human skin to sun without tanning indicates that the ultraviolet light is attenuated also. The slow evening cooling after the 40°C summer day in the Indian or Rajputana Desert indicates that outward infrared radiation from the ground is also reduced.

Measured Effects of Dust over India

Instrumental measurements verify the sensory evidence. Of particular importance to the climate of the area is the radiation variation with height, for the vertical divergence of the radiation is a measure of the diabatic or direct radiative cooling of the atmosphere which is necessary to maintain the mean subsident motion that is characteristic of deserts. Over wet northeastern India the mean motion is upwards, but west of New Delhi it is downwards. Downward mean motion implies mean compressional warming of the air and in turn an average *in situ* warming which is greater than that observed.

Das (1962) calculated the sinking motion and its relation to the temperature change and found that in mid-troposphere, perhaps 5 km above the ground, diabatic cooling of 2.4°C/day was necessary to sustain the subsidence. Calculating the infrared diabatic cooling that would occur in the air over northwestern India, with the observed temperature and moisture distribution, Das found that he could account for only 1.8°C/day. He had, of course, made the calculation in the standard way, assuming that water vapor, carbon dioxide, and ozone were the significant radiating gases. Without data on the distribution of particulate matter in suspension, and a much more complicated and less certain calculation, he could not have allowed for the effect of dust.

Shortly after Das's paper was published, a series of balloon-sonde measurements of the infrared radiation divergence was started on the fringes of the Rajputana Desert, in New Delhi, in

Poona, and later in the desert at Jodhpur (Bryson et al., 1964; Mani et al., 1965). These measurements showed that the discrepancy between observed and calculated cooling-rates was very nearly the same as the discrepancy found by Das (1962) between required and calculated cooling rates—and that discrepancy depended on the dustiness of the air!

To the uninitiated, the difference between 1.8 and 2.4°C/day may seem small; but it means a difference of 33 percent in the sinking rate of the desert air—and it is sinking air primarily that makes aridity—leading one inevitably to the conclusion that the dusty desert air makes the area more desertic. Subsidence increases air stability (usually) and decreases relative humidity, thus inhibiting precipitation.

In short, the radiative effect of dust in the air over northwest India and West Pakistan is enormous when compared with the effect of particulates in North America and Europe. But then the amount of dust in the air is enormously greater in the former region, too. In order to put numbers on the sensory evidence of greatly reduced visibility in the Rajputana Desert, a series of research flights were made in 1966 (Peterson & Bryson, 1968). The effect of dust content of the air on the infrared radiation divergence, as measured during these flights, is shown in Figure 1.

The measurements of dust concentration made in the air over northwest India suggest that even on the average, away from the cities, there are 300-800 micrograms of dust per cubic meter in the lowest 5 to 10 km of air. Compare that with 150-200 micrograms per cubic meter in the lowest 1 to 2 km over Chicago, and one sees that if Chicago air is turbid, Rajputana Desert air is *very* turbid and covers a much larger area. During pollution episodes, city air is as turbid as the average air over the Rajputana Desert, but to lesser heights.

Instrumental observations are not available to assess the areal extent of this turbid air, but visual observations indicate maximum dust density over the Rajputana Desert area. In all directions from there the dust density diminishes—especially to the east and south—though remnant layers may be traced to Cambodia, northern Malaya, and at least as far south as Madras. To the west, the dust density decreases somewhat along the Mekran Coast and then may increase over Arabia and the Sahara in some seasons. Over the mountains to the north and northwest of the Rajputana Desert the air appears to be less turbid.

If one uses the homely principle that the smoke is densest near

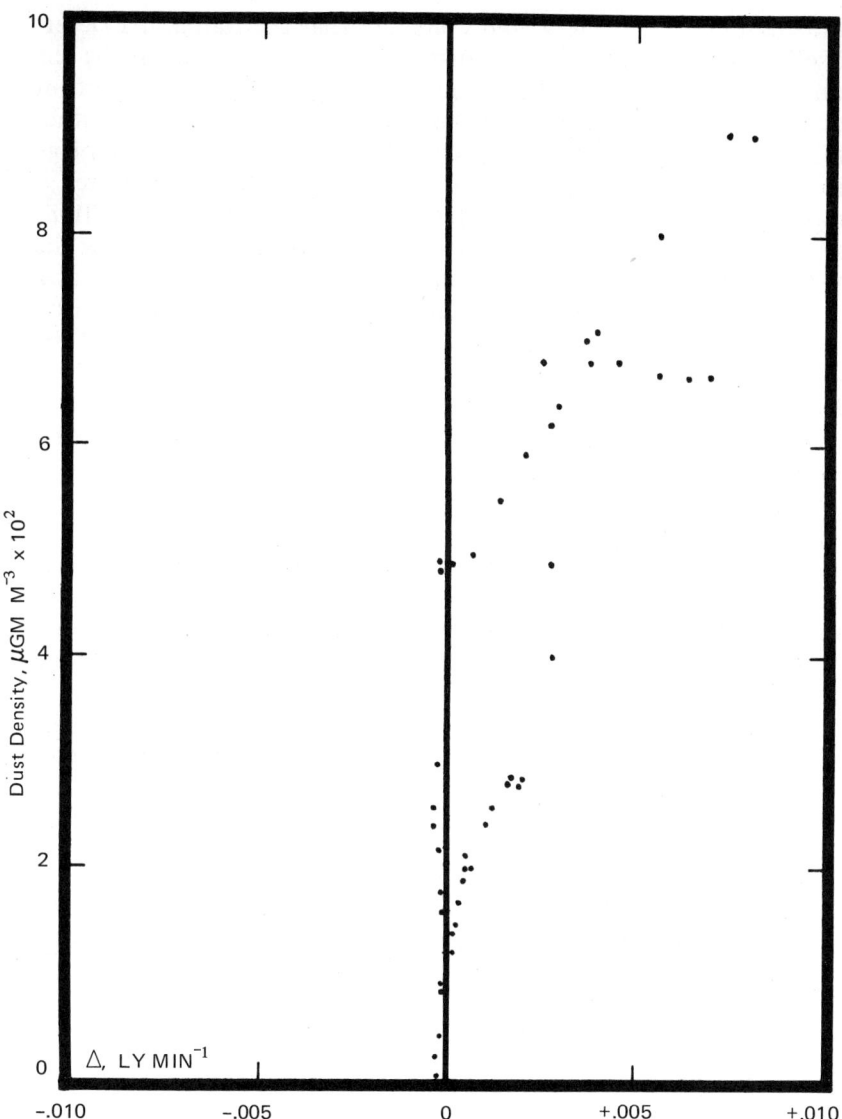

Figure 1. Infrared radiation effect of atmospheric dust as measured over northwest India. The ordinate is in hundreds of micrograms per cubic meter. The abscissa is in the difference between observed and calculated upward radiation flux-gradient, corrected for variation in ambient temperatures and downward fluxes. This figure shows that the upward radiation flux decreases less rapidly in dusty air than in clean air. (After Peterson and Bryson, 1968.)

the chimney, then one would conclude that the Rajputana Desert itself was the source of the dust. Indeed, this appears to be the case. Samples of the airborne dust were composed of 36 percent quartz and 20 to 22 percent each of carbonate, mica, and either feldspar or clay minerals. Of particular interest in this assemblage is the clay. A major clay mineral in the Rajputana Desert area is montmorillonite. E. D. Goldberg's (personal communication) studies of wind-borne dust on the ocean floor near India show the major clay to be montmorillonite, which is rather effective as a nucleating agent for cloud-seeding. With very large characteristic dust-loads one would suspect that the few clouds would be typically over-seeded and that the region would be a poor prospect for artificial rain-enhancement.

The thick blanket of turbid air affects the incoming solar radiation as well as the outgoing, absorbing or scattering a considerable portion of the radiation before it reaches the ground. This diminishes the daytime instability that might promote convective showers if moisture were present (incidentally, the air of the Rajputana Desert is surprisingly moist for a desert—there just isn't much rain). At night, the large radiative cooling of the upper part of the dusty blanket and suppressed radiative cooling near the ground increase the instability and keep the ground warmer than it otherwise would be. This reminds the scientific traveler of the nighttime difference in near-surface temperatures between the city and the countryside in less desertic, better vegetated, areas. The reason is different, but there *is* a relationship, as will be seen hereafter.

Several comments in the preceding paragraphs may have alerted the reader to some rather special circumstances of the particular desert under discussion. Thus he may have asked some questions, such as why this particular desert is so dusty, and why it is a desert if the air is so moist.

The Rajputana Desert does have some unusual features:
1. It is probably the dustiest of all deserts,
2. The air over the part of this desert having the least rainfall has a total water content that is comparable with that of some very rainy tropical forest areas,
3. The dew point of the air is quite high during the summer, and there are quite a few clouds,
4. It is the most densely-populated desert in the world,
5. Much of the desert area was once occupied by a high culture with an agricultural base—the Indus civilization, and

6. The region appears to be more barren than the measured rainfall would lead one to expect.

The remains of the Indus civilization suggest that the region was not always as desertic as it is now, and paleobotanical studies verify this suggestion. Gurdip Singh (1970), studying the pollen in the bed of Sambhar Salt Lake near Jaipur, in Rajasthan, found that during the time of the Indus people, the lake held fresh water and the vegetation of the surrounding land was indicative of much moister conditions. Then the lake became salty as the culture disappeared, and about 1000 B.C. the lake dried up entirely. After a long dry interval, scattered settlements reappeared to be replaced by the extensive Rangmahal culture by the fourth century A.D. Then extensive nomadism developed. Many dust storms in southwest Rajasthan were reported in the seventh century A.D., and it appears that by 1000 A.D. considerable spread of the desert had occurred, to be accentuated in the recent past.

One acquires, at this point in the investigation, a nagging suspicion that perhaps this region should not be desertic at all but rather some sort of savanna. This suspicion is compounded by awareness of cities buried under sand dunes, great castles in areas that are now too poor to have provided the excess captial to have built them, and ancient paintings which depict a lusher, wetter land.

If the Rajputana Desert is really anomalous, how might one change it back to a more productive land? We know that the air over this desert subsides, more than it otherwise would do, because of the high dust content and its effect on radiative cooling. Without the dust there would be less subsidence—and more frequent showers! If there were more rain there would be more grass, and as the dust is from the surface soil, the grass would reduce the deflation from the surface and result in there being less dust carried into the atmosphere (Bryson & Baerreis, 1967).

Now think of the contrast between the concrete, brick, and asphalt of the city on one hand and the grass and loose soil of the countryside on the other. Just as the nighttime temperatures fall lower in the country than they do in the city, so lower nocturnal temperatures would prevail over the desert grass than over the compact, bare desert soil—and the night temperatures would be still lower if the dust blanket were not there. With the high dew point of the Rajputana Desert, heavy dew would form on the grass (and does to some extent in the areas which are grassy, even now).

With dew to help the grass grow and hold down the desert soil, there would be less dust in the air. In turn, there would be less radiative cooling of the top of the dusty layer of air, less subsidence, and more frequent showers to make the grass grow, etc., etc.

Obviously, more grass, if it improved the climate and so made possible the growth of still more grass, would be a good thing for the goatherds of the desert—except that it appears that their herds are the reason why the grass is sparse, the air dusty, and the land unproductive in a semidesert climate!

The evidence for this influence of goats is to be found in a simple experiment that was performed at the Central Arid Zone Research Institute in Jodhpur. There, a barren plot of land was fenced to exclude goats, sheep, and other grazers. Nothing was done inside the fence except to let nature have its way—and in less than two years there was tall, rich native grass except within one goat-neck distance from the fence.

This gets at the heart of the whole matter, for it indicates that *a significant fraction of the dust over the Rajputana Desert is there because of human activity*. Such dust falls within the definition of air pollution as used in this paper*, and it thus appears that overuse of the land can change regional climates even without the "blessings" of mechanization and industrialization.

References

Bryson, R. A., and Baerreis, D. A. 1967. Possibilities of major climatic modification and their implications: northwest India, a case for study. *Bulletin of the American Meteorological Society* 48:136-42.

Bryson, R. A., Wilson, C. W., III, and Kuhn, P. M. 1964. Some preliminary results from radiation sonde ascents over India. In *Symposium on tropical meteorology, Rotorua, New Zealand, Nov., 1963,* pp. 501-16. Wellington: New Zealand Meteorological Service.

Das, P. K. 1962. Mean vertical motion and non-adiabatic heat sources over India during the monsoon. *Tellus* 14:212-20.

Mani, A., Sreedharan, C. R., and Srinivasan, V. 1965. Measurements of infrared radiative fluxes over India. *Journal of Geophysical Research* 70:4529-36.

Peterson, J. T., and Bryson, R. A. 1968. Influence of atmospheric particulates on the infrared radiation balance of northwest India. In *Proceedings of*

---

*For the purpose of the discussion in this paper, the materials put into the air by man and his machines, and those which enter the air in increased amounts due to a variety of human activities, will be regarded as atmospheric pollutants.

1st National Conference on Weather Modification, Albany, N. Y., pp. 153-62. Boston: American Meteorological Society.

Singh, G. 1970. *History of post-glacial vegetation and climate of the Rajasthan Desert.* Birbal Sahni Institute of Paleobotany, Lucknow, unpublished report.

3 When Will the Present Interglacial End?

G. J. Kukla and R. K. Matthews

George J. Kukla is senior research associate at the Lamont-Doherty Geological Observatory of Columbia University. R. K. Matthews is the chairman of the Department of Geological Sciences at Brown University.

A group of scientists interested in Quaternary research gathered recently to review the possibility that their data concerning climates of the past might be valuable for long-term global climatic forecasting. They met at a working conference entitled "The Present Interglacial, How and When Will it End?" held at Brown University, Providence, Rhode Island, on 26 and 27 January 1972 (1). The discussion was divided into five sections: (i) environmental changes in the historical period (independent of man), (ii) the pattern of change within the last 10,000 years, (iii) the last interglacial and its end, (iv) comparison of the last interglacial with the present warm interval and projection of future change, and (v) consideration of the causes of global climatic change.

The present global cooling, which reversed the warm trend of the 1940s, is still under way. Even though man-made pollution may have contributed to the observed fluctuations, the bulk of the change is probably of natural origin (Mitchell). The present cooling is especially demonstrable in certain key regions in arctic and subarctic latitudes. Thus, snowbanks today cover areas of Baffin Island that were seasonally free of snow for the 30 or 40 years preceding the present summer cooling (Andrews, Barry, Bradley, Miller, and Williams); pack ice around Iceland is once again becoming a serious hindrance to navigation (2); and warmth-loving animals, such as armadillos, which expanded northward into the American Midwest in the first half of the century, are now retreating southward (Schultz).

Periods of cooling more severe than the present one are known to have occurred in the past 5000 years. They are recognized not only in deep-sea sediments (Burckle and others) but also in advances of mountain glaciers (Denton and Karlen). The climatic shifts recorded in pollen-rich lake beds in northern mid-latitudes or in the sequences of stream alluviation and downcutting are closely related in time to the social disorders that ended or severely castigated flourishing civilizations in Egypt, Mesopotamia, and the Indus Valley about 4000 years ago and the lowland Mayas in A.D. 770. This suggests that former human civilizations may have been severely affected by (3) "failure in the rain supply, without which neither man, nor beast, nor growing vegetation can survive" (4). It is hard to envision how our modern economy and social structure would react to widespread droughts several decades long, should they occur in the near future.

On the geologic time scale, the general warmth and basic bipartite pattern of the last 10,000 years of the earth's history (the elapsed part of the Holocene), which are characteristic of interglacials, were underlined by several workers (Fairbridge, Wright, and others). It has long been recognized that the climatic optimum passed 6000 to 7000 years ago and was succeeded by slow, oscillatory cooling, interrupted by milder episodes like the one in the 10th and 11th centuries (Burckle, Fairbridge). In some places, the present fauna and flora can be compared to those of the early portion of the Holocene (Absolon, Wright). The warmth-loving species of the climatic optimum have migrated south (Lozek).

One conclusion reached at the session was that there is no qualitative difference between the climatic fluctuations in the 20th century and the climatic oscillations that occurred before the industrial era. The present climatic trends appear to have entirely natural causes, and no firm evidence supports the opposite view.

The next group of contributions dealt with the structure of the last interglacial and its end. Participants considered the periodic occurrence of interglacials within the stratigraphic record to be sufficiently well established to warrant comparison with the present interglacial. There are discrepancies concerning the time-stratigraphic boundaries of interglacials (McIntyre, Mörner, Ruddiman, Terasmae), but at least for the purpose of the meeting the interglacial was tacitly defined as one uninterrupted warm interval in which the environment on a global scale reached present or even warmer climatic conditions (5).

These discussions focused on two points: the length of interglacials and the environmental change that marked their end. Out of more than 800 determinations of ^{18}O in fossil planktonic foraminifera (the ^{18}O content is a function of the temperature and salinity of the surface waters) covering about the last 0.5 million years, only about 10 percent indicate conditions similar to or warmer than those of today (Emiliani). In lake beds of Germany and England the length of an interglacial was found to be about 10,000 years (Shakleton, Wright). Pollen diagrams of interglacial lake beds so closely parallel the Holocene records in composition and thickness that basically the same duration must be expected for both intervals. However, recent soils in the American West are weaker than those believed to be of the last interglacial age (Morrison, Richmond), an occurrence consistent with views that the last interglacial was somewhat warmer and wetter than the Holocene (Fairbridge, Lozek).

Sea level is related inversely to the volume of continental glaciers. Thus, information concerning the duration of an interglacial high stand of the sea can be directly correlated with the ice volume of continental glaciers. On Barbados, sedimentological considerations suggest that the high stand associated with the last interglacial (terrace III, 124,000 years ago) lasted no longer than about 5000 years. Further, preliminary evidence suggests that the interglacial high stand was immediately followed by a drop in sea level of more tens of meters within 10,000 to 15,000 years (Matthews). For this same time interval, data from deep-sea cores show that the cold subarctic waters of the North Atlantic extended to latitudes about 15 degrees south of where they are today, or roughly two-thirds of their maximum full glacial southward displacement in the late Wisconsin (McIntyre, Ruddiman). Summer temperatures at the sea surface dropped by 7°C at 50°N latitude in the Atlantic (Imbrie). During this same time interval, fine sand and dust were blown from Africa into the central Atlantic, which indicates a time of considerable disruption of the vegetation cover on the continent (Hays).

Some data indicate how rapid the cooling could have been near the end of an interglacial. In the Greenland ice core (Camp Century) a spectacular drop in ^{18}O values appears to have occurred within a time interval only about 100 years long. The event is considered to have happened around 90,000 years ago. A similar event could have happened 20,000 years earlier, but a critical segment of ice core is missing (Clausen, Dansgaard, Johnsen, Langway). A rapid cooling event is also indicated at about 90,000 years ago by the temporary complete disappearance of the warmth-loving *Globorotalia menardii* group from the southern Gulf of Mexico, an event completed within less than 500 years (Kennett). In the vicinity of Prague and Brno in Czechoslovakia, where mixed broadleaf forests flourished in past interglacials as they do today, the end of an interglacial is marked by the replacement of forests with grassland. Eolian dust of distant origin then buried the vegetation, and torrential rains turned the countryside into badlands. Woolly rhinoceros and the land snail *Puppilla loessica*, cold-resistant species of Pleistocene fauna, lived there at this time, about 110,000 years ago. The date is supported by a magnetic event interpreted as Blake (Kukla, Koci). At Tenaghi Phillipon in Greece, the interglacial forest was succeeded by grassland within a few centuries (6), and in the Netherlands and Denmark subarctic tundra with heath and birch replaced the temperate forests (7).

When comparing the present with previous interglacials, several investigators showed that the present interglacial is in its final phase (Emiliani, Imbrie, Lozek, Mörner, Wright) and that if nature were allowed to run its course unaltered by man, events similar to those which ended the last interglacial should be expected to occur perhaps as soon as the next few centuries.

The possible causes of past climatic changes were discussed in the last section. The ice-age preconditioning of the present globe (Fairbridge), the instability of atmospheric and oceanic circulation (Broecker, Flohn, Hendy, MacCracken, Mitchell, Shaw, Van Donk, Weyl), and the possibility of rapid antarctic ice "surges" (Hollin) were stressed. Theoretical considerations and some empirical data suggest that climatic change is closely related to the earth's precessional torques and thereby to the earth's magnetic field, episodic volcanism, and so forth, and to elements of the earth's orbit (Emiliani, Kukla, Stuiver). It was speculated that the astronomical motions of the earth may have led to stresses within the lithosphere with a maximum every 40,000 years. Enhanced volcanism, tectonic activity, and changes in magnetic parameters would be expected to follow this periodicity, contributing to glaciations and speeding evolution (Emiliani). Artificial heating, and production of dust and CO_2 by man's activities were shown to have diverging effects on global temperatures (Mitchell, Schneider), at present subordinate to natural processes. However, with continuing human input these effects might eventually trigger or speed climatic change. The general conclusion of this section of the conference was that knowledge necessary for understanding the mechanism of climatic change is lamentably inadequate, and that the ultimate causes remain unknown.

At the end of the working conference, the majority of the participants agreed to the following points:

> The global environments of the last several millennia is in sharp contrast with climates that existed during most of the past million years. Warm intervals like the present one have been short-lived and the natural end of our warm epoch is undoubtedly near when considered on a geological time scale. Global cooling and related rapid changes of environment, substantially exceeding the fluctuations experienced by man in historical times, must be expected within the next few millennia or even centuries. In man's quest to utilize global resources, and to produce an adequate supply of food, global climatic change constitutes a first order environmental hazard which must be thoroughly understood well in advance of the first global indications of deteriorating climate. Interdisciplinary attacks on these problems must be internationally organized and encouraged to develop at a rate substantially exceeding the present pace.

In the view of the majority of participants, further investigation is especially needed in the following fields: (i) detailed reconstruction of the history of intervals of rapid environmental change, especially of the termination of the last interglacial, as well as those periods of cold or dry "events," or both, in historical times; (ii) periodicity in climatic change on all time scales; (iii) records of past climatic change contained in stratigraphic sequences of the deep-sea sediments, of continental basins in loess areas, in ice sheets, and in mountain glaciers; (iv) computer modeling of past climatic systems based on boundary conditions prescribed by the stratigraphic data; and (v) the possible interrelationships between solar radiation, solar magnetics, earth magnetics, episodic volcanism, and global climatic change.

References and Notes

1. Participants present at the working conference were R. G. Barry (Institute of Alpine and Arctic Research, University of Colorado, Boulder); L. H. Burckle, J. D. Hays, C. Hendy, and A. McIntyre (Lamont-Doherty Geological Observatory, Palisades, New York); G. Denton and J. Hollin

(University of Maine, Orono); C. Emiliani (Institute of Marine Sciences, Miami, Florida); R. W. Fairbridge (Columbia University, New York); J. Imbrie and R. K. Matthews (Brown University, Providence, Rhode Island); W. Karlen (University of Stockholm, Stockholm, Sweden); J. P. Kennett and D. W. Shaw (University of Rhode Island, Kingston); G. J. Kukla (Czechoslovakian Academy of Sciences, Prague); J. M. Mitchell, Jr. (Environmental Data Service, National Oceanic and Atmospheric Administration, Silver Spring, Maryland); G. M. Richmond (U. S. Geological Survey, Denver, Colorado); W. Ruddiman (Office of Naval Research, Washington, D. C.); C. B. Schultz (University of Nebraska, Lincoln); N. J. Shackleton (University of Cambridge, Cambridge, England); T. W. Webb III (University of Michigan, Ann Arbor); and P. W. Weyl (State University of New York, Stony Brook).

Participants who sent contributions were A. Absolon, A. Koci, and V. Lozek (Czechoslovakian Academy of Sciences); J. T. Andrews, R. S. Bradley, and G. H. Miller (Institute of Alpine and Arctic Research); W. S. Broecker and J. Van Donk (Lamont-Doherty Geological Observatory); H. B. Clausen, W. Dansgaard, and S. J. Johnsen (University of Copenhagen, Copenhagen, Denmark); H. Flohn (Meteorologisches Institut, Bonn, West Germany); C. C. Langway (U. S. Cold Regions Research Laboratory, Hanover, New Hampshire); M. C. MacCracken (University of California, Livermore); N. A. Mörner (University of Stockholm); R. Morrison (U. S. Geological Survey, Denver); S. H. Schneider (Institute for Space Sciences, National Aeronautics and Space Administration, New York); M. Stuiver (University of Washington, Seattle); J. Terasmae (Brock University, Quebec, Canada); L. D. Williams (University of Colorado, Boulder); H. E. Wright, Jr. (University of Minnesota, Minneapolis).

Papers resulting from this working conference have been accepted for publication in *Quaternary Research*.
2. Eimarsson, N. A. 1969. *Hafisinn*. Reykjavik, Iceland: Almenna Bokafelagia.
3. Carpenter, R. 1966. *Discontinuity in Greek Civilization*. Cambridge, England: Cambridge University Press.
4. See also Bell, B. 1971. *American Journal of Archeology* 75:1.
5. The cold fluctuations of the historical period are considered to be within the range of present general climates and environments.
6. Wijmstra, T. A. 1969. *Acta Botan. Neer.* 18:511.
7. Van der Hammen, T., Maarleveld, G. C., Vogel, J. C., and Zagwijn, W. H. 1967. *Geol. Mijnb.* 45:79.

4

Climatic Effects of Man's Activities

Study of Critical Environmental Problems

The Study of Critical Environmental Problems represents the conclusions of a group of fifty participants from many fields after a one-month examination of the global climatic and ecological effects of man's activities. The Study was sponsored by Massachusetts Institute of Technology at Williams College in Williamstown, Massachusetts. The director was Professor Carroll L. Wilson and the associate director was Professor William H. Matthews, both of MIT.

Introduction

There is geological evidence that there have been five or six glacial periods (ice ages); the most recent (the Pleistocene) lasted 1 to 1.5 million years. In the past century there has been a general warming of the atmosphere of about 0.4°C up to 1940, followed by a few tenths degree cooling. It seems clear that our climate is subject to a wide variety of fluctuations, with periods ranging from decades to millennia, and that it is changing now.

We know that the atmosphere is a relatively stable system. The solar radiation that is absorbed by the planet and heats it must be almost exactly balanced by the emitted terrestrial infrared radiation that cools it; otherwise the mean temperature would change much more rapidly than just noted. This nearly perfect balance is the key to the changes that do occur, since a reduction of only about 2 percent in the available energy can in theory lower the mean temperature by 2°C and produce an ice age.

That there have not been wider fluctuations in climate is our best evidence that the complex system of ocean and air currents, evaporation and precipitation, surface and cloud reflection, and absorption form a complex feedback system for keeping the global energy balance nearly constant. Nonetheless, the delicacy of this balance and the consequences of disturbing it make it very important that we attempt to assess the present and prospective impact of man's activities on this system.

The total mass of the atmosphere and the energy involved in even such a minor disturbance as a thunderstorm (releasing the energy equivalent to many hydrogen bombs) should convince us immediately that man cannot possibly hope to intervene in such a gigantic arena. However, in reality man does intervene, because he can—without intending to do so—reach some leverage points in the system.

All the important leverage points that this Study has identified control the radiation balance of the atmosphere in one way or another, and most of them control it by changing the composition of the atmosphere. For example, man can change the temperature of the atmosphere by introducing a gas such as CO_2 or a cloud of particles that absorbs and emits solar and terrestrial infrared radiation, thereby altering the delicate balance we have described. He can also affect the heat balance by changing the face of the earth or by adding heat as a result of rising energy demands.

A thorough understanding and reliable prediction of the influence of atmospheric pollutants on climate requires the mathematical simulation of atmosphere-ocean systems, including the pollutants. At present, computer models successfully simulate many observed characteristics of the climate and have significantly advanced our knowledge of atmospheric phenomena. They have, however, a number of drawbacks that become serious when modeling new states of equilibrium or changes of climate in its transition toward these new states. Unless these limitations are overcome, it will be difficult, if not impossible, to predict inadvertent climate modifications that might be caused by man.

Recommendations

1. We recommend that current computer models be improved by including more realistic simulations of clouds and air-sea interaction and that attempts be made to include particles when their properties become better known. Such models should be run for periods of at least several simulated years. The effects of potential

global pollutants on the climate and on phenomena such as cloud formation should be studied with these models.

2. We recommend that possibilities be investigated for simplifying existing models to provide a better understanding of climatic changes. Simultaneously, a search should be made for alternative types of models which are more suitable for handling problems of climatic change.

Carbon Dioxide from Fossil Fuels

All combustion of fossil fuels produces carbon dioxide (CO_2), which has been steadily increasing in the atmosphere at 0.2 percent per year since 1958. Half of the amount man puts into the atmosphere stays and produces this rise in concentration. The other half goes into the biosphere and the oceans, but we are not certain how it is divided between these two reservoirs. CO_2 from fossil fuels is a small part of the natural CO_2 that is constantly being exchanged between the atmosphere/oceans and the atmosphere/forests.

A projected 18 percent increase resulting from fossil fuel combustion to the year 2000 (from 320 ppm to 379 ppm) might increase the surface temperature of the earth 0.5°C; a doubling of the CO_2 might increase mean annual surface temperatures 2°C. This latter change could lead to long-term warming of the planet. These estimates are based on a relatively primitive computer model, with no consideration of important motions in the atmosphere, and hence are very uncertain. However, these are the only estimates available today.

Should man ever be compelled to stop producing CO_2, no coal, oil, or gas could be burned and all industrial societies would be drastically affected. The only possible alternative for energy for industrial and commercial use is nuclear energy, whose by-products may also cause serious environmental effects. There are at present no electric motor vehicles that could be used on the wide scale our society demands.

Although we conclude that the probability of direct climate change in this century resulting from CO_2 is small, we stress that the long-term potential consequences of CO_2 effects on the climate or of societal reaction to such threats are so serious that much more must be learned about future trends of climate change. Only through these measures can societies hope to have time to adjust to changes that may ultimately be necessary.

Recommendations

1. We recommend the improvement of present estimates of future combustion of fossil fuels and the resulting emissions.
2. We recommend study of changes in the mass of living matter and decaying products.
3. We recommend continuous measurement and study of the carbon dioxide content of the atmosphere in a few areas remote from known sources for the purpose of determining trends. Specifically, four stations and some aircraft flights are required.
4. We recommend systematic scientific study of the partition of carbon dioxide among the atmosphere, the oceans, and the biomass. Such research might require up to twelve stations.

Particles in the Atmosphere

Fine particles change the heat balance of the earth because they both reflect and absorb radiation from the sun and the earth. Large amounts of such particles enter the troposphere (the zone up to about 12 km or 40,000 feet) from natural sources such as sea spray, windblown dust, volcanoes, and from the conversion of naturally occurring gases into particles.

Man introduces fewer particles into the atmosphere than enter from natural sources; however, he does introduce significant quantities of sulfates, nitrates, and hydrocarbons. The largest single artificial source is the production of sulfur dioxide from the burning of fossil fuel that subsequently is converted to sulfates by oxidation. Particle levels have been increasing over the years as observed at stations in Europe, North America, and the North Atlantic but not over the Central Pacific.

In the troposphere, the residence times of particles range from six days to two weeks, but in the lower stratosphere micron-size particles or smaller may remain for one to three years. This long residence time in the stratosphere and also the photochemical process occurring there make the stratosphere more sensitive to injection of particles than the troposphere.

Particles in the troposphere can produce changes in the earth's reflectivity, cloud reflectivity, and cloud formation. The magnitudes of these effects are unknown, and in general it is not possible to determine whether such changes would result in a warming or cooling of the earth's surface. The area of greatest uncertainty in connection with the effects of particles on the heat balance of the atmosphere is our current lack of knowledge of their optical properties in scattering or absorbing radiation from the sun or the earth.

Particles also act as nuclei for condensation or freezing of water vapor. Precipitation processes can certainly be affected by changing nuclei concentrations, but we do not believe that the effect of man-made nuclei will actually be significant on a global scale.

Recommendations

1. We recommend studies to determine optical properties of fine particles, their sources, transport processes, nature, size distributions, and concentrations in both the troposphere and stratosphere, and their effects on cloud reflectivity.
2. We recommend that the effects of particles on radiative transfer be studied and that the results be incorporated in mathematical models to determine the influence of particles on planetary circulation patterns.
3. We recommend extending and improving solar radiation measurements.
4. We recommend beginning measurements by lidar (optical radar) methods of the vertical distribution of particles in the atmosphere.
5. We recommend the study of the scientific and economic feasibility of initiating satellite measurements of the albedo (reflectivity) of the whole earth, capable of detecting trends of the order of 1 percent per ten years.
6. We recommend beginning a continuing survey, with ground and aircraft sampling, of the atmosphere's content of particles and of those trace gases that form particles by chemical reactions in the atmosphere. For relatively long-lived constituents about ten fixed stations will be required, for short-lived constituents, about 100.
7. We recommend monitoring several specific particles and gases by chemical means. About 100 measurement sites will be required.

The Role of Clouds

The importance of clouds in the atmosphere stems from their

relatively high reflectivity for solar radiation and their central role in the various processes involved in the heat budget of the earth-atmosphere system.

Recommendations

1. We recommend that there be global observations of cloud distribution and temporal variations. High spatial resolution satellite observations are required to give "correct" cloud population counts and to establish the existence of long-term trends in cloudiness (if there are any).
2. We recommend studies of the optical (visible and infrared) properties of clouds as functions of the various relevant cloud and impinging radiation parameters. These studies should include the effect of particles on the reflectivity of clouds and a determination of the infrared "blackness" of clouds.

Cirrus Clouds from Jet Aircraft

Contrail (condensation trail) formation, which is common near the world's air routes, is more likely to occur when jets fly in the upper troposphere than in the lower troposphere because of the different meteorological conditions in these two regions.

There are very few, if any, statistics that permit us to determine whether the advent of commercial jet aircraft has altered the frequency of occurrence or the properties of cirrus clouds. We do not know whether the projected increase in the operation of subsonic jets will have any climate effects.

Two weather effects from enhanced cirrus cloudiness are possible. First, the radiation balance may be slightly upset, and, second, cloud seeding by falling ice crystals might initiate precipitation sooner than it would otherwise occur.

Recommendations

1. We recommend that the magnitude and distribution of increased cirrus cloudiness from subsonic jet operations in the upper troposphere be determined. A study of the phenomenon should be conducted by examining cloud observations at many weather stations, both near and remote from air routes.
2. We recommend that the radiative properties of representative contrails and contrail-produced cirrus clouds be determined.

3. We recommend that the significance, if any, of ice crystals falling from contrail clouds as a source of freezing nuclei for lower clouds be determined.

Supersonic Transports (SSTs) in the Stratosphere

The stratosphere where SSTs will fly at 20 km (65,000 feet) is a very rarefied region with little vertical mixing. Gases and particles produced by jet exhausts may remain for one to three years before disappearing.

We have estimated the steady-state amounts of combustion products that would be introduced into the stratosphere by the Federal Aviation Agency projection of 500 SSTs operating in 1985-1990 mostly in the Northern Hemisphere, flying seven hours a day, at 20 km (65,000 feet), at a speed of Mach 2.7, propelled by 1700 engines like the GE-4 being developed for the Boeing 2707-300. We have used General Electric (GE) calculations of the amount of combustion products because no test measurements exist. In our calculations we used jet fuel of 0.05 percent sulfur. We have been told that a specification of 0.01 percent sulfur could be met in the future at higher cost.

We have compared the amounts that would be introduced on a steady-state basis with the natural levels of water vapor, sulfates, nitrates, hydrocarbons, and soot in the stratosphere. We have also compared these levels with the amounts of particles put into the atmosphere by the volcano eruption of Mount Agung in Bali in 1963.

Based on these calculations, we have concluded that no problems should arise from the introduction of carbon dioxide and that the reduction of ozone due to interaction with water vapor or other exhaust gases should be insignificant. Global water vapor in the stratosphere may increase 10 percent, and increases in regions of dense traffic may be 60 percent.

Very little is known about the way particles will form from SST-exhaust products. Depending upon the actual particle formation, particles from these 500 SSTs (from SO_2, hydrocarbons, and soot) could double the pre-Agung eruption global averages and peak at ten times those levels where there is dense traffic. The effects of these particles could range from a small, widespread, continuous "Agung" effect to one as big as that which followed the Agung eruption. (The analogy between the SST input and that by the Mount Agung eruption is not exact.) The temperature of the equatorial stratosphere (a belt around the earth) increased 6°

to 7°C after the eruption and remained at 2° to 3°C above the pre-Agung level for several years. No apparent temperature change was found in the lower troposphere.

Clouds are known to form in the winter polar stratosphere. Two factors will increase the future likelihood of greater cloudiness in the stratosphere because of moisture added by the SSTs: the increased stratospheric cooling due to the increasing CO_2 content of the atmosphere and the closer approach to saturation indicated by the observed increase of stratospheric moisture. Such an increase in cloudiness could affect the climate. The introduction of particles into the stratosphere could also produce climatic effects by increasing temperatures in the stratosphere, with possible changes in surface temperatures.

A feeling of genuine concern has emerged from these conclusions. The projected SSTs can have a clearly measurable effect in a large region of the world and quite possibly on a global scale. We must, however, emphasize that we cannot be certain about the magnitude of the various consequences.

Recommendations

1. We recommend that uncertainties about SST contamination and its effects be resolved before large-scale operation of SSTs begins.
2. We recommend that the following program of action be initiated as soon as possible:
   a. Begin now to monitor the lower stratosphere for water vapor, cloudiness, oxides of nitrogen and sulfur, hydrocarbons, and particles (including the latter's composition and size distribution).
   b. Determine whether additional cloudiness or persistent contrails will occur in the stratosphere as a result of SST operations, particularly in certain cold areas, *and* the consequences of such changes.
   c. Obtain better estimates of contaminant emissions, especially those leading to particles, under simulated flight conditions and under real flight conditions, at the earliest opportunity.
   d. Using the data obtained in carrying out the preceding three recommendations, estimate the change in particle concentration in the stratosphere attributable to future SSTs *and* its impact on weather and climate.
3. We recommend implementation now of a special monitoring

program for the lower stratosphere (about 20 km or 60,000 to 70,000 feet) to include the following activities:
a. Measurement by aircraft and balloon of the water vapor content of the lower stratosphere. The area coverage required is global, but with special emphasis on areas where it is proposed that the SST should fly.
b. Sampling by aircraft of stratospheric particles, with subsequent physical and chemical analysis.
c. Monitoring by lidar (optical radar) of optical scattering in the lower stratosphere, again with emphasis on the region in which heavy traffic is planned.
d. Monitoring of tropospheric carbon monoxide concentration because of its potential effects on the chemical composition of the lower stratosphere.

Atmospheric Oxygen: Nonproblem

Atmospheric oxygen is practically constant. It varies neither over time (since 1910) nor regionally and is always very close to 20.946 percent. Calculations show that depletion of oxygen by burning all the recoverable fossil fuels in the world would reduce it only to 20.800 percent. It should probably be measured every 10 years to be certain that it is remaining constant.

Surface Changes and the Climate

The most important properties of the earth's surface that have a bearing on climate and are likely to be affected by human activity are reflectivity, heat capacity and conductivity, availability of water and dust, aerodynamic roughness, emissivity in the infrared band, and heat released to the ground.

Since the amount of carbon dioxide in the atmosphere is dependent on the biomass of forest lands which serves as a reservoir, widespread destruction of forests could have serious climatic effects. Population growth or overgrazing that increases the arid or desert areas of the earth creates conditions that allow the introduction of dust particles to the atmosphere.

Other important surface changes are from man's activities that modify snow and ice cover, particularly in polar regions, and from some possible projects involving the production of new, very large water bodies. Increased urbanization is of possible global importance only as it produces extended areas of contiguous cities. Still, it is not certain whether effects of urbanization extend far beyond the general region occupied by the cities.

Recommendation

We recommend that before actions are taken which result in some of the very extensive surface changes described mathematical models be constructed which simulate their effects on the climate of a region or, possibly, of the earth.

Thermal Pollution

Although by the year 2000 global thermal power output may be as much as six times the present level, we do not expect it to affect global climate. Over cities it does already create "heat islands," and as these grow larger they may have regional climatic effects. We recommend that these potential effects be studied with computer models.

5

Air Pollution in the Inner City

Council on Environmental Quality

The Council on Environmental Quality was established by the National Environmental Policy Act of 1969, and the Office of Environmental Quality, which provides staff for the Council, was subsequently established in 1970. The Council develops and recommends to the President national policies which promote environmental quality, performs a continuing analysis of changes or trends in the national environment, and assists the President in the preparation of the annual environmental quality report to the Congress.

Air pollution generally hangs more heavily over the inner city than the rest of the urban area, and far more heavily than over most suburban and rural areas. In some cities the central business district absorbs the most severe air pollution; in other cities close-in industrial areas bear the heaviest pollution loads. The largest concentrations of the urban poor often live near these two areas. A 1969 study conducted for the National Air Pollution Control Administration (NAPCA, now a part of the Environmental Protection Agency) confirmed that concentrations of particulate matter, carbon monoxide, and sulfur oxides decline steadily out from urban areas (urban measuring points are all located in downtown areas). This may be seen in Table 1.

Based on samples collected over a 2-year period, the study shows that average concentrations of particulates in nonurban areas are between 10 and 50 percent of the average in urban areas. Moreover, within metropolitan regions, similar variations hold between suburbia and the central city.

TABLE 1. SELECTED PARTICULATE CONSTITUENTS AS PERCENTAGE OF GROSS SUSPENDED PARTICULATES, 1966-67

	Urban		Nonurban					
	(217 stations)		Proximate (5)[1]		Inter-mediate (15)[2]		Remote (10)[3]	
	$\mu g/m^3$	Percent	$\mu g/m^3$	Percent	$\mu g/m^3$	Percent	$\mu g/m^3$	Percent
Suspended particulates	102.0	-----	45.0	-----	40.0	-----	21.0	-----
Benzene soluble organics	6.7	6.6	2.5	5.6	2.2	5.4	1.1	5.1
Ammonium ion	0.9	0.9	1.22	2.7	0.28	0.7	0.15	0.7
Nitrate ion	2.4	2.4	1.40	3.1	0.85	2.1	0.46	2.2
Sulfate ion	10.1	9.9	10.0	22.2	5.29	13.1	2.51	1.8
Copper	0.16	0.15	0.16	0.36	0.078	0.19	0.060	0.28
Iron	0.16	1.38	0.56	1.24	0.27	0.67	0.15	0.71
Manganese	0.073	0.07	0.026	0.06	0.012	0.03	0.005	0.02
Nickel	0.017	0.02	0.008	0.02	0.004	0.01	0.002	0.01
Lead	1.11	1.07	0.21	0.47	0.096	0.24	0.22	0.10

[1] Technically in nonurban areas, but conspicuously influenced by proximity to city, i.e., Cape Vincent, N.Y.; Kent County, Del.; Washington County, Miss.
[2] Closer to urban areas, usually with agricultural activity, i.e., Jackson County, Miss.; Humboldt, Calif.
[3] Farthest from large population center, i.e., Glacier National Park; White Pine County, Nev.

Source: Environmental Protection Agency.

A study of the St. Louis area made in 1966 by NAPCA reported that suspended particulates, dust fall, and concentrations of sulfur dioxide were higher in the predominantly poor black neighborhoods of St. Louis and East St. Louis than elsewhere in the metropolitan area.

A recent air pollution computer model of the Chicago region, when correlated with census data, indicates that the lowest income neighborhoods are in the areas of highest pollution concentrations (see Figures 1 and 2). Similar conclusions surface in data drawn from Kansas City, Mo., St. Louis, and Washington D. C. (see Table 2).

Another survey conducted by NAPCA of 22 other metropolitan areas, including Cincinnati, Dallas-Fort Worth, Denver, Indianapolis, Louisville, New York City-Northern New Jersey, Pittsburgh, Providence, San Antonio, and Seattle, substantiates unequal geographical distribution of certain pollutants within metropolitan areas. Emissions of carbon monoxide, sulfur oxides,

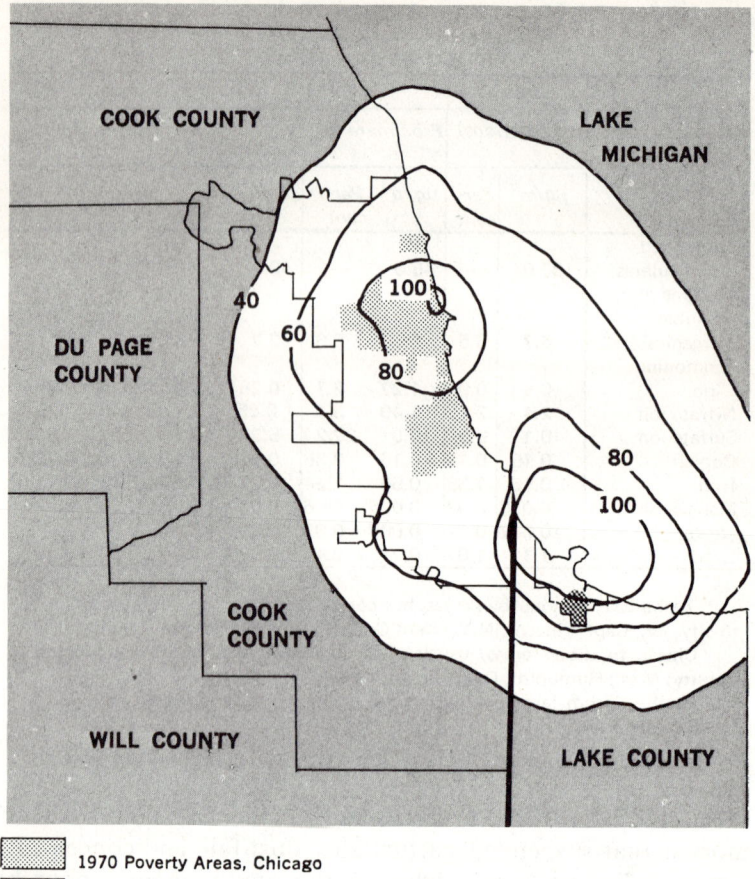

[Figure: Map of Chicago/Gary area showing sulfur dioxide concentrations with contour lines at 40, 60, 80, and 100 micrograms per cubic meter across Cook County, Du Page County, Will County, and Lake County, with Lake Michigan to the northeast.]

1970 Poverty Areas, Chicago

1960 Poverty Areas, Gary (1970 census statistics not available)

Source: Based on computer simulation model of data from Atomic Energy Commission, Argonne National Laboratory, and on U. S. Department of Commerce, Bureau of the Census data.

Figure 1. Current expected mean concentrations of sulfur dioxide, Chicago, Ill., and Gary, Ind. (In micrograms per cubic meter)

and particulate matter in these areas were measured by source and proximity to heavy population centers. Emissions of carbon monoxide result from vehicles; sulfur oxides and particulate emissions come from residential and commercial burning of coal and oil. For each of the three pollutants, the pattern is the same: Emission densities are highest in the core city and diminish with

1970 Poverty Areas, Chicago

1960 Poverty Areas, Gary (1970 census statistics not available)

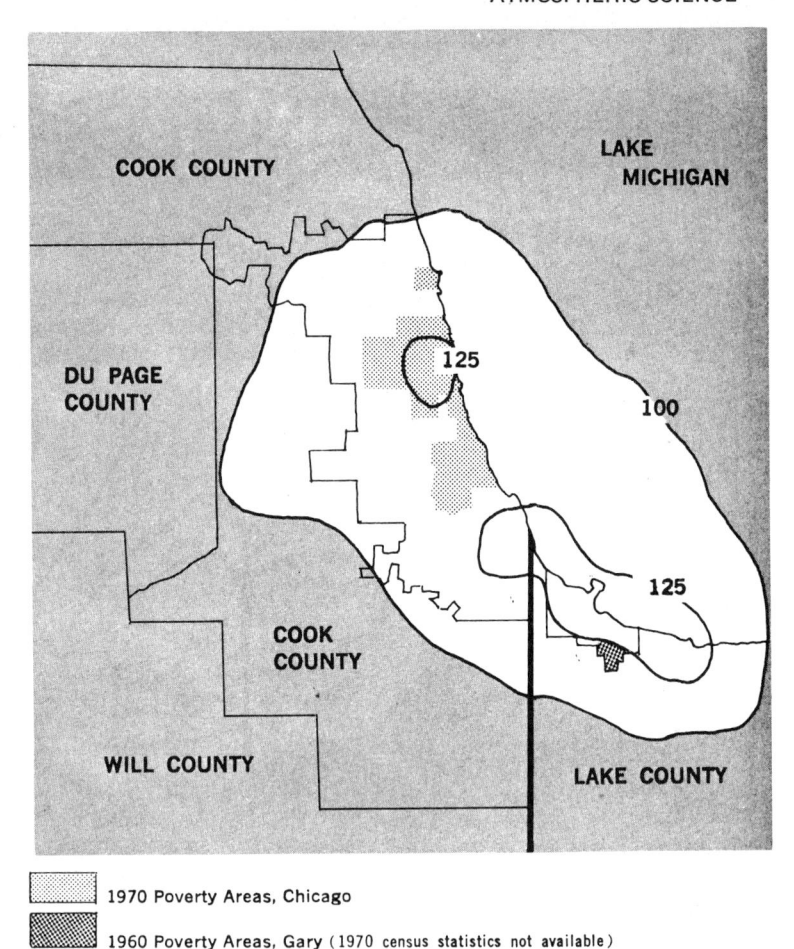

Source: Based on computer simulation model of data from Atomic Energy Commission, Argonne National Laboratory, and on U. S. Department of Commerce, Bureau of Census data.

Figure 2. Current expected particulate concentrations for Chicago, Ill.; and Gary, Ind. (In micrograms per cubic meter)

distance outward. Overall, the areas of highest emissions within the tested cities coincide with the areas of highest population density.

Other pollutants confirm this pattern. Lead, for example, is a potentially poisonous metal commonly found in the atmosphere and in the soil near heavily traveled roadways. Data vary widely among sampling sites, but lead levels in the blood appear particu-

TABLE 2. AIR POLLUTION EXPOSURE INDICES BY INCOME

Income	Suspended particulates	Sulfation	Mean
Kansas City			
0-2999	76.7	0.22	1.16
3000-4999	72.4	.20	1.09
5000-6999	66.5	.18	.98
7000-9999	63.5	.17	.93
10,000-14,999	60.1	.15	.86
15,000-24,999	57.6	.14	.80
25,000-over	58.1	.12	.76
St. Louis			
0-2999	91.3	.97	1.19
3000-4999	85.3	.88	1.10
5000-6999	79.2	.78	1.00
7000-9999	75.4	.72	.93
10,000-14,999	73.0	.68	.89
15,000-24,999	68.8	.60	.82
25,000-over	64.9	.52	.74
Washington, D.C.			
0-2999	64.6	.82	1.19
3000-4999	61.7	.82	1.16
5000-6999	53.9	.75	1.04
7000-9999	49.7	.69	.96
10,000-14,999	45.5	.64	.88
15,000-24,999	43.2	.58	.82
25,000-over	42.0	.53	.77

Source: Freeman, A. M., based on U. S. Department of Commerce, Bureau of the Census, and U. S. Department of Health, Education, and Welfare, Public Health Service data.

larly high among urban area dwellers. A study of major cities by the Environmental Protection Agency reported that atmospheric lead had increased significantly in the seven years ending in 1969. In Cincinnati, the increase ranged from 13 to 33 percent; in Los Angeles, from 33 to 64 percent; and in Philadelphia, from 2 to 36 percent.

Blood specimens of three groups of persons in the Philadelphia area were taken to determine the amount of lead in their systems. The groups were divided into those who had lived and worked within a 25-block radius of City Hall for 5 years, those who commuted regularly from the suburbs to work in the downtown area, and those who lived in the same neighborhood as the suburban commuters but also worked in the suburbs. For both men and women, lead was significantly highest in the city dwellers. The suburbanites who worked in the city showed higher lead concentrations than those who lived and worked in the suburbs.

Additional Readings

American Association for the Advancement of Science. 1965. *Air conservation*. AAAS Publication 80.
Arnold, K. 1961. An investigation into methods of accelerating the melting of ice and snow by artificial dusting. In *Geology of the Arctic*, ed. G. O. Raasch, pp. 989-1012. Toronto: University of Toronto Press.
Budyko, M. I. 1971. *Climate and life*. Leningrad: Hydrological Publishing House.
———. 1972. The future climate. *Trans. American Geophysical Union* 53(10):868-74.
Callender, G. S. 1961. Temperature fluctuations and trends over the earth. *Quarterly Journal of the Royal Meteorological Society* 87:1-12.
Campbell, W. J., and Martin, S. 1973. Oil and ice in the Arctic Ocean: possible large-scale interactions. *Science* 181:56-68.
Chang, J. H. 1968. *Climate and agriculture: an ecological survey*. Chicago: Aldine.
Detwyler, T. R., and Marcus, M. G. 1972. *Urbanization and environment*. Belmont, Calif.: Duxbury Press.
Donn, W., and Shaw, D. 1966. The heat budgets of an ice-free and ice-covered Arctic Ocean. *Journal of Geophysical Research* 71:1087-93.
Esposito, J. C. 1970. *Vanishing air: the Ralph Nader Study Group report on air pollution*. New York: Grossman.
Ewing, M., and Donn, W. L. 1956. A theory of ice ages, I. *Science* 123:1061-66.
———. 1958. A theory of the ice ages, II. *Science* 127:1159-62.
Fairbridge, R. W., ed. 1967. *Encyclopedia of the atmospheric sciences and astrogeology*. Encyclopedia of Earth Sciences Series, vol. 2. New York: Reinhold.
Fletcher, J. O. 1966. The arctic heat budget and atmospheric circulation. In *Proceedings of the symposium on the heat budget and atmospheric circulation*. Santa Monica, Calif.: RAND Corporation.
———. 1969. Controlling the planet's climate. *Impact of Science on Society* 19:151-68.
———. 1970. Polar ice and global climate machine. *Bulletin of the Atomic Scientists* 26:40-47.
Hidore, J. J. 1972. *A geography of the atmosphere*. 2nd ed. Dubuque, Iowa: Wm. C. Brown.

Ladurie, E. L. 1972. *Times of feast, times of famine: a history of climate since the year 1000*. Translated by B. Bray. London: George Allen and Unwin.

Landsberg, H. E. 1970. Climates and urban planning. In *Urban climates*. Geneva: World Meteorological Organization, no. 254, Tech. Paper 141, pp. 364-74.

Lave, L. B., and Seskin, E. P. 1970. Air pollution and human health. *Science* 169:723-33.

Leighton, P. A. 1966. Geographical aspects of air pollution. *Geographical Review* 56:151-74.

MacDonald, G. J. F. 1968. How to wreck the environment. In *Unless peace comes: a scientific forecast of new weapons*, ed. N. Calder. New York: Viking Press.

Malone, T. P. 1967. Weather modification: implications of the new horizons in research. *Science* 156:897-901.

Manabe, S., and Bryan, K. 1969. Climate circulation with a combined ocean-atmosphere model. *Journal of Atmospheric Sciences* 26:786-89.

Matthews, W. H., Smith, F. E., and Goldberg, E. D., eds. 1971. *Man's impact on terrestrial and oceanic ecosystems*. Cambridge, Mass.: MIT Press.

Miller, A. 1971. *Meteorology*. 2nd ed. Columbus, Ohio: Charles E. Merrill.

Petterssen, S. 1969. *Introduction to meteorology*. 3rd ed. New York: McGraw-Hill.

Rasool, S. R., and Schneider, S. H. 1971. Atmospheric carbon dioxide and aerosols: effects of large increases on global climate. *Science* 173:138-41.

Riehl, H. 1972. *Introduction to the atmosphere*. 2nd ed. New York: McGraw-Hill.

Sawyer, J. S., ed. 1966. World climate from 8000 to 0 B.C. In *Proceedings of International Symposium at Imperial College, London, 18 and 19 April, 1966*. London: Royal Meteorological Society.

Sellers, W. D. 1965. *Physical climatology*. Chicago: University of Chicago Press.

Study of Man's Impact on Climate (SMIC). 1971. *Inadvertent climate modifications*. Cambridge, Mass.: MIT Press.

Part Two

Oceanography

Each time certain properties of the sea floor were postulated on theoretical grounds, investigations in situ have upset the picture.

P. H. Kuenen, 1958

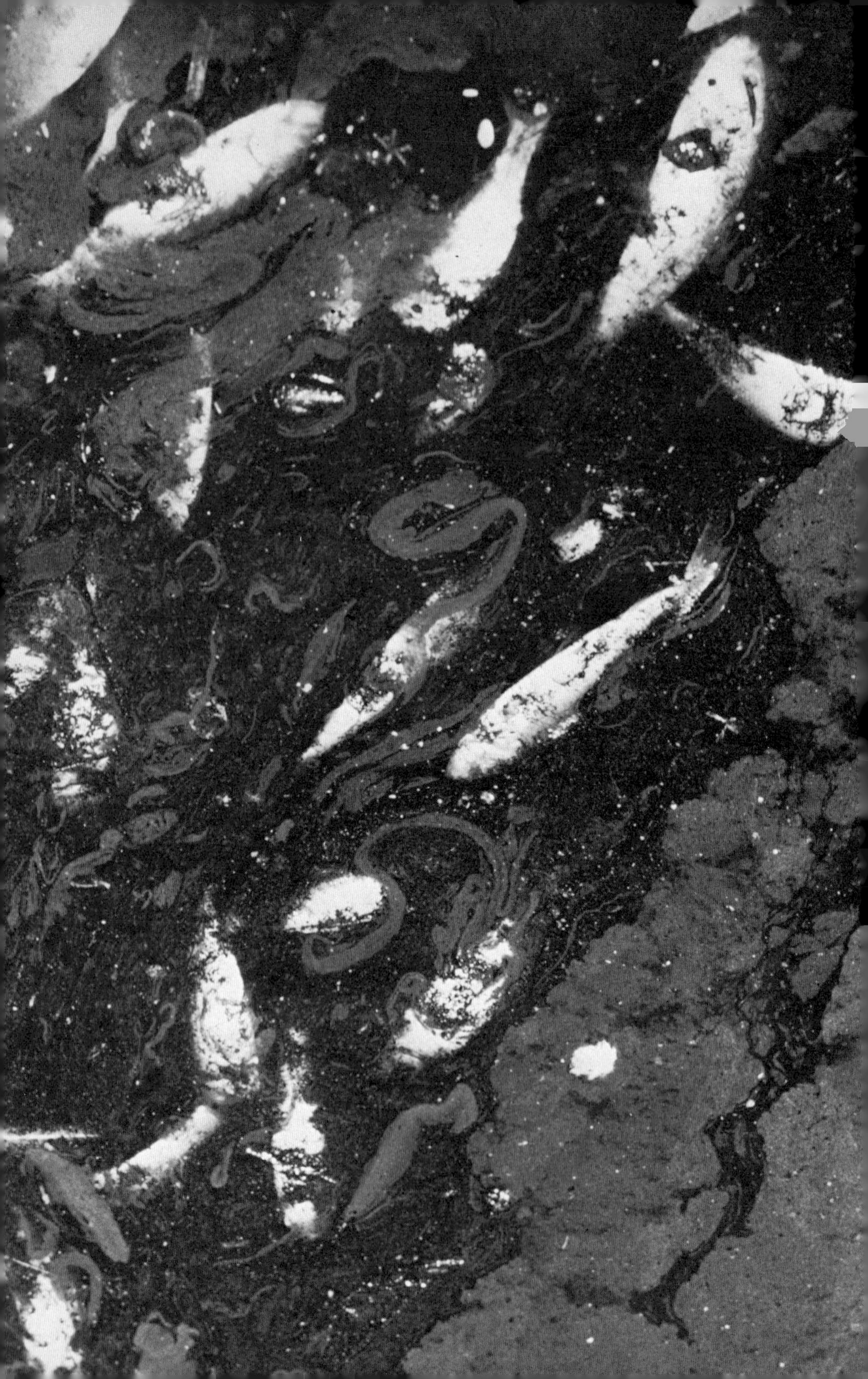

*O*ceanography is a field that is currently attracting much interest from students in many different areas of study. It is a multidisciplinary field encompassing disciplines such as geology, physics, chemistry, and biology. Public relations for the oceanic industry may have oversold the joys of ocean cruises to students or it may just be that this glamorous and exciting field is attracting the interest of many students. Some students even may want to know if the oceans will really be the cornucopia for our mineral and food resources.

The first two readings on chemical oceanography and marine sediments are not as recent as other articles in this book, and thus their inclusion does not strictly follow the main purpose of this book. They do, however, provide an encyclopedic overview of two major aspects of oceanography that might not be well-covered in most introductory earth science textbooks. The third article provides an exciting look at discoveries in two unique seas.

The birth of oceanography—the study of the oceans—is often considered to be the voyage of the HMS Challenger, which in 1873 began soundings and dredgings in the Atlantic. This voyage covered almost 70,000 miles in three years, and scientists recovered numerous specimens of sediments and strange creatures. Many of the biological specimens from this and later voyages were used to study the principles of biochemistry, physiology, and biological evolution.

Studies in oceanography prior to World War II included shore processes and, to some degree, the ocean atmosphere and sea floor. During the early forties, much interest developed and much research effort went into studying the physical aspects of the oceans for military purposes. After World War II came the application of many of these developments for scientific purposes through funding by the Office of Naval Research and the National Science Foundation. One major project of the past decade— Project Mohole—failed to reach its goal of drilling through the ocean floor and MOHO (Mohorovičić Discontinuity) into the mantle. Another project which has produced results of great benefit to many disciplines and has helped to provide data for investigations of the patterns of sea-floor spreading is the Deep Sea Drilling Project. This project was initiated in 1968 by an organization of leading oceanographic institutions known as Joint

Oceanographic Institutions for Deep-Earth Sampling, or JOIDES. Drilling and sampling is done from a specially adapted vessel, the Glomar Challenger, which is capable of drilling in as much as 20,000 feet of water. Like most oceanographic ventures, it is expensive—$24,000 per day—to operate. Though expensive, this project has advanced the techniques of oceanography which are now being used by industry to search for marine resources, in addition to providing many data of purely scientific nature.

Large industrial organizations are now moving into the field of oceanography, and new designs for large ocean structures capable of drilling for oil and gas in depths of over 300 meters, submersible vehicles for servicing wells and mining the sea floor, or deep-water drilling ships appear almost daily. In about two years the capability of well completions at depths of 1000 meters is expected. Exploitation of the oceans will accelerate as the need for resources increases, and numerous technological advances can be expected.

For some resources the oceans provide an inexhaustible reserve. One of the best examples is magnesium which occurs in seawater in concentrations of 1272 ppm or 10.9 pounds per 1000 gallons. The magnesium plant at Freeport, Texas, has been producing magnesium by the electrolytic process for over 30 years and has yet to produce magnesium equivalent to that found in a cubic mile of seawater.

Two problems centered on the oceans loom in the future. One problem, which first became apparent with the controversy over fishing rights and now includes rights to seabed and sub-sea resources, is the ownership of ocean resources. Canada is one nation that is pushing for development of international laws governing the allocation of marine resources and has issued oil and gas exploration permits for areas more than 400 miles off shore and in water over 3500 meters deep. Lesser developed countries and wealthy nations are equally interested in the outcome of planned world conferences to decide the ownership of the mineral resources of the oceans.

The other problem is the possibly serious pollution of the oceans and estuaries of the world. The cry that the oceans are rapidly becoming polluted has been heard for several years from such distinguished ocean travelers as Jacques Cousteau and Thor

Heyerdahl, and a scenario describing the death of the oceans by 1979 has been produced (Ehrlich, 1969). Although the oceans may be immense and can accept large quantities of wastes, some wastes being utilized as nutrients, they are not infinite, particularly with respect to disposal of some of the chemicals and plastics being produced by man. Recent surveys in the Atlantic Ocean have found 665,000 square miles over the American continental shelf and around the Caribbean that are littered with tar, oil, and chunks of plastic. Similar degradation has been observed in the Pacific Ocean 600 miles from land where, in an 8-hour period, 22 plastic fragments, six plastic bottles, four glass bottles, three pieces of paper, one coffee can, one finished piece of wood, a rubber sandal, and a shoebrush were seen!

6 Chemical Oceanography

Francis A. Richards

Francis A. Richards is a professor and assistant chairman for research in the Department of Oceanography at the University of Washington.

Introduction

(1) Definition and Role of Chemical Oceanography

(a) General. Oceanography can be defined as the scientific study of the seas, and chemical oceanography is the application of chemical techniques, laws and principles to oceanography. It is devoted to the accurate description of the chemical nature of the seas, to the elucidation of the processes that produce and alter the distribution of chemical species in the oceans (in both time and space), and to the consequences of the chemical nature of seawater on biological, geological, and physical processes in the marine environment. The chemical nature of seawater may control, limit, or help define these processes, which in turn may alter the chemical make-up of the environment.

In many respects, chemical oceanography serves to elucidate biological, physical, and geological processes in the ocean more than it serves to elucidate chemical processes, and it tends to be analytical and diagnostic. In general, knowledge of the chemistry of seawater is of more application to biological, geological, and physical problems than it is to problems of chemistry *per se.*

(b) Marine Chemical Engineering. The recovery of chemical products of value from seawater and the corrosive effects of

seawater on man-made objects are primarily chemical problems, but they have little to do with the study of the oceans themselves, and therefore are of more interest to chemical engineers than to chemical oceanographers. In general, such problems require a knowledge of the composition of seawater and its variability, but their solution is concerned only to a limited degree with the processes by which the properties of seawater were acquired or are altered.

(c) The Chemical Environment. Chemical oceanography is primarily an environmental science and is concerned with describing, explaining, and predicting the environment in which biological, geological, and physical events take place or have taken place, and the ways in which the environment may control or limit such events. The process of describing this environment is an analytical one, so the analysis of seawater and the development of adequate analytical methods have taken up a large fraction of the efforts of chemical oceanographers. Many effects of the chemical environment on the marine biological system have been reviewed by Harvey (1957).

Historical

(1) The Beginnings of Chemical Oceanography

Man's early interests in the chemical nature of seawater were no doubt highly practical ones, primarily concerned with the recovery of table salt. Robert Boyle (1627-1691), however, turned his attention to matters of scientific chemical oceanography as defined above. Boyle was interested in many aspects of oceanography, and T. G. Thompson has referred to him as the father of modern chemical oceanography. He devised the silver nitrate test for seawater and observed that, when silver nitrate was added to seawater, a heavy precipitate was formed but only a cloudiness formed with river or lake water, and he concluded that the sea got its saltiness from the leaching of the land.

(2) Early Analyses of Seawater

Goldberg (1961) has credited Torbern Bergman (1735-1784) and Antoine Lavoisier (1743-1794) with the first quantitative analyses of seawater, but these investigations shed little light on the chemical nature of the oceans as a whole. Lavoisier's interest in seawater was primarily because of its use as a medicinal.

(3) The Concept of Uniform Composition of Seawater

The general chemical nature of seawater began to be understood as a result of analyses made by Forchhammer and published in 1865. Even though major solutes were omitted from these analyses, they led Forchhammer to conclude that the composition of sea salts was always much the same, regardless of the source of the seawater. The uniformity of the composition of seawater was emphasized by M. F. Maury in his book "Physical Geography of the Sea," published in 1855, where he pointed out that the total salt content of seawater did not depart widely from 3.5% except in regions of excessive evaporation, such as the Red Sea, or excessive river runoff, such as the outfall of the Amazon River. He appears also to have realized that the composition of the sea salts was similarly uniform, and he concluded that the reason for the uniformity was that the oceans were well mixed—"well shaken together," in his words. The work of Dittmar, published in 1884, placed the concept of the uniformity of the ratios of the major constituents of seawater on a firm footing that has received substantial confirmation by many subsequent analyses. Real and significant variations in the ratios of the major constituents of seawater to each other do exist, especially in seawaters highly influenced by land runoff, such as the Baltic Sea, and probably in waters highly influenced by the freezing and thawing of sea ice, but the variations are generally small.

(4) Dissolved Gases

Attempts were made to determine the dissolved oxygen content of seawater as early as 1869, and these attempts were continued during the scientific cruise of the *Challenger* (1873-1876), but the analytical methods were inadequate until the Winkler titrimetric method, described in 1888, was introduced to oceanography in the first decade of the twentieth century. Since that time, a large body of data on the distribution of dissolved oxygen has been acquired, and dissolved oxygen has been more widely observed than any other constituent of seawater except the chlorinity. The subject of oxygen in the ocean was reviewed by Richards (1957a).

(5) Solubility of Dissolved Gases

C. J. J. Fox made careful determinations of the solubilities, published in 1909, of oxygen, carbon dioxide, and nitrogen (including the other inert gases, principally argon) in seawater equilibrated with air under varying conditions of chlorinity and temperature. The validity of Fox' data came into question in 1955, but more recent work suggests that his data were generally highly accurate, so far as O_2 solubility is concerned, although the values of these constants await definitive determination.

(6) Concept of Limiting Nutrients

Near the end of the nineteenth century, K. Brandt suggested that depletion of phosphorus and nitrogen compounds might limit plant production in the sea, but analytical difficulties prevented the satisfactory investigation of this idea until sensitive colorimetric methods were applied to the analysis of seawater in the early 1920s, primarily by W. R. G. Atkins of the Laboratory of the Marine Biological Association, Plymouth, England. These methods have permitted extensive studies of the nutrient distributions to be carried out in the intervening years and have led to our understanding of some of the biochemical circulations in the sea referred to below.

H. Wattenberg, chemist aboard the German research vessel *Meteor*, developed the concept of a modern shipboard chemical laboratory, and his observations of dissolved oxygen, nutrients, pH, and alkalinity during the investigations of the South Atlantic Ocean in 1925-1927 greatly extended the concepts of the biochemical circulation of nutrients and their importance in limiting primary production by phytoplankton organisms.

(7) Recent Advances in Radiochemistry and Isotope Distributions

Since World War II, there have been marked advances in the development of methods for the analysis of seawater and in the fields of radiochemistry and isotope distributions in the ocean. Although the radioactivity of seawater was first reported by R. J. Strutt in 1906, its exact determination was difficult and its distribution essentially unknown until the introduction of modern methods of radiochemistry. Y. Miyake successfully followed the distribution of artifically introduced radionuclides by United States bomb tests in the Pacific Ocean and used his observations to estimate rates of oceanic circulation. Artificial radioactivity provided the first tracer that man has been able to introduce into the ocean in sufficient quantity to permit its being followed over vast expanses. The natural fractionation and distribution of stable isotopes of several of the elements, particularly of dissolved oxygen and of hydrogen and oxygen combined as water, have been observed in recent years.

Processes Leading to and Altering the Distribution of Chemical Variables in the Oceans

(1) General

Solutes, solids and gases enter and leave the oceans across its boundaries between solid earth and the atmosphere. The composition of solutes introduced by rivers and streams differs markedly from those in seawater, being generally dominated by the alkaline earth metal cations and the anions of carbonic acid, while seawater is dominated by the ions of sodium chloride. Other solutes can be introduced directly with rainfall on the oceans; some nitric acid, formed during lightning discharges, is supposed to enter by this source. Volcanism, both submarine and subaerial, introduces other solutes. Dust, the particulate burdens of rivers and streams, volcanism and meteorites provide solid matter to the oceans. Bottom materials may be redissolved or resuspended through the agencies of wave action, biological activity, or changing physicochemical conditions.

(2) Gas Exchanges

The surface waters of the oceans are nearly in equilibrium with the atmosphere, i.e., they are saturated, or nearly so, with the air gases. There are net exchanges of dissolved gases across the sea surface in response to changes in temperature and changes in dissolved gas contents resulting from biological activity. The circulation of the oceans may bring waters into the surface layers that are oversaturated or undersaturated with respect to the atmosphere. Seawater saturated with air gases at high latitudes and low temperatures will give up some of its gas content to the air if brought into contact with the atmosphere at lower latitudes and higher temperatures, but subsurface waters introduced into the surface layers will generally be undersaturated with oxygen, because of its consumption during respiration and decomposition of organic matter.

(3) Changes in Chemical Species and Physical State

Within the body of the ocean, chemical species will be altered by inorganic and by biological-biochemical reactions. These reactions include the life processes, by which relatively simple inorganic substances are transformed into complicated living forms, but most of the inorganic materials are eventually returned to solution through the processes of excretion, respiration, and decomposition, although some of the materials combined into the bodies of plants and animals will fall to the ocean floor and be incorporated in the sediments.

(4) Aggregates and Precipitates

Inorganic and organic aggregates and precipitates can be formed from soluble or more finely divided materials. The formation of organic aggregates from soluble organic compounds at the surfaces of bubbles and other interfaces has recently been demonstrated. The formation of pelagic sediments is not well understood, but they apparently represent solids that were introduced to the sea as inorganic matter or by life processes and were subsequently altered by physicochemical exchanges with seawater solutes. The direct inorganic precipitation of insoluble salts is relatively rare in the ocean, but $CaCO_3$ is known to precipitate under conditions of high temperature and low carbon dioxide concentrations, for example on the Bahama Banks. The formation of insoluble sulfides in anoxic, sulfide-bearing waters may take place—it appears that copper is precipitated under these conditions and the Black Sea is probably saturated with ferrous sulfide. The ocean may be supersaturated with other inorganic salts, notably calcium carbonate.

(5) Body or *in situ* Processes

In the body of the oceans, chemical species can be altered by chemical or biochemical processes, and atoms and molecules can pass into and out of the liquid, solid, and gaseous (under limited circumstances near the surface) states. The distribution of solutes will be determined by the hydrodynamics of the system, but matter in the particulate form, which can be living or not, will also be subject to the swimming of organisms and to sinking (or rising, in the case of bubbles) under the influence of gravity. Colloid-sized particles will behave more like true solutes, but they may be gathered into aggregates or sorbates and precipitated.

(6) Conservative and Non-Conservative Solutes

There can be considered to be two major classes of solutes in the oceans, and their distributions accordingly differ widely. The first class, the conservative solutes, are those which are present in such abundance or are so chemically and biochemically inert that only minute fractions of their total amounts enter the particulate state. These elements make up over 99.9% of the solutes and are distributed as (and make up) the solutes measured as the *salinity*. Except at the boundaries of the ocean, changes in the concentration of these solutes in a unit volume are brought about by mixing or diffusion with adjacent units of water and by advection of water into the unit. The magnitude of the concentration changes will depend on the amount of mixing or diffusion and advection, and the concentration gradients. Dynamic or molecular diffusion is generally too slow to bring about major changes in the distribution of chemical variables, so eddy diffusion assumes the major role. Molecular diffusion coefficients for solutes have values of about $2 \times 10^{-5} \cdot g/cm^2$, depending somewhat on the nature of the solute. Eddy diffusion coefficients can be expected to be several orders of magnitude larger and to vary widely.

Most of the major constituents of seawater (Na^+, Mg^{++}, Ca^{++}, K^+, Sr^{++}, Cl^-, $SO_4^=$, Br^-, F^-, and H_3BO_3) are generally considered to be conservative solutes, although biological activity may make significant changes in the ratios of the alkaline earth metal ions to the chloride ion, and about 20% of the HCO_3^- ions can be involved in biological cycles. The ratios of boric acid and fluoride ions to the other major constituents may also vary considerably, and appreciable fractions of the sulfate ions may be reduced to sulfides under anoxic conditions such as occur in the Black Sea and other stagnant basins and fjords. Of the dissolved gases, nitrogen, argon, and the other rare gases are conservative, or nearly so, although there is some evidence that nitrogen fixation may occur fairly extensively in the surface layers of the tropics. Nitrogen has been shown to be produced in oxygen-free marine environments, but this process is quantitatively of little importance in considerations of the entire oceans.

The non-conservative solutes are generally present in low concentrations, and *major fractions* of their total amounts enter and leave the particular phase, where they are redistributed by swimming motions and under the influence of gravity in a way different from the more abundant, conservative major solutes. Although these constituents form a small fraction of the total matter dissolved in seawater, they include many that are essential to life processes (e.g., phosphorus, nitrogen, and manganese), others that are useful as tracers of water masses (dissolved oxygen and silicates are examples), and others that are of particular geochemical interest. The deposition of sediments may remove large fractions of some of these materials from the sea, but it is generally assumed that equal amounts are being added from extraoceanic sources so that an overall steady state exists.

(7) Biochemical Circulation

Large fractions of the non-conservative constituents are involved in biological cycles in the sea . . . . The biological aspects (Redfield, Ketchum, and Richards, 1963) of this redistribution involve the incorporation of inorganic ions into the bodies of plants and animals, ultimately by means of photosynthesis and therefore primarily in the upper levels that visible light penetrates. While in the particulate form, there will tend to be a net downward transport of large fractions of these materials which may later be either returned to solution by decomposition (remineralization) or incorporated into the sediments. Thus, there are large reservoirs of the nutrients, for example nitrogen and phosphorus compounds, at depths below the photosynthetic zone, and they can be returned to the photosynthetic zone only by the physical circulation of the water. This return is accomplished on a worldwide scale by the general circulation of the oceans, in regions of upwelling along certain coastal regions and along the equator, and by annual vertical mixing in the temperate zone and high latitudes. Thus, the biochemical circulation produces a net downward flux of non-conservative constituents in the particulate form that is essentially balanced, in the long run, by a net upward flux of these constituents in solution, as a result of water motions.

(8) Geochemical Circulations

Some constituents are introduced to the oceans from outside sources, such as the leaching of the land, volcanism, meteoritic bombardment, and the radioactive decay of elements in the atmosphere and lithosphere. Such materials can accumulate in the oceans, be removed to the atmosphere, or be incorporated in the sediments. The latter eventually may be incorporated in sedimentary rock which may, in geologic time, be returned to the land and again cycled through the weathering process. Geochemical estimates of the residence times of many of the elements in the sea have been made. These are based on the assumption that the sea is in a steady state, that is, that the rate at which a given element is being introduced to the oceans is balanced by the rate at which it is being deposited in the sediments or otherwise lost (such as by volatilization or as sea spray). Goldberg (1963) has listed a set of such residence times for most of the elements whose abundance in seawater has been estimated. They range from 100 years for aluminum to 2.6×10^8 years for sodium. The long residence times for sodium, Li (2.0×10^7 years), Mg (4.5×10^7), K (1.1×10^7), Ca (8.0×10^6), and Sr (1.9×10^7) in comparison with estimates of the age of the ocean (10^9 years) is associated with the lack of reactivity of these elements in the marine environment. Conversely, it is evident that such elements as aluminum and silicon (residence time 8.0×10^3 years) enter the oceans and are then deposited in the sediments relatively rapidly.

The Methods and Materials of Chemical Oceanography

(1) Seawater Analyses

The description of the distribution of chemical variables in the oceans depends on the accumulation of accurate analyses of seawater samples taken within appropriate networks in time and in horizontal and vertical space. The most frequent analyses have been made of (1) the salinity, estimated from the chlorinity, the electrical conductivity, or some other physicochemical property of the water, (2) the dissolved oxygen content of the water, (3) the nutrient ions in the water, principally phosphate, nitrite, nitrate, and silicate, and (4) the pH and alkalinity of the water. Other chemical properties of seawater have been observed less frequently or rarely, and the observations have often been confined to surface waters or carried out using methods of dubious reliability (Goldberg, 1961).

In 1932, Thompson and Robinson reviewed the then extant knowledge on the occurrence of the various elements in the sea and discussed the physical-chemical properties of seawater and their precise determination. This review also contains information on analytical methods. There are more recent reviews by Sverdrup, Johnson, and Fleming (1942), Richards (1957b), and Hood (1963), that contain many references to the distributions of various constituents of seawater, the oceanographic processes that contribute to the chemical environment, and the reaction chemistry of seawater.

(2) Discrete Sampling Programs

Oceanographic observations are usually designed to give a synopsis of the distribution of a variable within a definite area of the ocean, or to show the changes of a variable in a limited area as a function of time. In the former case, attempts have been made to describe the variable in horizontal or vertical space, requiring observations at various depths at a network of oceanographic stations. Changes in time have been described by making repeated observations (sometimes at a selection of depths) at a station or series of stations. Some of these time-dependent series have been carried out over many years, such as the long series of nutrient observations that have been made at specific locations near the laboratory of the Marine Biological Association at Plymouth, England. The Woods Hole Oceanographic Institution has made repetitive observations of chemical, physical, and biological variables along a section between Cape Cod, Massachusetts, and Bermuda. These series and similar ones permit the examination of temporal changes in the chemical properties of the waters and facilitate attempts to relate these changes to the factors causing them and to evaluate their effects on biological and other oceanographic events in the region. Other programs have been carried out in attempts to detect longer-term secular changes. One of the objectives of the oceanographic program of the International Geophysical Year (1957-1958) was to detect and evaluate secular changes in the properties of the South Atlantic Ocean by comparing the distribution of these properties observed during the IGY with those observed during the *Meteor* expedition in 1925-1927. However, uncertainties in the validity and comparability of the respective analyses are major problems in such comparative studies. This has proved particularly true of dissolved oxygen analyses, and it is difficult to judge whether apparent changes are due to real differences or to variations in the analytical methods used by the various observers.

(3) Continuous Sensing Devices

As chemical observations in the oceans have become closer, both in time and space, more and more detailed structure has been revealed in the chemical distributions. It has been possible to make continuously sensing devices to record chemical variations as a function of distance from a continuously moving vessel, or as a function of time at a fixed point. So far, these devices have been successfully used to record the temperature, electrical conductivity, and the dissolved oxygen content. In all cases, they have revealed details of the variations in these properties that would have been missed by more conventional spot sampling at discrete depths at an array of stations arranged in horizontal space.

(4) Analytical Problems

In most cases, chemical observations at sea still require the collection of water samples and their subsequent analysis either aboard ship or at a land-based laboratory. For many of the more usual chemical constituents, this is a relatively routine matter, but the observation of some of the more transitory or very dilute constituents becomes difficult and uncertain. Some of the constituents may be rapidly and radically altered by increased biological activity in the water after the sample is taken, requiring either immediate analysis under frequently difficult shipboard conditions or the careful preservation of the samples by methods that have proved to be effective. Many of the procedures for the analysis of seawater constituents are given by Barnes (1959). The observation of highly dilute constituents, such as many trace metals and organic compounds, requires rigid precautions to prevent contamination of the samples. An example is copper, for which no valid subsurface observations were made until sampling devices that prevented any contact of the sample with metal were devised. Only surface samples, taken directly in glass or other nonmetallic samplers, could be considered to be valid.

(5) Analytical Advances

Advances in analytical and radiochemistry have made possible the detection and determination of a variety of chemical constituents of seawater that could not readily be observed formerly. Many of the older methods of analysis, particularly for the trace metals, required the separation and concentration of the element from very large volumes, and these pretreatments frequently resulted in losses or contamination or both. Modern methods, depending on activation analysis, spectrometric methods, various forms of polarography, probes for the direct *in situ* observation of radionuclides, etc., avoid many of the former difficulties and should result in a more rapid increase in our knowledge of the distributions of some of these elements than was previously possible. It should be emphasized that investigations of the processes significant in controlling these distributions and the specific effects of these distributions has to await the accurate description of the variations in time and space.

References

Barnes, H. 1959. *Apparatus and methods of oceanography, part one: chemical.* London: George Allen and Unwin Ltd.
Goldberg, E. D. 1961. Chemistry in the oceans. In *Oceanography*, ed. M. Sears, pp. 583-97. Washington, D. C.: American Association for the Advancement of Science.
———. 1963. The oceans as a chemical system. In *The sea—ideas and observations on progress in the study of the seas,* vol. 2, ed. M. N. Hill, pp. 3-25. New York and London: Interscience Publishers.
Harvey, H. W. 1957. *The chemistry and fertility of sea waters.* 2nd ed. Cambridge: Cambridge University Press.
Hood, D. W. 1963. Chemical oceanography. *Oceanography and Marine Biology Annual Review* 1:129-55.
Redfield, A. C., Ketchum, B. H., and Richards, F. A. 1963. The influence of organisms on the composition of sea-water. In *The sea—ideas and observations on progress in the study of the seas,* vol. 2, ed. M. N. Hill, pp. 26-77. New York and London: Interscience Publishers.
Richards, F. A. 1957a. Oxygen in the ocean. In *Treatise on marine ecology and paleoecology*, vol. 1, ed. J. W. Hedgpeth, pp. 185-238. *Geological Society of America Memoir* 67 (1).
———. 1957b. Some current aspects of chemical oceanography. In *Physics and chemistry of the earth*, vol. 2, ch. 4, eds. L. H. Ahrens, F. Press, K. Rankama, and S. K. Runcorn, pp. 77-128. London: Pergamon Press.
Sverdrup, H. U., Johnson, M. W., and Fleming, R. H. 1942. *The oceans.* New York: Prentice-Hall.
Thompson, T. G. 1932. The physical properties of sea water. In *Physics of the earth. Bulletin of the National Research Council 85(5):63-94.*
Thompson, T. G., and Robinson, R. J. 1932. Chemistry of the sea. In *Physics of the earth. Bulletin of the National Research Council* 85(5):95-203.

7 Marine Sediments

Rhodes W. Fairbridge

Rhodes W. Fairbridge has been a professor of geology at Columbia University since 1955. He edits the series entitled Encyclopedia of Earth Sciences.

The first systematic classification of marine sediments was offered by Murray and Renard (1884, 1891) on the basis of their studies of the worldwide samples collected by the *Challenger* Expedition (1873-76). Many of these samples are still available for study at the British Museum. Two fundamental categories were recognized: *pelagic* (pertaining to the open ocean, i.e., fine-grained material capable of being carried in suspension of fine organic debris, that fell "as a gentle rain" to the floor of the deep sea) (Tables 1, 2) and *terrigenous* (material of all dimensions and compositions, derived directly from continental or island sources). The following subdivisions were made:
 (a) Pelagic
   (i) Inorganic: red clay;
   (ii) Organic (or Biogenic): calcareous oozes (globigerina, pteropod, coccolith oozes); siliceous oozes (radiolarian, diatom oozes).
 (b) Terrigenous
   (i) Blue, green and red muds;
   (ii) Volcanic mud;
   (iii) Coral sand and mud.

Strictly speaking, Murray and Renard employed this mixed classification only for deep-sea deposits, and not for neritic and shallow oceanic environments. Unfortunately some of their successors

TABLE 1. OCEANIC AREAS COVERED BY PELAGIC SEDIMENTS*

Type of Sediment	Atlantic Ocean Area (10^6 km^2)	Pacific Ocean Area (10^6 km^2)	Indian Ocean Area (10^6 km^2)	Total Area (10^6 km^2)
Calcareous oozes				
Globigerina	40.1	51.9	34.4	
Pteropod	1.5			
Total	41.6	51.9	34.4	127.9
Siliceous oozes				
Diatom	4.1	14.4	12.6	
Radiolarian		6.6	0.3	
Total	4.1	21.0	12.9	38.0
Red clay	15.9	70.3	16.0	102.2
Grand total	61.6	143.2	63.3	268.1

TABLE 2. DISTRIBUTION IN PERCENTAGES OF AREA COVERED BY PELAGIC SEDIMENTS*

	Indian Ocean (%)	Pacific Ocean (%)	Atlantic Ocean (%)
Calcareous ooze	54.3	36.2	67.5
Siliceous ooze	20.4	14.7	6.7
Red clay	25.3	49.1	25.8
Total	100.0	100.0	100.0

	Calcareous Ooze (%)	Siliceous Ooze (%)	Red Clay (%)
Indian Ocean	26.9	33.9	15.7
Pacific Ocean	40.6	55.3	68.7
Atlantic Ocean	32.5	10.8	15.6
Total	100.0	100.0	100.0

*H. U. Sverdrup, Martin W. Johnson, and Richard H. Fleming, *The Oceans, Their Physics, Chemistry, and General Biology,* © 1942, Renewed 1970. Reprinted by permission of Prentice-Hall, Inc. Englewood Cliffs, New Jersey.

assumed that it was an all-embracing system for marine sediment classification. Geikie (1903) expanded it so that the terrigenous group would take in: (A) shore deposits, and (B) infra-littoral or deeper water (shelf) deposits.

Revelle (1944) found that oxidation-state and color changes in any case rendered the terrigenous group difficult to handle and suggested that, like the pelagic, it should simply be divided into two main types: *(i)* organic muds (i.e., reef derivatives, etc.) and *(ii)* inorganic muds. Striving further toward completeness, Shepard (1948) added a *Glacial* (glacigene) *class*, and restricted the terms terrigenous and pelagic to allow red clay and the volcanic sediments to stand by themselves: *(i)* red clay, *(ii)* pelagic oozes, *(iii)* terrigenous muds, *(iv)* glacial marine sediments, and *(v)* volcanic sediments.

A different approach was made by Kuenen (1950), who grouped marine sediments according to location (i.e., environment) in primary division. Thus, he considered *shelf sediments and mixed environments* (littoral, delta, lagoon and estuarine) as well as the *deep-sea group,* sometimes called *thalassic* or *thalatogenic*, in which, beside the usual *pelagic* class, he recognized a class of *"hemipelagic and terrigenous"* that included the terrigenous muds, volcanic, coral and carbonate muds. Considerable virtue is seen in the environmental approach, which is especially useful in consideration of facies. However, it might be improved by the recognition of special pigeon holes for reef environments, marine volcanic associations and the marine glacial sediments.

Genetic-Geochemical Approach—The Sedimentary Cycle

In 1942, Sverdrup et al. offered a wholly genetic approach to marine sediments, in place of the partly descriptive, partly genetic approach of most of the above schemes. In 1964, Goldberg came back to the problem, viewing the whole sedimentary cycle as a closed (balanced) geochemical system, which, with minor modifications, can be summarized as follows:

(1) Source Materials

(a) Meteorites (and other extraterrestrial matter);
(b) Volcanic and continental weathering products.

(2) Transport and Phase

Through the air or by water (deflation, rivers, littoral erosion, glaciers, etc.), these products are carried into the world ocean, as

(a) *Dissolved phases* (i.e., the salts, etc. of seawater);
(b) *Particulate phases* (in suspension, turbidity flow or traction).

(3) Marine Sediment Components

(a) *Pore solutions* (connate water; that may become systematically modified and concentrated during diagenesis);

(b) *Hydrogenous components* (authigenic minerals, "halmyrolytic" or "halmeic" as Arrhenius calls them), i.e., inorganic precipitates or metasomatic products, derived entirely from seawater solution, or partially, by ion exchange or other process. Examples include ferromanganese minerals, glauconite, phosphorite, barite, phillipsite, montmorillonite and hydromagnesite.

(c) *Biogenous components* (all types of benthonic skeletal debris, coral and calcareous algal material, "biohermal" and "biostromal" assemblages, both in biocoenotic and thanatocoenotic associations. These may be massive [i.e., relatively undamaged], or detrital [wave-worked and sorted]. The benthonic material is normally calcareous, but sometimes siliceous, as with sponge spicule deposits [spongolite], etc. Also the pelagic oozes— *calcareous* [globigerina, coccolith, pteropod ooze]; *siliceous* [radiolaria, diatom]).

(d) *Lithogenous components* (clastic debris of all grain sizes [including flocculates of extremely fine-grained colloids] derived from preexisting rocks of all sorts, siliceous, carbonate, ferruginous, etc. Included are quartz, feldspars, micas, hornblende [and other "heavy minerals"], laterite [largely as grains or hematite], clay particles [of many species], and the volcanic glasses [those unaltered by seawater, i.e., those which do not fall in class (b)]. Arrhenius [1963] believes that there are two genetic components in the deep-sea *red clay*—*chthonic* [i.e., lithogenous], and *halmeic* [i.e., hydrogenous], the former dominant in the northern hemisphere and the latter in the southern).

(e) *Cosmogenous components* (all sorts of meteoritic material, tektites, cosmic dust, magnetic spherules; these have been found in ancient sediments back to Cambrian in age, and although not bulking large, they are universally present).

Criteria for Classification

Goldberg (1964) has offered the following observation about distinguishing criteria: chemical composition of the mineral; size distribution (see Figure 1); isotopic analyses of an element or elements; geographic occurrence; association with other minerals; and the form or habit. For any specific case, a combination of several of the cited approaches is usually necessary. As an example, the oolitic calcium carbonates, deposited inorganically in coastal waters, such as on the Great Bahama Bank, can be distinguished from organically deposited calcium carbonates by means of differences in the minor element abundances (Tatsumoto and Goldberg, 1959), the isotopic compositions of the carbon and oxygen (Lowenstam and Epstein, 1957), geographic occurrences, and mineral habits and forms in the two groups of minerals (Newell, Purdy, and Imbrie, 1960). Feldspars in East Pacific Rise deposits are characterized as volcanic in origin on the basis of their size distributions as a function of distance from sources, their chemical compositions, and associations with other minerals (Peterson and Goldberg, 1962).

Although this genetic classification of sedimentary components has proved especially useful, certain ambiguities do arise. Griffin and Goldberg (1963) have indicated that in the North Pacific, ordered illites and chlorites are forming from a lithogenous montmorillonite-like clay mineral. To typify the partially converted starting substances as either lithogenous or hydrogenous is clearly inappropriate. Volcanic debris in various states of alter-

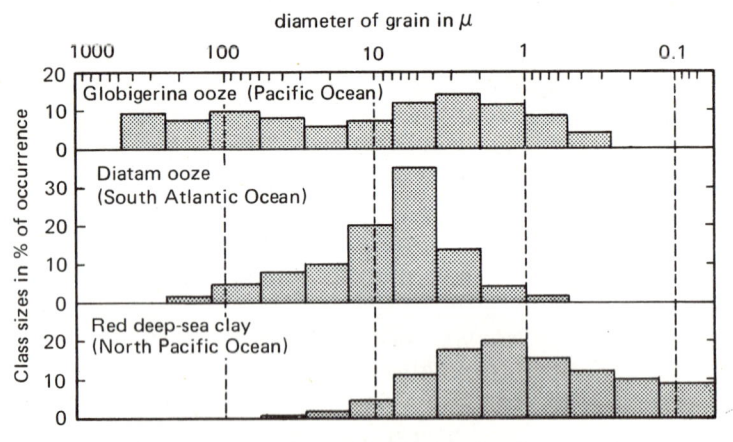

Figure 1. Average distribution of grain sizes in the most important pelagic bottom sediments (H. U. Sverdrup, Martin W. Johnson, and Richard H. Fleming, *The Oceans, Their Physics, Chemistry, and General Biology.* © 1942, Renewed 1970. Reprinted by permission of Prentice-Hall, Inc. Englewood Cliffs, New Jersey.)

ation is equally difficult to categorize in a clear-cut way. However the number of minerals not subject to a rigid classification is relatively small.

Transportation

(1) River Systems

Kuenen calculated that the total annual particulate and dissolved matter delivered by the world's rivers amounted to 12 km^3/yr. Livingston (1963) gives, for example, the figure of 5700 g/sec of dissolved minerals in the discharge of the Amazon. The location of *submarine fans or cones* opposite river deltas clearly correlates the latter with fluvial transport; the size of the Ganges cone is particularly impressive with 5 x 10^6 km^3, representing an annual transport (by eastern Himalayan rivers) of something like 0.3 km^3/yr.

(2) Ocean Current Systems

Dissolved phases are rapidly diffused and thus difficult to trace definitely, although certainly some more localized components can be identified, e.g., silica solutions. Biogenous particulate material is largely related to the patterns of the principal current systems and the zones of high organic productivity (related to *planktonic photosynthesis* and *upwelling*; see also Ryther, 1963). Arrhenius (1963) in particular has demonstrated the coincidence of high calcareous ooze accumulation along the belt of the intertropical convergence right across the central Pacific. Areas of low productivity and little accumulation occur in the anticyclonic gyres like the Sargasso Sea (low nutrient supply) or in the ice-covered Arctic (low nutrients, low temperature and limited light penetration). Heezen (personal communication) has pointed out that the great geostrophic currents of the world are frequently matched by deep countercurrents; at their boundary zones (sites of low energy) along the continental slope and rise, suspended sediment must settle to the bottom, leading to the curious phenomenon of thick sediments accumulating on slopes (and not necessarily, strangely enough, in trough-shaped "geosynclines").

(3) Turbidity Current Systems

Associated with major river sources, most submarine canyons seem to be the sources of gravitationally activated turbidity flows, the velocities of which are one or two orders of magnitude above the rates in category (b). Accordingly, lithogenous components of all sizes are transported down to the abyssal plains. One may note that the mean direction of the initial turbidity flow may well be at 90° to the mean geostrophic current direction; the latter current is semicontinuous, while the turbidity flow is aperiodic, perhaps only once a decade or once a century.

(4) Wind Transportation

Charles Darwin first pointed out the probability of widespread eolian transport of desert materials to the ocean, especially in an area such as that in the eastern Atlantic, off the Sahara Desert, where the hot harmattan winds blow dusts up to 1000 miles out. Radczewski (1937) studying the *Meteor* samples found hematite-coated desert quartz grains over a broad area here; in the 5-10μ range this desert material represented up to 22% of the whole sample and in the 10-50μ range up to 39%. Goldberg (1964) suggested that fine fractions of both quartz and illite, found far out in the Pacific, away from obvious ocean current sources, could well be explained by the patterns of the upper atmospheric jet streams. Volcanic eruptions shower ash and dust over broad areas, generally limited to about 500 km in radius (Kuenen and Neeb, 1943), but may reach further downwind. The finest dust goes up into the stratosphere and achieves almost worldwide distribution as shown also by the radioactive tracers from atomic bomb testing.

(5) Glacial Transportation

Ice floe and iceberg rafting are the most important transport agencies in all high latitudes; today the usual distribution range is to 55° or, in regions of strong equatorward-setting currents, as far as 45°N and S. During Pleistocene maxima these limits were generally extended to 35°, and locally to about 25°. Because of the large size of glacial boulders and the thin layer of postglacial sediment in many areas of the world ocean today, oceanographic rock trawls bring up the Pleistocene boulders in many regions where ice does not reach today (Lisitzin, 1960).

(6) Biologic Transportation

Floating vegetation is a major agency of transport, from the size of pebbles and gravel adhering to the roots of trees, carried seaward by river floods, to all sort of terrestrial organisms, from vertebrates down to gastropods, which are carried thus into highly unusual environments for fossilization (thanatocoenoses).

Sedimentation Rates

In regions of very high bottom scour such as Formosa Strait, parts of the English Channel and the Straits of Dover, and parts of Florida Straits, there is no accumulation and the chart is marked "hardground" or "rocks." In other areas near delta fronts, the sedimentation rate may exceed 1mm/yr. In the deep sea, the rates vary from the turbidite accumulations (*abyssal plains*) where rates may *average* 0.5 mm/yr, to regions of exclusively red clay with a very slow accumulation rate, below the "calcium carbonate compensation depth," and to regions of the carbonate oozes, at less than 4500-5000 m, where rates of sedimentation depend largely on primary productivity. Mean pelagic rates are in the range 1-10 mm per 1000 years (Menard, 1964). In iceberg melting areas, e.g., off the Newfoundland Banks, a very high dump rate is maintained over a relatively small area. The first picture of sedimentation in depth in the deep ocean has come from the first experimental Mohole boring off Guadalupe Island in the eastern Pacific (Murata and Erd, 1964).

References

Arrhenius, G. 1963. Pelagic sediments. In *The sea,* vol. 3, pp. 655-727. New York: Interscience Publishers.
Darwin, C. 1846. An account of the fine dust which falls on vessels in the Atlantic Ocean. *Quarterly Journal of the Geological Society of London* 2:26.
Fairbridge, R. W. 1955. Warm marine carbonate environments and dolomitization. *Geological Society Digest* 23:39-48.
―――. 1964. The importance of limestone and its Ca/Mg content to paleoclimatology. In *Problems in paleoclimatology,* pp. 431-77, 521-30. London: Interscience Publishers.
Geikie, A. 1903. *Text book of geology.* 4th ed. London: Macmillan.
Goldberg, E. D. 1964. The oceans as a geological system. *Trans. N. Y. Acad. Sci. Ser.* 11,27:7-19.
Goldberg, E. D., and Griffin, J. J. 1964. Sedimentation rates and mineralogy in the South Atlantic. *Journal of Geophysical Research* 69:4293-4309.
Griffin, J. J., and Goldberg, E. D. 1963. Clay mineral distribution in the Pacific Ocean. In *The sea*, vol. 3, pp. 728-41. New York: Interscience Publishers.
Kuenen, P. H. 1950. *Marine geology.* New York: John Wiley & Sons.
Kuenen, P. H., and Neeb, G. A. 1943. *The Snellius expedition (bottom samples),* vol. 5, part 3. Leiden: E. J. Brill.

Lisitzin, A. P. 1960. Bottom sediments of the Eastern Antarctic and the Southern Indian Ocean. *Deep-Sea Research* 7:89-99.

Livingston, D. A. 1963. *Chemical composition of rivers and lakes*. U. S. Geological Survey Professional Papers 440-G: 1-64.

Lowenstam, H. A., and Epstein, S. 1957. On the origin of sedimentary aragonite needles of the Great Bahama Bank. *Journal of Geology* 65:364-75.

Menard, H. W. 1964. *Marine geology of the Pacific*. New York: McGraw-Hill.

Murata, K. J., and Erd, R. C. 1964. Composition of sediments in cores from the experimental Mohole Project (Guadalupe site). *Journal Sediment. Petrol.* 34:633-55.

Murray, J., and Renard, A. F. 1884. On the nomenclature, origin and distribution of deep-sea deposits. *Proc. Roy. Soc. Edinburgh* 12:495-529.

———. 1891. Deep-sea deposits. *Report of the H. M. S.* Challenger *1873-1876*.

Newell, N. D., Purdy, E. G., and Imbrie, J. 1960. Bahamian oolite sand. *Journal of Geology* 68:481-97.

Peterson, M. N. A., and Goldberg, E. D. 1962. Feldspar distributions in South Pacific pelagic sediments. *Journal of Geophysical Research* 67:3477-92.

Radczewski, O. E. 1937. Eolian deposits in marine sediments. In *Recent marine sediments*, ed. P. Trask, pp. 496-502. American Association of Petroleum Geologists.

Revelle, R. R. 1944. Marine bottom samples collected in the Pacific Ocean by the *Carnegie* on its seventh cruise. *Carnegie Institute Washington Publication* 556:part 1.

Ryther, J. H. 1963. Geographical variations in productivity. In *The sea*, vol. 2, pp. 347-80. New York: Interscience Publishers.

Sverdrup, H. U., Johnson, M. W., and Fleming, R. H. 1942. *The oceans: their physics, chemistry, and general biology*. New York: Prentice-Hall.

Tatsumoto, M., and Goldberg, E. D. 1959. Some aspects of the marine geochemistry of uranium. *Geochim. Cosmochim. Acta.* 17:201-8.

8   The Red and the Black Seas

David A. Ross

David A. Ross is an associate scientist in the Geology and Geophysics Department of the Woods Hole Oceanographic Institution, Woods Hole, Massachusetts.

Because of the uniqueness and mystery associated with their names and geographical locations, the Red and the Black seas have always aroused man's curiosity. The Red Sea, an important Egyptian highway of trade as early as 2000 B.C., received its name because of a blue-green alga called *Trichodesmium erythraeum*. This alga can sometimes undergo rapid growth, or bloom, and because of an accessory red pigment it contains, it colors the water red after its death. The Black Sea is thought to have received its name from the Turks, who feared this occasionally stormy body of water. They called it Karadeniz, which was their word for black. To others, who were perhaps better sailors, it was called Pontus Euxinus, or the hospitable sea.

These two seas also have a fascination for present-day oceanographers for several reasons. One is their intermediate geological position between the continents and the ocean; another is the response they have had to the recent changes in sea level associated with the Pleistocene glaciation. For these and other reasons, the two seas have been the subject of recent expeditions by the Woods Hole Oceanographic Institution.

In many aspects, the geology of the Red and the Black seas differs, although some interesting similarities exist. Probably the main difference between the two is their age. The Black Sea is geologically moderately old, having been formed over

100,000,000 years ago. Since that time it has been receiving sediments from the surrounding areas, resulting in an accumulation that may be as thick as 16 km in the central part of the basin. The crustal structure of the Black Sea is different from that of the typical structure of continents or oceans (Figure 1). It has a crustal thickness similar to that of continental areas, but the seismic velocity of the deeper material is similar to that of basalt, the typical volcanic rock of the ocean. These two facts have caused considerable controversy over the origin of the Black Sea—is it an oceanic area being converted into a continental area or vice versa?

The Red Sea, on the other hand, is geologically relatively young, its central portion being only a few million years old. This sea was formed as a result of the spreading or rifting apart of the Arabian peninsula from the African continent. Its crustal structure is similar to that of oceanic areas (Figure 1). The sediment

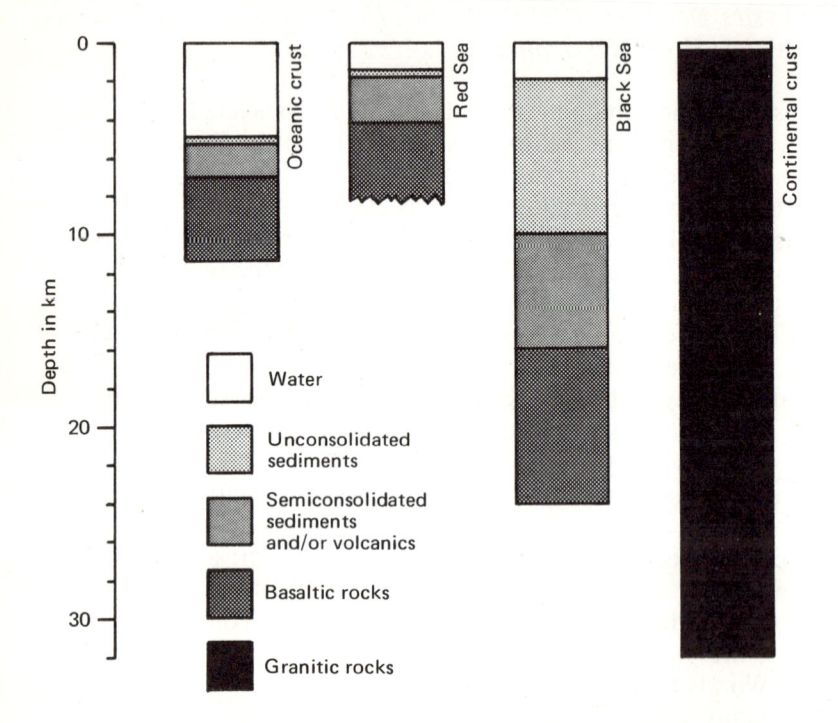

Figure 1. Typical crustal sections obtained by seismic refraction techniques. The different rock types are determined on the basis of their observed compressional sound velocity. (Adapted from data in Girdler [15] and Neprochnov [16].)

accumulation in the Red Sea has been small for two reasons, the relative youth of the area and the small supply of sediment from land being carried into the Red Sea. Most of the sediment in the Red Sea comes either from chemical precipitation from seawater or from biological deposition (shells of dead organisms).

The difference in geological structure of the upper few kilometers can be seen from seismic reflection profiles from the areas (Figure 2). A seismic profile is obtained by discharging energy in the form of compressed air or an electrical spark into the water and recording the energy reflected back to the ship from the surface and subsurface layers. In the Black Sea the upper part of the sedimentary record shows up as a horizontal sequence of uniform layers. This sequence attests to the relative tectonic inactivity in the central part of the Black Sea during the last several million years. In the Red Sea, however, the sedimentary sequence, where observed, is usually moderately deformed. In the

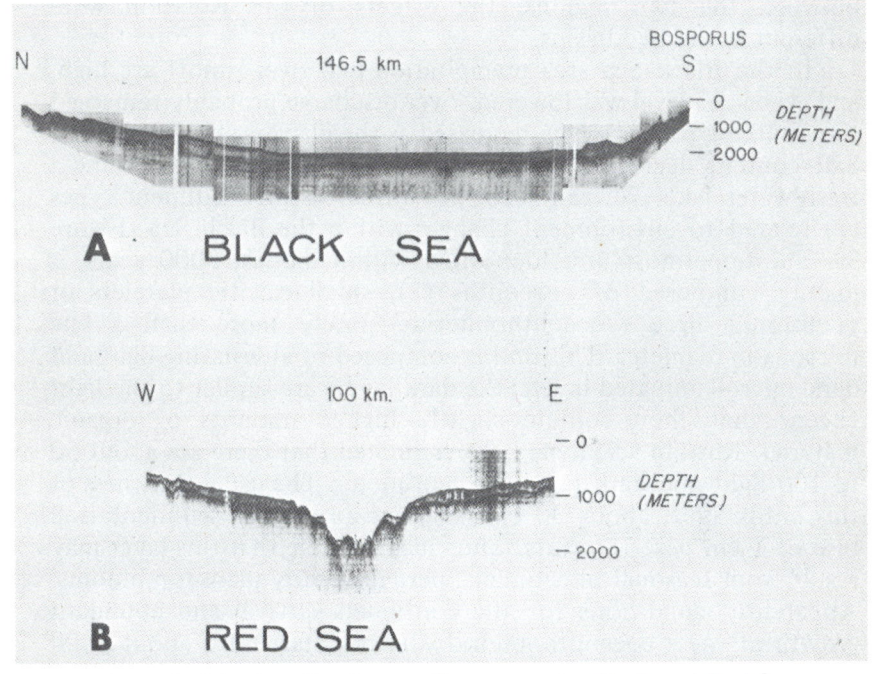

Figure 2. Seismic reflection profiles across the Black and Red Seas. The Black Sea profile starts in the north about 72 miles east of the Romanian-Bulgarian border and trends essentially south to 20 miles off the Bosporus. The Red Sea profile runs approximately west to east across the sea at about 19°N.

central region little or no sediment was observed, indicating the recent origin of this rift feature.

One important similarity between these two seas is that they are both connected to the ocean by relatively shallow and narrow channels. The Black Sea is connected to the Mediterranean by the Bosporus, which, at maximum, is about 50 meters deep. The Red Sea, at its southern end, is connected to the Indian Ocean by the Straits of Bab el Mandeb, which are about 100 meters deep. These narrow channels restrict the exchange of seawater between the seas and the main ocean. Any effect of this restricted circulation was even greater in the past, when sea level was lowered by the Pleistocene glaciation (Figure 4). We know from many studies that sea level was as low as -130 meters about 15,000 years ago and since that time has risen relatively rapidly until about 7000 years ago. Subsequently it has risen more gradually. Thus for many years the Red and Black seas were essentially isolated from their sources of salt water. Partly because of the climate difference between the two regions, the effects of this isolation were different in the two basins.

In the Black Sea area precipitation and river runoff are high, and when sea level was lowered, river discharge probably remained the same, or perhaps even increased as the glaciers started to melt. Salt content decreased in the Black Sea until it almost became a fresh-water lake. We can recognize three distinct sediment types (1) related to environment changes within the Black Sea (Figure 3). The uppermost unit, deposited within the last 3000 years, is mainly composed of coccoliths (2), small calcitic platelets of planktonic algae (coccolithophorids) rarely more than a few microns in diameter. This unit is composed of alternating light and dark micro-laminated layers; the dark layers are similar to the light except that they contain slightly higher amounts of organic material. Russian scientists have estimated that there are about 50 to 100 light and dark pairs per centimeter. The usual thickness of this entire unit, about 30 cm, suggests an average sedimentation rate of 1 cm per 100 years. Thus it is possible that this layer may result from seasonal phenomena, perhaps yearly plankton blooms. Apparently conditions for the continued survival and abundant growth of these coccolithophorids did not start until about 3000 years ago, although they may have made some earlier attempts at populating the Black Sea. The record of these prior attempts is preserved in the middle unit as individual, thin (less than 1 mm) coccolith layers.

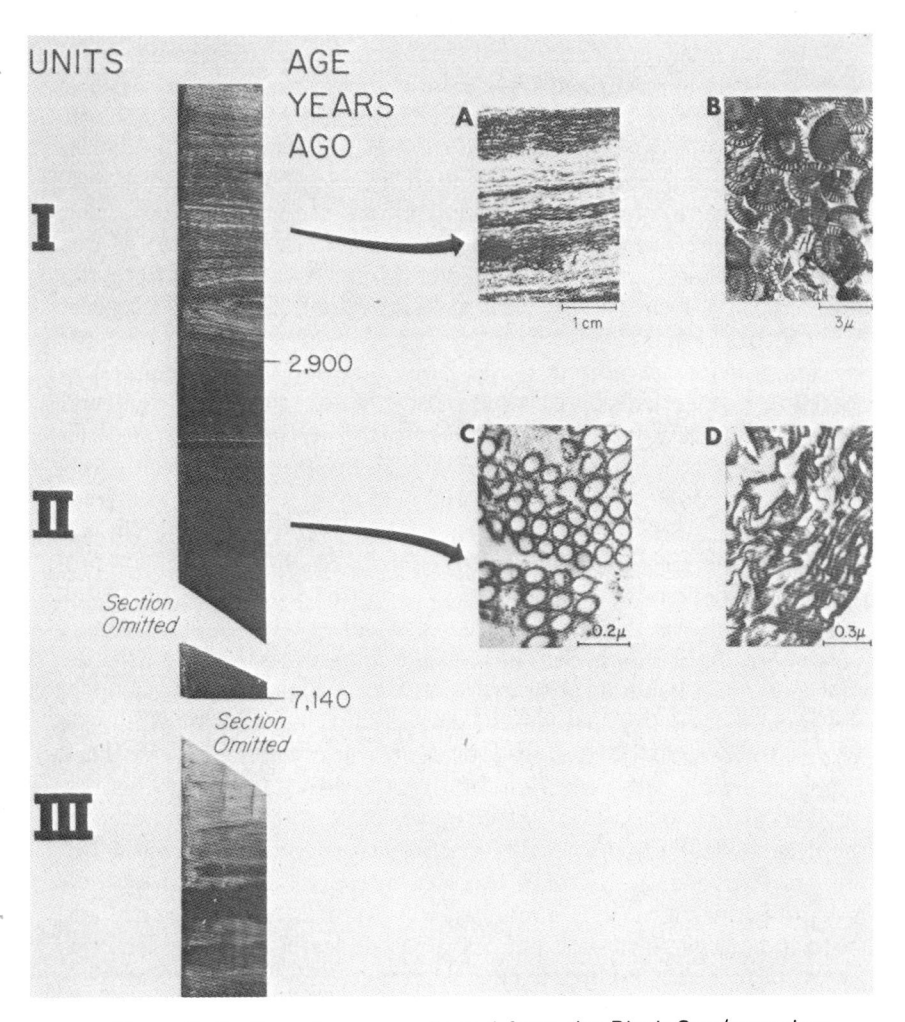

Figure 3. Portion of a core collected from the Black Sea (note that parts have been left out). The three distinct units are indicated respectively by the numbers 1, 2, 3, for the upper coccolith unit, the middle organic-rich unit, and the lower light and dark mud unit. The ages are based on carbon-14 dating determined on correlative layers in other cores. Insert A shows the alternating light and dark micro-laminated layers; both layers are mainly composed of the coccolith *Emiliania huxleyi*, which is shown enlarged in insert B (18). Insert C shows some large tubular membranes having a diameter of about 700 to 800 Å. Insert D shows branched tubular membranes aggregated together and bound by a limiting membrane (3).

The middle unit, deposited between about 3000 and 7000 years ago, is characterized by a relatively high content of organic matter (between 10 to 20% organic carbon). Electron microscope studies (3) of this material show numerous structures similar to biological membranes, although their exact identity is unknown.

The lowermost unit is a monotonous sequence of light and dark mud. The base of this unit was not penetrated by any of our cores, the longest of which was over 11 m. A carbon-14 date at a depth of 1,120 cm of this core gave an age of 22,830 ± 800 years B.P. (before present).

These three sediment types from bottom to top apparently correlate respectively with the period when the Black Sea was essentially a fresh-water lake (lowermost unit), with the transition from fresh to salt water (middle unit), and with present conditions (uppermost unit). The mixing rate of the Black Sea is relatively slow (about 2000 years). Thus it would take some time for the effects of the reestablishment of the Black Sea-Mediterranean connection to be felt.

This close dependence of sediment types with the hydrologic history of the Black Sea has also been confirmed by isotope studies (4). In general, with evaporation, and thus higher salinity, the heavier forms of an element will remain in the water while the lighter forms will be removed by the evaporation process. Thus fresh waters are relatively richer in light isotopes than salty waters.

The effect of sea-level lowering in the Red Sea as compared to the Black Sea was quite different because of the climatic difference between the two. In the Red Sea, evaporation far exceeds precipitation (rainfall is generally less than 2 cm per year), and there is no river inflow (5). Loss of fresh water from the Red Sea by evaporation accounts for its relatively high salinity—40 parts per 1000 vs. about 35 for most of the oceans. Estimates of evaporation range between about 180 to 215 cm per year over the entire sea (6,7); this water must be replaced by inflow from the Indian Ocean through the Straits of Bab el Mandeb. Thus, when the sea level was lowered even slightly, the amount of inflow was restricted and the Red Sea became even saltier. Inflow can also be influenced by tectonic movement of the Straits of Bab el Mandeb, or of the Red Sea itself; however, the likelihood of this having happened during the last 20,000 years is probably small.

Much of the sediment of the Red Sea is of biological origin— mainly shells of planktonic foraminifera. The percentages of the various species systematically fluctuate with depth in the sediment, and the fluctuations are related to changes in water salinity and temperature, which in turn are caused by changes in sea level (8). Isotope data likewise show the systematic changes resulting from periods of evaporation (9)—these data indicate that there were about four periods of high evaporation during the last 70,000 years (see Figure 5). The most recent was about 13,500 years ago, and it fits nicely with the generally accepted sea-level curve (see Figure 4). The other periods of high evaporation correlate closely with those detected by similar studies from the Mediterranean and other areas.

Our main reason for studying the Red Sea was to learn more about the hot brine region situated in the central rift valley of the sea at about 21°N latitude. Three small deeps containing hot (maximum temperature 59°C) salty (maximum salinity 256 parts per 1000; not salinity in the usual sense of the word as used by oceanographers, since the ratios of the major ions are considerably different from those of normal seawater) waters are known. These

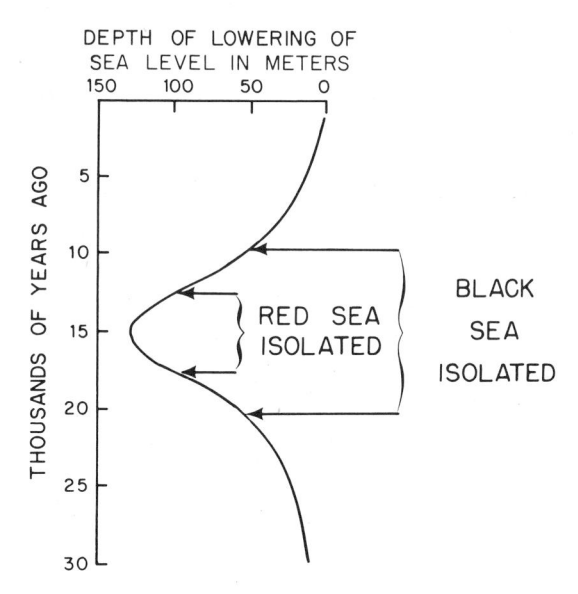

Figure 4. Recent sea-level curve and times of isolation of the Red and Black Seas from the Indian Ocean and Mediterranean Sea respectively. The sea-level curve is drawn on the basis of material collected from the continental shelf whose original depth of formation is known and whose age can be dated by carbon-14. (From Milliman and Emery [17].)

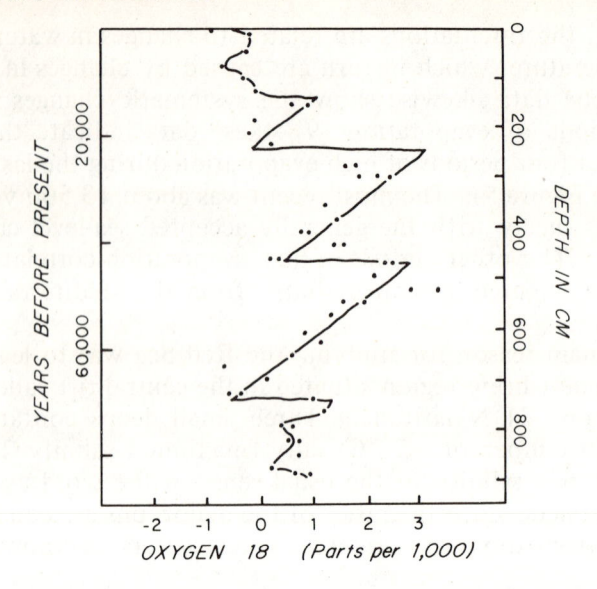

Figure 5. Changes in the oxygen-18 content of fossil foraminiferas found at various depths in the bottom sediments of the Red Sea. The oxygen-18 values shown represent deviation with respect to an ocean-water standard in parts per 1000 (the standard carbonate = 0). The four major peaks of increasing oxygen-18 indicate periods of increasing salinity (probably due to lowered sea level). These four peaks are followed by abrupt declines, perhaps related to an increase in the inflow of Indian Ocean water into the Red Sea (data from Deuser and Degens [9]).

deeps are apparently related to the recent rifting apart of the Red Sea. Underlying the brines we found sediments rich in many heavy metals (Table 1), as were the overlying waters (10, 11). One sedimentary unit contained as much as 11% zinc oxide, 4-5% copper oxide, and 0.01% silver oxide. The *in-situ* mineral value of the top 10 meters of sediment is conservatively estimated to be over 2 billon dollars (12), but one must exercise a little caution when considering this number; it is somewhat like considering the value of the elements in a cubic mile of seawater. Clearly the cost of recovering the elements both from the sea floor and, later, metallurgically must be evaluated. Another difficulty is establishing the ownership of these deposits; they lie about equidistant between Saudi Arabia and Sudan at a depth of 2000 meters. Economic potential aside, these deposits are of great interest

TABLE 1. METAL CONTENT IN SOME OF THE SEDIMENTARY UNITS UNDERLYING THE RED BRINE (12)

Sedimentary unit	Fe_2O_3	Mn_3O_4	ZnO	CuO	PbO	AgO
Goethite	49	2.8	1.0	0.1	0.1	0.0033
Iron-montmorillonite	35	20	4.7	0.5	0.1	0.0062
Sulfides	33	1.3	11.1	4.6	0.2	0.013

Content in Percent

because here we may be seeing the beginning of a hydrothermal ore deposit, whereas previously we have only been able to study ones that were laid down millions of years ago. We returned to the Red Sea in early 1971 to continue our study of these deposits, in particular to determine their thickness and to see if any other brine areas exist; preliminary results indicate that old or fossil brine deposits are present.

The Black Sea likewise has an interesting economic potential. Our seismic profiling records show many structures, mainly on the slope, that could be favorable for oil accumulation. Another possible deposit is related to a peculiar oceanographic characteristic of the Black Sea—an anoxic environment. The uppermost 100 to 200 meters of water are oxygenated, while the deeper waters have no oxygen and are reducing. Water samples from the transition zone between the waters indicate that manganese is being precipitated as particles at this interface (13). When manganese precipitates in the oceans, other ions such as cobalt and nickel tend to precipitate also; this may be happening here. Conceivably, where this transition zone intersects the shelf or slope of the Black Sea, the particles should form a bottom deposit, somewhat like a bath tub ring, around the entire basin.

Unfortunately, because of diplomatic difficulties in getting permission to work in nearshore waters we have been unable to explore this interesting hypothesis. I did mention the idea to some Russian scientists whom I met when we visited a Russian port, and they acknowledged that some manganese encrustations have been found. Upon checking the literature, it appears that the first Russian Black Sea expedition (14) found manganese concretions around *Modiolus* shells from depths where the transition zone intersects the bottom. The results of our seismic reflection survey interested our Russian colleagues. Within a few weeks they had a research vessel in the Black Sea making their own survey.

In our studies of the Red and Black seas we were fortunate to have the cooperation and participation of scientists from many countries, including those of neighboring countries. By this participation we were able to able to exchange knowledge about the geology and other aspects of the surrounding region as well as about the oceans.

References

1. Ross, D. A., Degens, E. T., and MacIlvaine, J. 1970. Black Sea: recent sedimentary history. *Science* 170:163-65.
2. Burky, D., Kling, S. A., Horn, M. K., and Manheim, F. T. 1970. Geological significance of coccoliths in fine-grained carbonate bands of post-glacial Black Sea sediments. *Nature* 226:156-58.
3. Degens, E. T., Watons, S. W., and Remsen, C. C. 1970. Fossil membranes in cell wall fragments from a 7,000-year-old Black Sea sediment. *Science* 168:1207-8.
4. Deuser, Werner. 1970 Personal Communication.
5. Seidler, G. 1969. General circulation of water masses in the Red Sea. In *Hot brines and recent heavy metal deposits of the Red Sea*, ed. E. T. Degens, and D. A. Ross, pp. 131-37. New York: Springer Verlag.
6. Privett, D. W. 1959. Monthly charts of evaporation from the Indian Ocean including the Red Sea and the Persian Gulf. *Quarterly Journal of the Royal Meteorological Society* 85:424.
7. Neumann, J. 1952. Evaporation from the Red Sea. *Israel Explor. Journal* 2:153.
8. Berggren, W. A., and Boersma, A. 1969. Late Pleistocene and Holocene planktonic foraminifera in the Red Sea. In *Hot brines*, ed. Degens and Ross, pp. 282-98.
9. Deuser, W. G., and Degens, E. T. 1969. O^{18}/O^{16} and C^{13}/C^{12} ratios of fossils from the hot brine deep area of the central Red Sea. In *Hot brines*, ed. Degens and Ross, pp. 336-47.
10. Brewer, P. G., and Spencer, D. W. 1969. A note on the chemical composition of the Red Sea brines . In *Hot brines*, ed. Degens and Ross, pp. 174-89.
11. Bischoff, J. L. 1969. Red Sea geothermal brine deposits: their mineralogy, chemistry, and genesis. In *Hot brines*, ed. Degens and Ross, pp. 368-401.
12. Bischoff, J. L., and Manheim, F. T. 1969. Economic potential of the Red Sea heavy metal deposits. In *Hot brines*, ed. Degens and Ross, pp. 535-41.
13. Brewer, P. G., Spencer, D. W., and Sachs, P. L. 1970. Trace metals in the Black Sea. *Oceanus* 4:23-25.
14. Andrussov, N. I. 1890. Predvarital nyi otchet ob uchastii v Chernomorskoi glubokomernoi ekspeditsii. *Isvest. Russk. Geografich. Obshch.* 26:398-409.

15. Girdler, R. W. 1969. The Red Sea—a geophysical background. In *Hot brines*, ed. Degens and Ross, pp. 38-70.
16. Neprochnov, Yu. T. 1968. Structure of the earth's crust of epi-continental seas: Caspian, Black and Mediterranean. *Can. J. Earth Sci.* 5:1037-43.
17. Milliman, J. D., and Emery, K. O. 1968. Sea levels during the past 35,000 years. *Science 162:1121-23*.
18. Degens, E. T., and Ross, D. A. 1970. Oceanographic expedition in the Black Sea. *Die Naturwissenschaften* 7: 349-51.

Additional Readings

Brooks, D. B. 1969. Ocean mining: political opportunities and economic consequences. *Technology Review* 71(9):22-29.

Carson, R. L. 1961. *The sea around us.* New York: Oxford University Press.

Costlow, J. D., Jr. 1971. *Fertility of the sea.* 2 vols. New York: Gordon and Breach.

Davis, R. A. 1972. *Principles of oceanography.* Reading, Mass.: Addison-Wesley.

Fairbridge, R. W., ed. 1966. *The encyclopedia of oceanography.* Encyclopedia of the Earth Sciences Series, vol. 1. New York: Reinhold.

Gross, M. G. 1971. *Oceanography.* 2nd ed. Columbus, Ohio: Charles E. Merrill.

Heezen, B. C., Tharp, M., and Ewing, M. 1959. *The floors of the oceans, I: the North Atlantic.* Geological Society of America Special Paper 65.

Horsfield, B., and Stone, P. B. 1972. *The great ocean business.* New York: Cowhard, McCann, and Geoghegan.

Idyll, C. P. 1973. The anchovy crisis. *Scientific American* 228(6):22-29.

Inman, D. L., and Brush, B. M. 1973. The coastal challenge. *Science* 181:20-32.

Keen, M. J. 1968. *An introduction to marine geology.* Oxford, England: Pergamon.

Manheim, F. T. 1972. *Mineral resources off the northeastern coast of the United States.* U. S. Geological Survey Circular 669.

Marx, W. 1967. *The frail ocean.* New York: Ballantine Books.

Matthews, W. H., Smith, F. E., and Goldberg, E. D., eds. 1971. *Man's impact on terrestrial and oceanic ecosystems.* Cambridge, Mass.: MIT Press.

McKelvey, V. E., Tracey, J. I., Stoertz, G. E., and Vedder, J. G. 1969. *Subsea mineral resources and problems related to their development.* U. S. Geological Survey Circular 619.

McKelvey, V. E., and Wang, F. H. 1970. World subsea mineral resources, preliminary maps. U. S. Geological Survey Map I-632.

Miller, R. L., ed. 1964. *Papers in marine geology: Shepard commemorative volume.* New York: Macmillan.

Moore, J. R., ed. 1971. *Oceanography: readings from Scientific American.* San Francisco: W. H. Freeman.

National Science Foundation, National Ocean Sediment Coring Program. 1969. *Initial reports of the deep-sea drilling project.* Washington, D. C.: U. S. Government Printing Office.
Scientific American. 1969. *The ocean.* San Francisco: W. H. Freeman.
Turekian, K. K. 1968. *Oceans.* Engelwood Cliffs, N. J.: Prentice-Hall.
Weyl, P. K. 1970. *Oceanography: an introduction to the marine environment.* New York: Wiley.

Part Three
Plate Tectonics

It is useful to be assured that the heavings of the earth are not the work of angry deities. These phenomena have causes of their own.

Seneca (4 B.C.-A.D. 65)

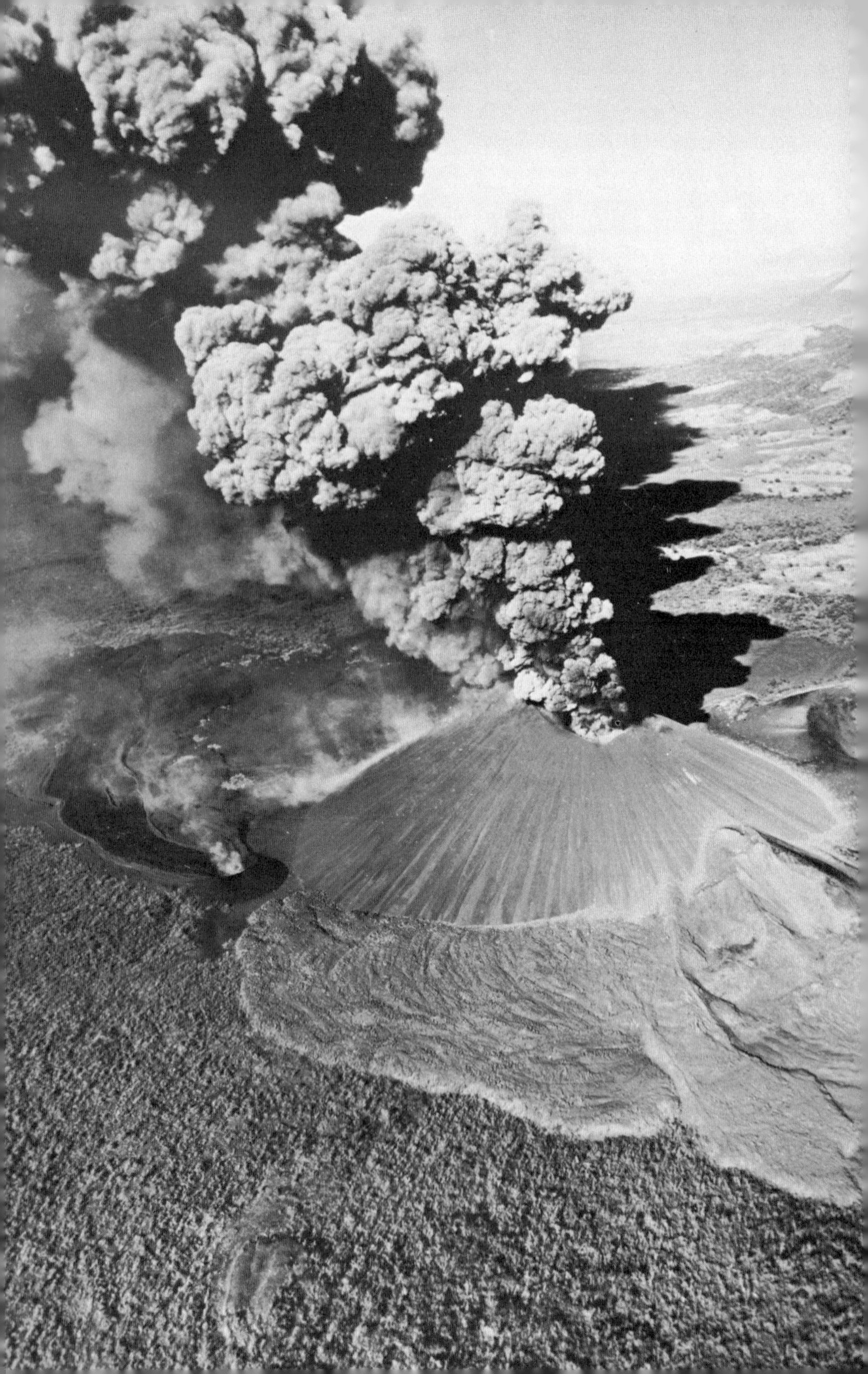

*T*he idea that we live on a dynamic earth with drifting continents, spreading ocean floors, and shifting crustal plates has become a much-publicized and accepted concept during the past decade. In this section we have included selections by two authors who have made significant contributions to the concept of sea-floor spreading: Robert Dietz (who coined the phrase sea-floor spreading following discussions with H. H. Hess who developed the concept) and F. J. Vine (who, with D. H. Matthews, related spreading to magnetic reversals and paleomagnetic imprints in the rocks on the sea floor). Interested students are urged to consult references listed in the suggested readings for this section.

The concept of drifting continents, now one aspect of plate tectonics which has caused a revolution in the earth sciences, can readily be traced back as far as Alfred Wegener, the German geographer who noticed that continents on either side of the Atlantic Ocean fitted very nicely together as if they had at one time in the past been one continent. In fact, he reconstructed all of the continents into a supercontinent that he called Pangaea, surrounded by a sea known as Panthalassa. Although the idea of drifting continents was not greeted by the scientific community of the day with open arms, and indeed a few people today regard it at best as an hypothesis with many shortcomings (Meyerhoff 1970, 1972), the idea persisted and was further developed by Du Toit (1937). Although it might be said that the disdain for the idea of moving continents is an example of the unscientific or closeminded attitude of a scientific community dominated by a few concepts and influential gentlemen, it was probably the result of insufficient data on which to judge the new concept. Geoscientists were trying to interpret the history of the earth through study of only 30 percent of the area. The oceans remained a blank.

Through the efforts of many scientists in the last fifteen years, the data needed to develop and support the plate tectonic story have been provided not only by seismologists and paleomagneticians, who no doubt made major contributions that converted nonbelievers to "drifters," but also paleontologists, geomorphologists, petrologists, and geochemists. A few scientists still remained sceptical of the geophysical evidence, but most were silenced when the fossil evidence—tetrapods discovered by members of Antarctic expeditions specifically planned to search *for such fossils—arrived. Plate tectonics, or the "new global tectonics" as it is sometimes called, has not explained everything in geology although there are some who appear to suggest this. It has sent many scientists scurrying back to the field or field notes to reinterpret findings of years past. This infusion of the geoscience community with new life has produced a new wave of research which has brought geology into the public eye.*

The latest application of the idea of mobile continents has come from archaeologists who have found that the Pyramid of Cheops is skewed a few minutes of arc from true north. It is suspected that this pyramid was originally aligned very close to true north and, in the motion of the continents since construction, has achieved its new position. More direct observations of the current rate of spreading are being tried using laser measurements from satellites to two continental plates and also long-range surface surveying instruments to measure the rate of spreading of Iceland. Measurements made in the past year will be repeated after a certain period and hopefully will be sufficiently different to determine within the limits of measurement the rate of spreading of those areas.

It may be that the application of the concept of plate tectonics will find its greatest usefulness in delineating areas for exploration of fossil fuels and metallic mineral deposits. From that standpoint, environmentalists would say that the maturation of the new global tectonics was none too soon.

9 Sea-Floor Spreading— New Evidence

F. J. Vine

Frederick J. Vine is a reader in the School of Environmental Sciences at the University of East Anglia, Norwich, England.

The concept of sea-floor spreading was first formulated in some detail by H. H. Hess of Princeton University in 1962; however, only since 1966 with the advent of new and rather compelling evidence, has it attracted widespread attention. In light of these new data it seems appropriate to try to reformulate the hypothesis and, starting with the current seismicity of the earth, develop it in terms of geologic time.

Seismicity of the Earth

Earthquakes are not randomly distributed over the surface of the earth but are largely restricted to the young fold mountains and trench systems of the Alpine-Himalayan and circum-Pacific belts, and to the crests of the mid-ocean ridges (Figure 1). By far the greatest number of earthquakes occur in the circum-Pacific belt in association with the trench systems. Earthquakes are less frequent in the Alpine-Himalayan belt and less frequent again on the mid-ocean ridge crests. The recent more accurate epicentral locations determined by the Environmental Sciences Services Administration—with data from the World Wide Standardized Seismograph Network—define these narrow belts in remarkable detail. Thus most of the current seismicity of the earth occurs in very restricted linear zones. This is where the action is.

The activity associated with ridge crests and strike-slip faults is confined to very shallow depths, probably not exceeding 10 or 20 km (Isacks et al., 1968). In the young fold mountains and trench systems, however, shallow focus earthquakes are present but in many of these settings intermediate and deep focus earthquakes, occurring to a maximum depth of 700 km, are also recorded (see Figure 1). According to the hypothesis of sea-floor spreading, these two seismic provinces—those in which only shallow earthquakes occur and those which are characterized by deeper focus earthquakes—reflect very different but complementary processes at work in the upper mantle and being accommodated in the earth's crust. The mid-ocean ridge crests are extensional features along which new oceanic crust is created, and the trench systems are regions in which oceanic crust is partly resorbed. The oceanic crust is thus considered to be a surface expression of the mantle,

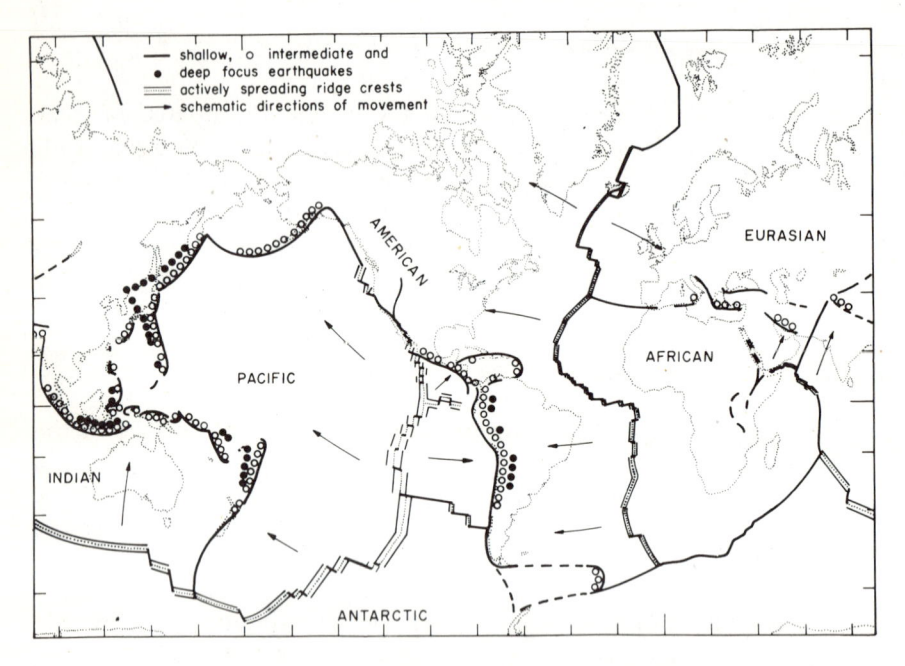

Figure 1. Summary of the seismicity of the earth (Gutenberg and Richter, 1954) and hence the extent of crustal plates bounded by active ridge crests, faults trench systems and zones of compression. The six major crustal blocks assumed by Le Pichon (1968) are named. Spreading rates at ridge crests are indicated schematically and vary from 1 cm per year in the vicinity of Iceland to 6 cm per year in the equatorial Pacific Ocean (Heirtzler et al., 1968).

derived from it by partial fusion and chemical modification beneath ridge crests, and in part returned to the mantle beneath the trench systems. Recent earthquake mechanism solutions have confirmed this picture of extension of the earth's crust at ridge crests in terms of normal faulting (Sykes, 1967), and compression and underthrusting of the crust landward of the trench systems (Stauder, 1968).

Movement of Crustal Plates

Within this basic framework of ridges and trenches—sources and sinks—several complications are produced by the distribution of continental crust. The simplest of these possibilities are illustrated in Figure 2. If an upwelling in the mantle is initiated beneath a continent, the continent will be rifted and passively drifted apart with the formation of a new ocean basin by lateral spreading from the original rift. The continent, coupled to a rigid conveyor belt of uppermantle material, drifts either with a trench system at its leading edge (Figures 2A and B), an island arc system ahead of it (Figure 2B), or until it encounters a trench. In the last case, because of the continent's lower density, it is unable to sink and overrides the trench, forming a new trench system (Figure 2A). Such an ocean basin, with a median ridge and surrounded by the recently undeformed trailing edges of drifting continents, is reminiscent of large parts of the Atlantic and Indian Oceans. Alternatively, if rifting is initiated within a former oceanic area, new oceanic crust is formed and older crust resorbed in marginal trenches (Figure 2B). With time the trench systems will encroach on the oceanic area. There is no reason why a ridge of this type should be median within the oceanic area. Such a picture is analogous to the situation in much of the Pacific at the present time. A third possibility, clearly, is that two continents may ultimately come together at a trench system, producing high mountain ranges such as the Alps and Himalayas which result from the collision of Africa and India with Eurasia (Figure 2C). Thus the distribution of the continents accounts for the marginal trench systems and island arcs of the Pacific and the median position of the Atlantic and Indian Ocean ridges.

The geometry of spreading on the real earth is a complex combination of these simple possibilities but it is, nonetheless, amenable to rigorous analysis in terms of the relative movement of rigid crustal plates bounded by the zones of current earthquake activity (Morgan, 1968). These plates or aseismic areas are out-

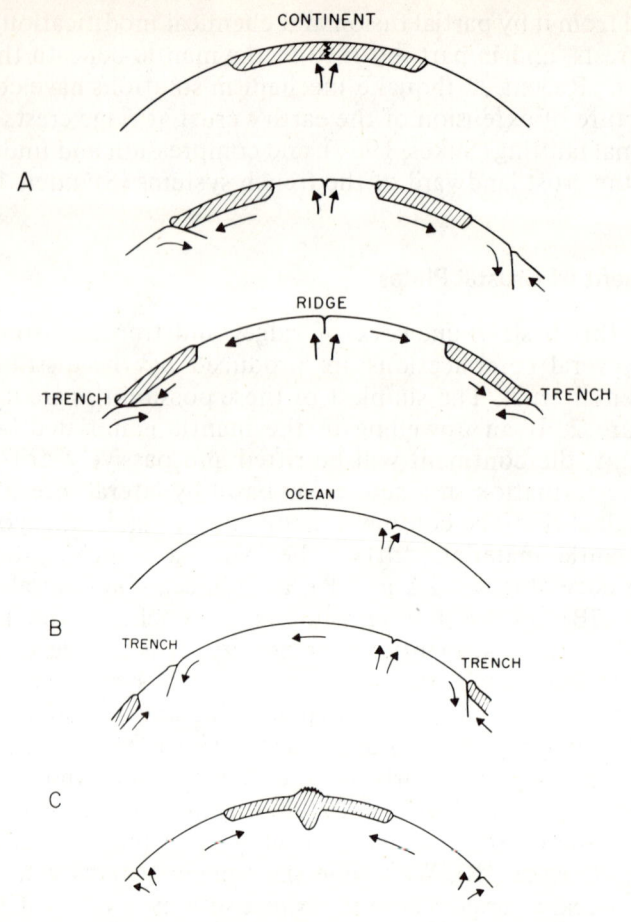

Figure 2. Simple geometric possibilities within the framework of sea-floor spreading, emphasizing the role of the continents. A and B illustrate the stages and possible configurations resulting from the initiation of rifting and spreading beneath continental and oceanic crust, respectively. C illustrates the possibility of two continental blocks coming together over a downcurrent.

lined, therefore, by the seismic belts and spreading ridge crests shown in Figure 1.

Transform Faults

Turning now to specific new evidence for spreading, one might first consider the distribution of earthquake epicenters and the nature of faulting on mid-ocean ridge crests. Greatly improved epicentral determinations for these areas, available only in the last few years, reveal that the seismicity is concentrated on the transverse fracture zones between the offset points of the ridge crest. Mechanism solutions obtained for a number of these earthquakes indicate a strike-slip movement on a vertical plane paralleling the trend of the fracture. The sense of movement on this plane is the opposite to that suggested by a simple offsetting of the ridge crest (Sykes, 1967). These points are illustrated diagrammatically in Figure 3A. Classically, the left-lateral offset of the ridge crest in Figure 3A would be interpreted as resulting from left-lateral movement along a transcurrent fault between two crustal plates as shown in Figure 3B. However, the implied sense of motion and distribution of earthquake activity along such a fault does not satisfy that which is observed. The sense of movement and distribution of epicenters summarized in Figure 3A is, however, compatible with the ridge-ridge type of transform fault formulated by Wilson (1965) within the framework of sea-floor spreading. This is illustrated in Figure 3C. An important corollary of this concept is that the offset of a ridge crest along a transform fault may well be an initial one in that the offset does not change with time as spreading occurs. In contrast, the offset of the ridge crest across a transcurrent fault would increase as renewed faulting occurs, as shown in Figure 3D. The faults and ridge crest in the equatorial Atlantic (Figure 1) parallel the continental margins of

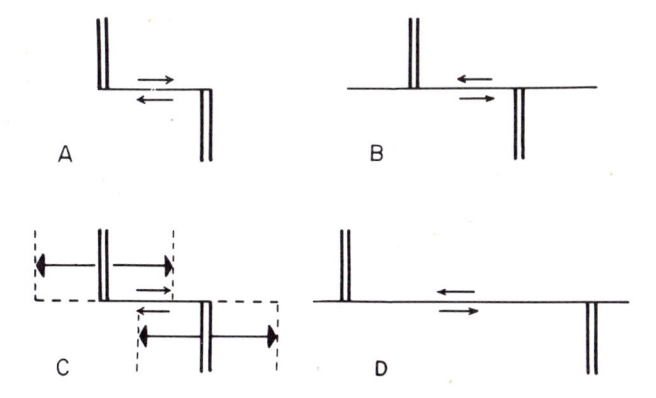

Figure 3. A right-lateral ridge-ridge type transform fault (A and C) contrasted with a left-lateral transcurrent fault (B and D). The ridge crest is indicated by the parallel bars, and the active trace of the faults by the thin solid line. (After Wilson, 1965.)

Africa and South America and may well represent the locus of initial rifting in this area. It was this point that led to the prediction of transform faults by Wilson (1965) prior to their verification by detailed epicentral determinations and focal mechanism solutions by Sykes (1967).

The Magnetic Record

Further evidence for spreading of the sea floor at ridge crests comes from studies of disturbances in the earth's magnetic field recorded at or above sea level. These anomalies are typically one or two percent of the total intensity in amplitude, and range from several kilometers to several tens of kilometers in wavelength. They are attributed to magnetization contrasts essentially within the earth's crust but potentially extending into the upper mantle beneath the ocean basins (i.e. to the depth of the Curie temperature isotherm—600-700°C—for the ferromagnetic rock forming minerals).

In 1963, Vine and Matthews suggested that if sea-floor spreading has occurred, it might be recorded by the remanent magnetization of the oceanic crust, especially within the basalt layer, because basalts have the highest intensities of remanent magnetization. It was postulated that if the earth's magnetic field reverses intermittently as spreading occurs, the resulting magnetization contrasts between normally and reversely magnetized material in the basalt layer should produce appreciable and predictable disturbances in the earth's magnetic field as measured at sea level. Such a crustal model for a ridge crest is shown schematically in Figure 4, and is discussed in some detail by Vine (1968).

At the time this suggestion was made the time scale for reversals was unknown (Cox, Doell and Dalrymple, 1963); it was still debatable as to whether the earth's magnetic field had reversed at all. However, since 1963, the reversal timescale for the last 3.5 million years has become increasingly well defined by paleomagnetic and radiogenic age measurements on young lava flows from many parts of the world (Cox, Dalrymple, and Doell, 1967), and has received striking confirmation from the paleomagnetic study of deep-sea sediment cores (Opdyke et al., 1966). This time-scale has been incorporated into Figure 4 by assuming a spreading rate of 1 cm per year per ridge flank.

We can now, therefore, test the Vine and Matthews hypoth-

RIDGE MODEL

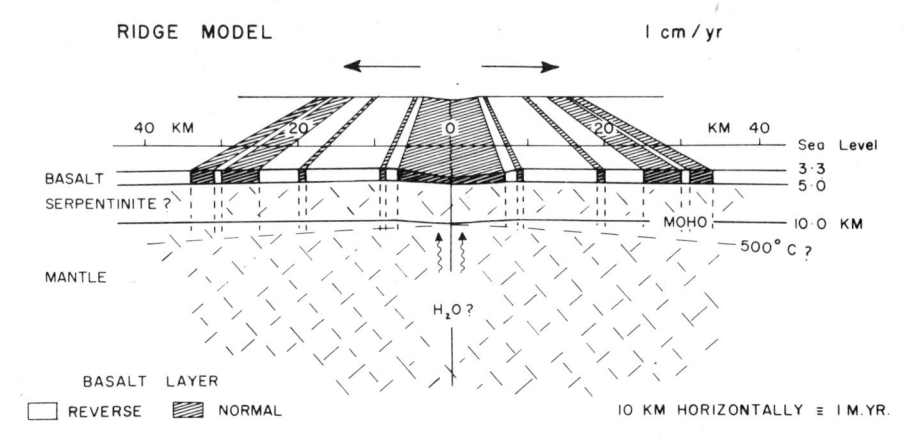

Figure 4. Diagrammatic representation of the oceanic crust at a mid-ocean ridge crest, assuming active spreading at a rate of 1 cm per year per ridge flank, and the geomagnetic reversal time scale of Cox et al. (1967).

esis. The central curve in Figure 5 shows the anomalies in the earth's magnetic field along a profile perpendicular to the crest of the East Pacific Rise at 51°S in the Pacific Ocean. The lower curve shows the predicted anomalies which result from assuming the above model and a rate of spreading of 4.4 cm per year per flank. The upper curve in Figure 5 is simply the central profile plotted in reverse about its mid-point to emphasize the symmetry of the observed anomalies about the ridge crest. The crustal model is perfectly symmetrical about the ridge axis; the asymmetry of the computed profile is due to the dipolar nature of the earth's magnetic field.

Rates and Relative Movements

This interpretation of the magnetic anomalies coupled with the paleomagnetic and dating results obtained from subaerial lava flows, enables one to deduce recent rates of spreading at ridge crests for which magnetic data are available. Rates deduced in this way vary from 1 cm per year per flank near Iceland, to 6 cm per year per flank in the equatorial Pacific (Vine, 1966; Heirtzler et al., 1968). These rates are summarized in Figure 1, and imply rates of separation of crustal blocks, i.e. rates of drift, varying from 2 to 12 cm per year.

EAST PACIFIC RISE   51° S

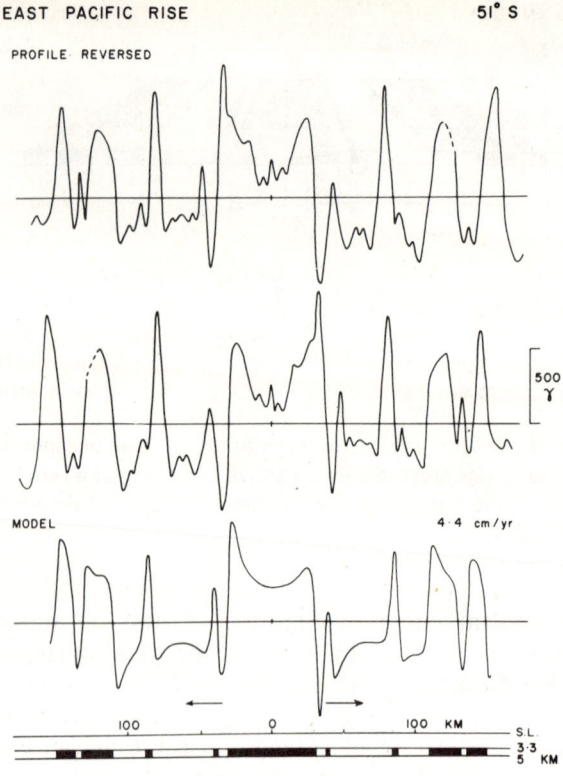

PROFILE REVERSED

MODEL   4.4 cm/yr

Figure 5. An observed magnetic anomaly profile across the East Pacific Rise (upper curves) compared with a computed profile for this area, assuming the model of Figure 4 and a spreading rate of 4.4 cm per year. The computation assumes an intensity and dip for the earth's magnetic field of 48,700 gamma and -63°, and a magnetic bearing for the profile of 102° (1 gamma=10^{-5} oersted). Normal or reverse magnetization is with respect to an axial dipole vector. Effective susceptibility assumed =±0.01, except for central block (+0.02). S.L.=seal level. (Profile: Pitman and Heirtzler, 1966.)

It is now possible, therefore, to assign rates of relative movement at certain of the boundaries between the crustal plates outlined in Figure 1. Having assumed these rates and a simple configuration of just six crustal plates, Le Pichon (1968) has deduced rates of compression and crustal shortening in the active mountain belts and trench systems by vectorially summing the relative movements between pairs of plates about their instantaneous centers of rotation. The rates of shortening deduced are highest in the Himalayas and in the trench systems. The rate for a trench system would appear to be directly proportional to the depth extent of the earthquake activity associated with it (Isacks et al., 1968). This agrees well with the interpretation by Oliver and Isacks (1967) of the attenuation of seismic waves (notably shear waves) traversing the crust and upper mantle in the vicinity of the Tonga Trench. These authors envisage a cold rigid plate of crust and upper-mantle material, the lithosphere, being pulled or thrust down beneath the trench system and into the hotter, perhaps partly molten, asthenosphere (Figure 6). The degree of partial melting, if present, must be very small since the asthenosphere only slightly attenuates seismic shear waves passing through it. The asthenosphere, however, correlates with the seismic low-velocity zone in the upper mantle, and convection within it may provide the necessary driving forces to move the lithospheric plates above. Presumably, stresses set up between and within the plates also determine their relative movements.

Current and recent poles of relative movement between plate pairs can be determined from slip vectors (deduced from large shallow focus earthquakes at plate boundaries), from the strike of transform faults, and from the variation in spreading rate along a ridge crest (Morgan, 1968; Le Pichon, 1968). Of these techniques clearly only the orientation of fossil transform faults and the variation in spreading rate at a given time (as revealed by the magnetic anomalies) can be used to determine past changes in the direction of spreading and relative movement (Vine, 1966; Menard and Atwater, 1968). The documentation of such changes is crucial

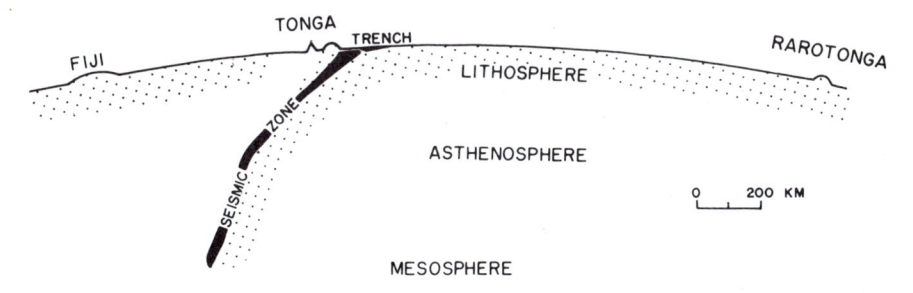

Figure 6. Postulated east-west section through the Tonga Trench, assuming that a relative lack of attenuation of seismic waves corresponds to rigidity. The lithosphere and mesosphere are believed to have appreciable strength, whereas the asthenosphere might flow more easily over geologic periods of time. (From Oliver and Isacks, 1967.)

to extending the plate theory back into the geologic past and gaining a clearer understanding of the detailed history of spreading and continental drift.

The Reversal Time Scale

The geomagnetic polarity time scale has only been defined for the last 3.5 million years by independent techniques; it is only possible, therefore, to deduce rates of spreading at recently active ridge crests. Nevertheless, the symmetry of the observed magnetic anomalies persists out beyond this central region. It seems reasonable, therefore, to make a provisional extrapolation of the reversal time scale on the basis of the magnetic anomalies, assuming that the Vine-Matthews hypothesis continues to apply and that the spreading rate has remained constant. Such an extrapolation has been made using a profile obtained at 51°S in the Pacific Ocean, the central part of which is shown in Figure 5. This profile shows a remarkable degree of symmetry, and many of the details of the reversal time scale are apparent because of the high spreading rate (Pitman and Heirtzler, 1966; Vine, 1966; 1968).

The reversal time scale for the last 11 million years obtained in this way has been used to predict the anomaly profile one might observe over the Reykjanes Ridge south of Iceland (Figure 7), where the spreading rate is thought to be 1 cm per year per flank on the basis of the central anomalies. Above the simulated profile is an observed anomaly profile obtained across this ridge. Although the time scale used is derived from the South Pacific Ocean, 15,000 km away, the simulation shows many similarities with the observed profile—one of 58 traverses flown by the U. S. Naval Oceanographic Office in 1963 in making a detailed aeromagnetic survey of the ridge (Heirtzler, Le Pichon, and Baron, 1966). The location of the Reykjanes Ridge and a summary map of the magnetic anomalies recorded over it are shown in Figure 8. The symmetrical pattern of anomalies provides striking evidence for sea-floor spreading and drift in this part of the North Atlantic Ocean (Vine, 1966).

Extrapolation Across the Ocean Basins

The magnetic anomalies and the extension of the reversal time scale enable us to map and date the ocean floor in the vicinity of

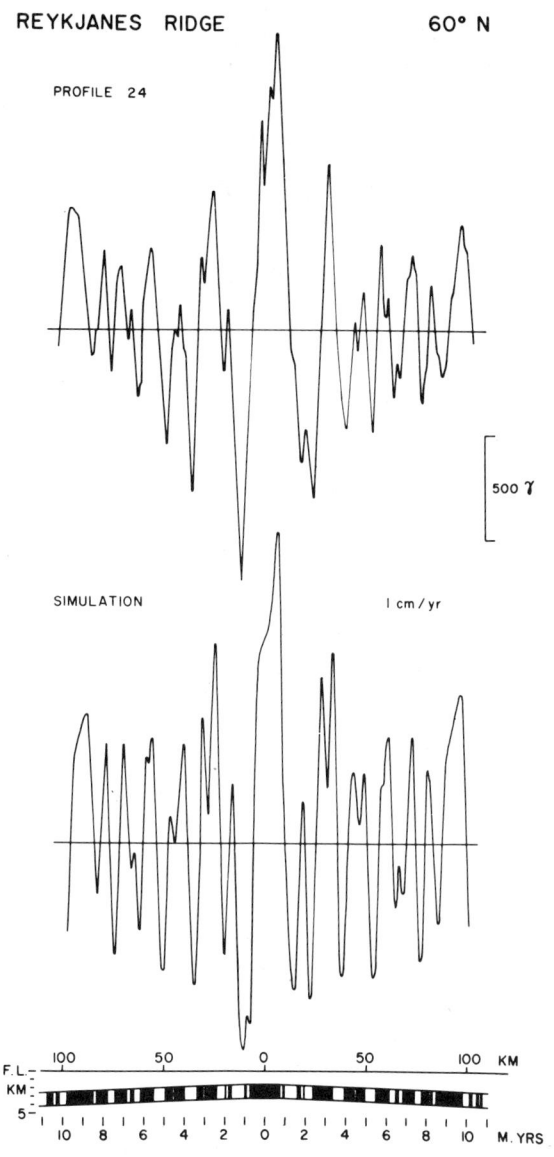

Figure 7. An observed aeromagnetic profile across the Reykjanes Ridge, southwest of Iceland, compared with a simulated profile, assuming a reversal time scale for the last 11 million years derived from a Pacific Ocean profile the central part of which is shown in Figure 5. Intensity and dip of the earth's magnetic field assumed to be 51,600 gamma and +74° respectively; magnetic bearing of profile 153°. (F.L.=flight level.) (Profile: Heirtzler, Le Pichon and Baron [1966].)

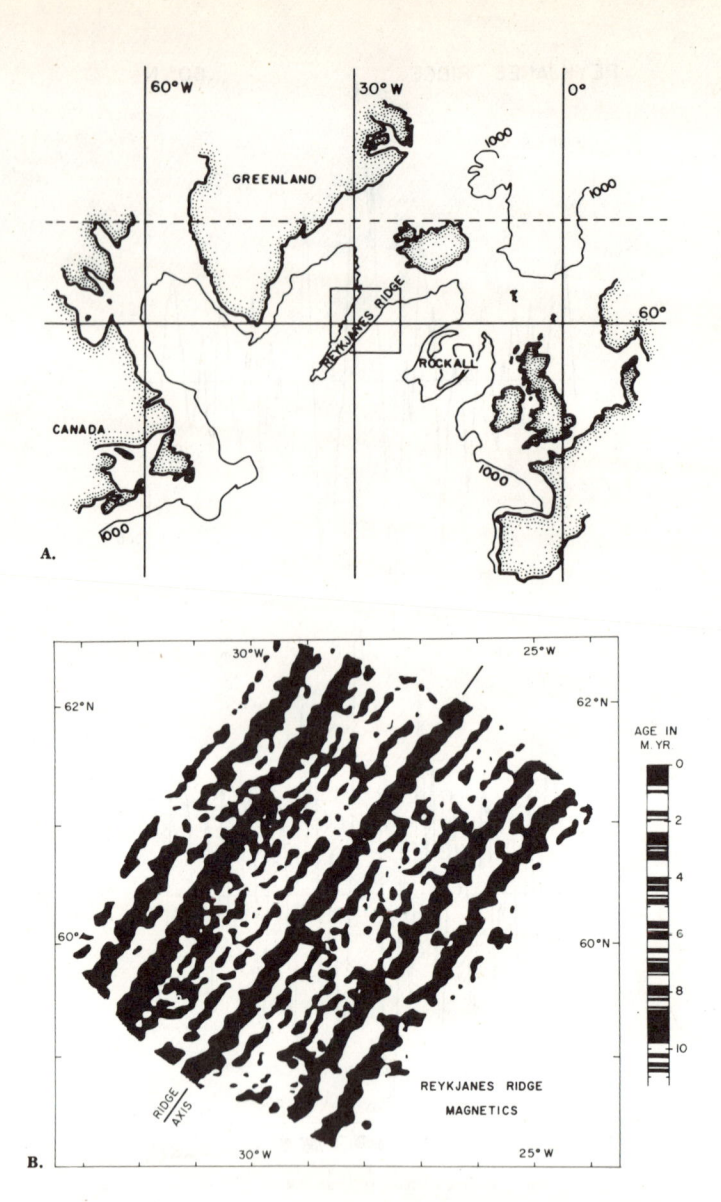

Figure 8. A. The location of the Reykjanes Ridge and the area of B. The 1000-fathom submarine contour is shown together with the 500-fathom contours for the Rockall Bank. B. Summary diagram of magnetic anomalies recorded over the Reykjanes Ridge. Areas of positive anomaly are shown in black. The central positive anomaly of Figure 7 correlates with the ridge axis. (After Heirtzler, Le Pichon and Baron [1966].)

active ridge crests. The question arises whether this hypothesis is also applicable to the remainder of the ocean basins. In that the same sequence of anomalies away from active ridge crests is reproduced in all the ocean basins, it seems highly probable that sea-floor spreading accounts for the formation of most, if not all, oceanic areas. Figure 9 shows this sequence of anomalies as observed across the Juan de Fuca Ridge and west of the Gorda Ridge in the northeast Pacific Ocean. It is reproduced on both sides of the East Pacific Rise in the South Pacific Ocean (Pitman et al., 1968), on both sides of the Mid-Atlantic Ridge in the South Atlantic Ocean (Dickson et al., 1968), and in part to the south of Australia and in the Indian Ocean (Le Pichon and Heirtzler, 1967). Those areas in which the sequence of anomalies shown in Figure 9 has been recognized have been summarized by Heirtzler et al. (1968) and are shown in Figure 10. Provisional attempts to assign

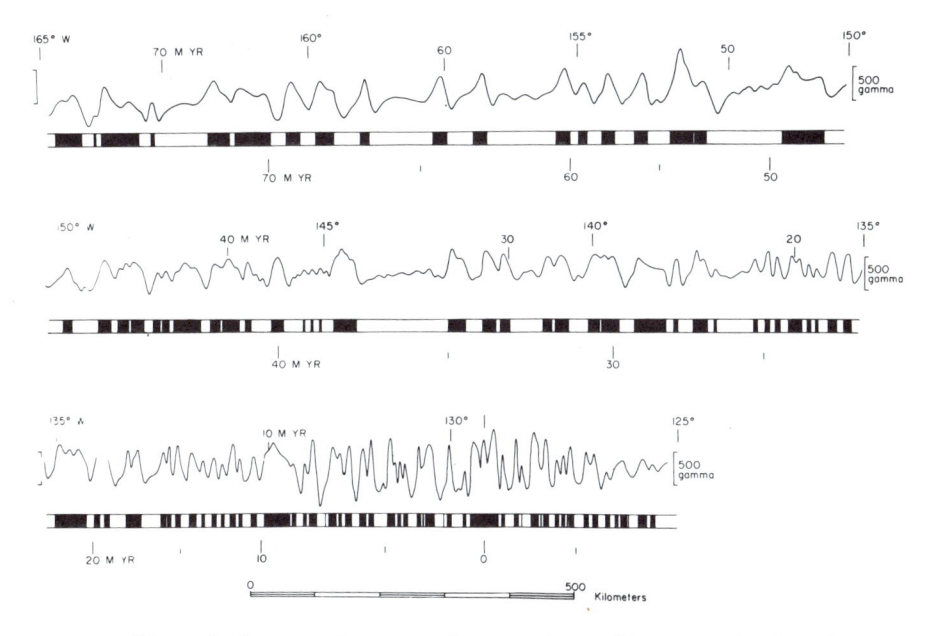

Figure 9. A composite magnetic anomaly profile across the Juan de Fuca Ridge, southwest of Vancouver Island, and to the west of the Gorda Ridge to the south, and immediately north of the Mendocino Fracture Zone (Vine, 1966). The time scales indicated are nonlinear. The upper one was suggested by Vine (1966) and the lower one by Heirtzler et al. (1968) and, for the last 11 million years, by Vine (1968).

106   MAN'S FINITE EARTH

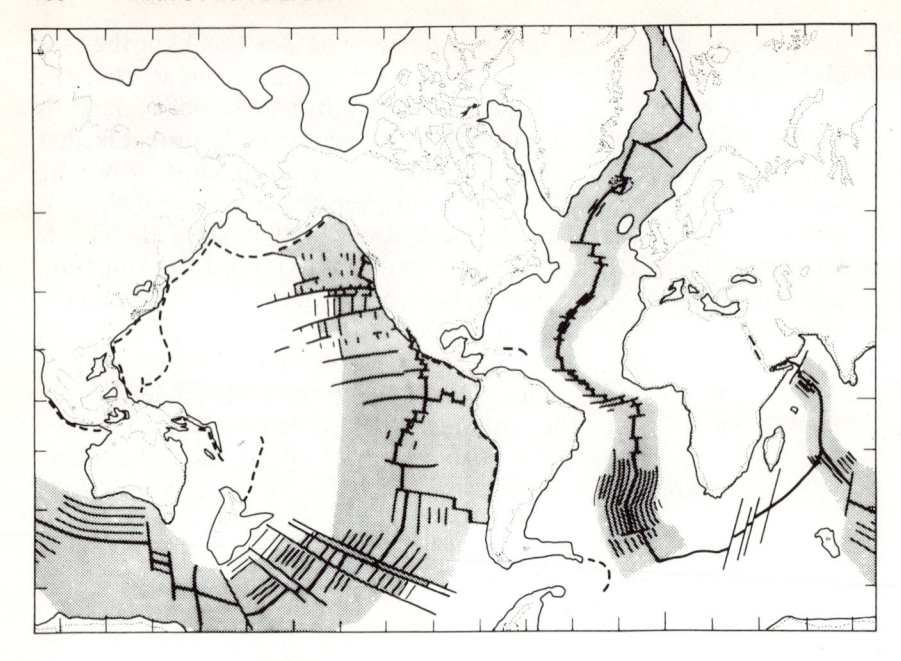

Figure 10. Provisional attempt to delineate areas of continental and oceanic crust. Within the ocean basins, trenches are indicated by thick dashed lines, ridge crests by thick solid lines, and fractures (transverse to the ridge crests) and correlatable linear magnetic anomalies (parallel to the ridge crests) by thin solid lines (Heirtzler et al., 1968). Oceanic crust thought to have been formed within the Cenozoic (i.e. the last 65 million years) is shaded.

ages to the anomalies, and hence dates for the underlying oceanic crust and implied geomagnetic reversals, have been made by Vine (1966), and by Heirtzler et al. (1968) as indicated in Figure 9. There is little difference between the two time scales.

Age of the Ocean Floor

The shaded area in Figure 10 indicates a preliminary estimate of the area of oceanic crust created within Cenozoic time, i. e. during the last 65 million years; the lines drawn parallel to the ridge crests are the ten-million-year "growth lines" suggested by Heirtzler et al. (1968). (A 65 m. yr. growth line is also included in the Pacific.) Whereas the shaded area in the Pacific and South Atlantic Oceans is thought to have been formed throughout the last 65 million years, in the Indian Ocean and south of Australia it has been formed within the last 40 million years or so. The earlier part of the sequence is missing between Australia and Antarctica, implying that the separation of these two continents was the last stage in the fragmentation of the southern continents which formerly constituted Gondwanaland.

This evidence for spreading and the age of the ocean floors is clearly readily compatible with "Continental Drift" as proposed by Wegener (1915), and Du Toit (1937). It is interesting and salutary to note that even such details as the post-Eocene separation of Australia and Antarctica, and some form of "subcontinental circulation," producing new ocean basins and the separation of continental fragments, were proposed by Wegener (1915) and Holmes (1928) respectively more than forty years ago. The implications of these earlier ideas and the new data regarding the age and ephemeral nature of the ocean basins are clearly revolutionary and far-reaching. However, it has been known for some time that the relatively thin blanket of sediments on the ocean floors can be accounted for, in terms of recent sedimentation rates, in 100-200 million years; less than five percent of geologic time. Similarly, marine geologists have never dredged rock or cored sediment more than 150 million years old from the ocean floor. The detailed pattern of sediment distribution and thickness now emerging from the results of seismic reflection profiling in no way conflicts with the age of the crust inferred from the magnetic anomalies. To some workers, however, it suggests pronounced discontinuities or irregularities in spreading in all areas (Ewing and Ewing, 1967; Le Pichon, 1968). The magnetic anomalies themselves suggest variations in spreading rates (Heirtzler et al., 1968) and discontinuities in spreading in the North Atlantic Ocean and northwest Indian Ocean (Le Pichon and Heirtzler, 1968); but the way in which the complete sequence of anomalies is reproduced in the remaining oceanic areas necessitates that any other lengthy stoppages must be worldwide.

Thick wedges of apparently undeformed sediments in the troughs associated with certain transform faults (Van Andel et al., 1967) and in parts of the trenches (Scholl et al., 1968) have led these observers to question the validity of spreading in these areas. However, it must be acknowledged that neither sedimentation rates nor the precise near-surface expression of faulting and under-thrusting in these settings is known. Other troughs and trenches are essentially devoid of sediment; this is difficult to account for without spreading, because they form such pronounced topographic lows and ideal sediment traps.

The one place where the mid-ocean ridge crest is exposed subaerially is Iceland. Although atypical because of its anomalous elevation above the sea floor, Iceland fulfills many requirements of a spreading ocean floor: its bedrock is entirely igneous (mainly basalt), the age of these rocks probably nowhere exceeds 20 million years (Moorbath et al., 1968), and there is evidence for crustal extension in the form of dike injection and normal faulting (Bodvarsson and Walker, 1964).

Summary

Earthquake activity throughout the world is largely confined to the young fold mountains and trench systems of the Alpine-Himalayan and circum-Pacific belts and to the crests of the mid-ocean ridges. The depth extent of this activity in the mountain and trench systems is very much greater than beneath the mid-ocean ridge crests. The hypothesis of sea-floor spreading proposes that the belts of shallow focus seismicity only, i.e. the mid-ocean ridge crests, are loci along which new oceanic crust is derived from the mantle, and that the trench systems are regions in which oceanic crust is partly resorbed. Focal mechanism solutions for shallow-focus earthquakes in these areas confirm that the ridges are tensional features producing extension of the earth's crust, and that the trenches are compressional features characterized by underthrusting.

The creation of new crust and the resulting spreading from the median ridges of the Atlantic and Indian Oceans appear to have been initiated within a former supercontinent producing the present ocean basins and separation of continental fragments. The Pacific Rise also exhibits spreading, but was presumably largely initiated within former oceanic crust. The sinks in the system—the trenches—occur at the leading edges of continents and are generally marginal to the Pacific.

Recent more accurate determinations of earthquake epicenters on mid-ocean ridge crests have revealed that the activity is concentrated along the transverse fractures or faults which offset the ridge crest. The distribution and nature of the earthquakes on these fractures indicate a transform type of faulting, which is only explicable in terms of spreading of the sea floor. Disturbances in the earth's magnetic field recorded at and above sea-level and due to the fossil magnetism of the oceanic crust also indicate that spreading occurs. The pattern of anomalies revealed in this way is strikingly symmetrical about ridge crests, and the anomalies correlate in every detail with reversals of the earth's magnetic field during the last few million years, as defined by paleomagnetic studies of terrestrial lava flows and deep-sea sediments. Such correlations with the reversal time scale, which has been dated by a refined radiogenic technique, enable one to determine recent rates of spreading at ridge crests. Rates deduced in this way vary from 1 to 6 cm per year per ridge flank.

The restricted belts of earthquake activity define essentially aseismic crustal plates bounded by active ridge crests, faults, trenches and mountain systems. If these plates are assumed to be perfectly rigid, then it is possible to calculate rates of shortening in the trench and mountain systems by assuming the rates of extension derived from the magnetic anomalies at the remaining boundaries. The predicted rates of crustal shortening correlate directly with the depth extent of the seismicity in these areas, particularly in the trench systems. Recent seismological studies of trench systems also suggest that a cold rigid plate of oceanic crust and uppermost mantle is being thrust down into the asthenosphere—the part of the mantle thought to be nearest its melting point.

The timetable for reversals of the earth's magnetic field has been accurately documented by independent techniques for only the last few million years. However, oceanic magnetic patterns appear to maintain their symmetry, parallelism and continuity beyond the ridge crests to the flanks and even across the deep ocean basins, implying that spreading has occurred earlier. The same sequence of anomalies away from ridge crests is reproduced in all the ocean basins, strongly supporting the concept that reversals of the earth's magnetic field are recorded in the oceanic crust as a result of sea-floor spreading. Recent spreading rates are consistent with the break-up of the continents and the formation of the present ocean basins within the last 200 million years, less than five percent of geologic time. Despite this gross extrapolation, ages assigned in this way to the anomalies, and hence to the underlying oceanic crust, are entirely consistent with all that is known of the age of the ocean floors from sediment thicknesses and the age of sediment cores and dredged rocks.

Thus, sea-floor spreading is capable of explaining many of the first-order structural features of the earth's surface and is compatible with the results of geological and geophysical investigations in the ocean basins. It also provides, perhaps for the first time, a plausible mechanism for continental drift.

References

Bodvarsson, G., and Walker, G. P. L. 1964. Crustal drift in Iceland. *Geophysical Journal* 8:285-300.
Cox, A., Doell, R. R., and Dalrymple, G. B. 1963. Geomagnetic polarity epochs and Pleistocene geochronometry. *Nature* 198:1049-51.
———. 1967. Reversals of the earth's magnetic field. *Scientific American* 216 (2):44-54.
Dickson, G. O., Pitman, W. C., and Heirtzler, J. R. 1968. Magnetic anomalies in the South Atlantic and ocean floor spreading. *Journal of Geophysical Research* 73:2087-2100.
Du Toit, A. L. 1937. *Our wandering continents.* Edinburgh: Oliver and Boyd.
Ewing, J., and Ewing, M. 1967. Sediment distribution on the mid-ocean ridges with respect to spreading the sea floor. *Science* 156:1590-92.
Gutenberg, B., and Richter, C. F. 1954. *Seismicity of the earth.* Princeton, N. J.: Princeton University Press.
Heirtzler, J. R., Dickson, G. O., Herron, E. M., Pitman, W. C., and Le Pichon, X. 1968. Marine magnetic anomalies, geomagnetic field reversals and motions of the ocean floor and continents. *Journal of Geophysical Research* 73:2119-36.
Heirtzler, J. R., Le Pichon, X., and Baron, J. G. 1966. Magnetic anomalies over the Reykjanes Ridges. *Deep-Sea Research* 13:427-43.
Hess, H. H. 1962. History of the ocean basins. In *Petrologic studies: a volume to honor A. F. Buddington,* pp. 599-620. Geological Society of America.
Holmes, A. 1928. Radioactivity and earth movements. *Trans. Geol. Soc. Glasgow* 18:559-606 (in *Principles of physical geology,* p. 1001. London/New York: Nelson/Ronald Press).
Isacks, B., Oliver, J., and Sykes, L. R. 1968. Seismology and the new global tectonics. *Journal of Geophysical Research* 73:5855-99.
Le Pichon, X. 1968. Sea-floor spreading and continental drift. *Journal of Geophysical Research* 73:3661-97.
Le Pichon, X., and Heirtzler, J. R. 1968. Magnetic anomalies in the Indian Ocean and sea-floor spreading. *Journal of Geophysical Research* 73:2101-17.
Menard, H. W., and Atwater, T. 1968. Changes in direction of sea-floor spreading. *Nature* 219:463-67.
Moorbath, S., Sigurdsson, H., and Goodwin, R. 1968. K-Ar ages of the oldest exposed rocks in Iceland. *Earth Planetary Science Letters* 4:197-205.
Morgan, W. J. 1968. Rises, trenches, great faults and crustal blocks. *Journal of Geophysical Research* 73:1959-82.
Oliver, J., and Isacks, B. 1967. Deep earthquake zones, anomalous structures

in the upper mantle, and the lithosphere. *Journal of Geophysical Research* 72:4259-5275.

Opdyke, N. D., Glass, B., Hays, J. D., and Foster, J. 1966. Paleomagnetic study of Antarctic deep-sea cores. *Science* 154:349-57.

Pitman, W. C., and Heirtzler, J. R. 1966. Magnetic anomalies over the Pacific-Antarctic Ridge. *Science* 154:1164-71.

Pitman, W. C., Herron, E. M., and Heirtzler, J. R. 1968. Magnetic anomalies in the Pacific and sea-floor spreading. *Journal of Geophysical Research* 73:2069-85.

Scholl, D. W., Von Huene, R., and Ridlon, J. B. 1968. Spreading of the ocean floor: undeformed sediments in the Peru-Chile Trench. *Science* 159:869-71.

Stauder, W. 1968. Tensional character of earthquake foci beneath the Aleutian trench with relation to sea-floor spreading. *Journal of Geophysical Research* 73:7693-7701.

Sykes, L. R. 1967. Mechanism of earthquakes and nature of faulting on the mid-oceanic ridges. *Journal of Geophysical Research* 72:2131-53.

Van Andel, Tj. H., Corliss, J. B., and Bowen, V. T. 1967. The intersection between the mid-Atlantic ridge and the Vema fracture zone in the North Atlantic. *Journal of Marine Research* 25:343-51.

Vine, F. J. 1966. Spreading of the ocean floor: new evidence. *Science* 154:1405-15.

―――. 1968. Magnetic anomalies associated with mid-ocean ridges. In *History of the earth's crust*, R. A. Phinney, ed., pp. 73-89. Princeton, N. J.: Princeton University Press.

Vine, F. J., and Matthews, D. H. 1963. Magnetic anomalies over oceanic ridges. *Nature* 199:947-49.

Wegener, A. 1915. *Die Entstehung der Kontinente und Ozeane*. Braunschweig: Viewag und Sohn. (1966, English translation of the fourth revised edition, 1929. New York: Dover.)

Wilson, J. T. 1965. A new class of faults and their bearing upon continental drift. *Nature* 207:343-47.

10 Plate Tectonics, Sea-Floor Spreading, and Continental Drift

Robert S. Dietz

Robert S. Dietz is a research oceanographer with the Atlantic Oceanographic and Meteorological Laboratories of the National Oceanic and Atmospheric Administration in Miami, Florida.

Over the past decade a geologic revolution has transpired. It can be summarized in the expression, "the new global tectonics," or more succinctly as *plate tectonics*. This, in turn, involves sea-floor spreading, descending lithospheric slabs, transform faulting, and continental drift. It has added a fourth dimension to classical geotectonics which was formerly concerned with uplift, subsidence, and time; but now we have added great horizontal shifts as well.

Of course, geologists have always accepted a limited amount of horizontal displacement by strike-slip faulting, but never before as the dominant dimension of tectonics. Consider the Bahama platform, where post-mid-Jurassic carbonates all deposited in shallow water have a thickness of about 6 km. This reveals a history of great subsidence. But during this time, according to the plate tectonic-continental drift solution, the Bahama platform has drifted 5000 km from an original geographic position near Ascension Island in the equatorial central Atlantic Ocean. Consider also the displacement of 6.5 cm/yr along the San Andreas fault, a transform fault separating the North American plate from the Pacific plate. This rate may seem ponderous, but it is a remarkably rapid geologic process compared to the upheaval of mountains or the peneplanation of continents. It is sufficient to shunt Los Angeles (on the Pacific plate) with respect to San Francisco (on the North American plate) entirely around the circumference of the earth in 600 million years, the time span of the Phanerozoic era.

Earth's Mosaic Carapace

The plate tectonics concept holds that the earth is divided into about eight rigid spherical caps 100 km thick, riding on the weak asthenosphere and in which the continents are embedded and drift as passive passengers. We can visualize the ideal plate as rectangular, although perhaps only the Indian plate attains this simplicity. Along one edge there is a subduction zone, usually marked by a trench, where the cold crustal plate dives steeply into the earth's mantle, reaching a depth of 700 km before being fully resorbed. Opposing the subduction zone is a mid-ocean rift, or

Figure 1. The African plate has drifted north and counterclockwise, closing up the Tethys seaway and leaving the Mediterranean Sea as a remnant. The northern margin of the plate has been subducted into the Tethyan trench; the mid-Atlantic ridge has been a spreading ridge; the Indian rift has acted mainly as a transform fault along which shearing has occurred.

pull-apart zone. As the rift opens, the gap is quickly healed by the inflow of liquid basalt and quasi-solid mantle rock. The other two antithetical sides of the plate, connecting the rifts to the trenches, are crust-piercing shears called transform faults. Three types of plate boundaries are possible: (1) divergent junctures, the mid-ocean rifts where new simatic crust is created; (2) shear junctures, the transform faults where plates slip laterally past one another so that crust is conserved; and (3) convergent junctures, the trenches where two plates overlap, with one plate descending and being consumed. This subducting plate is not fully digestible by the mantle, so that the sialic hyperfusible rocks rise to the surface as the andesitic volcanics and granodiorite plutons of island arcs.

All plates, except the great Pacific plate, may be identified by the particular continent embedded within the plate. Thus, in addition to the Pacific plate, there is the North American plate, the South American plate, the Eurasian plate, the African plate, the Indian plate, the Australian plate, and the Antarctic plate. Additionally, there are many small plates: e.g., Nazca, Cocos, Farallon, Caribbean, etc. There is even one plate, the Kula plate (from an Indian word meaning "all gone"), which has disappeared down the Aleutian trench. We know about its demise through the so-called Great Magnetic Bight in the northeast Pacific. Within the Pacific Ocean there probably was originally a West Pacific plate, an East Pacific plate, and a North Pacific plate (the Kula plate). The Antarctic plate is curious in that it contains no subduction zone along its perimeter, which in turn suggests that it is the most stationary of all the plates, having no place to go.

The South American and North American plates plus the Australian and Indian plates seem to be moving in unison and joined today so that, in terms of the major plates, the earth may be described as a highly distorted, spherical hexahedron or "cube." A few years ago S. W. Carey reviewed the various symmetries of the earth, including a listing of many historical comparisons that have been made about the earth with respect to various, regular polyhedrons and other solid forms. Of these, the tetrahedral hypothesis of Lowthian Green in the 1880s is especially well known. It is curious that a comparison with a hexahedron apparently has never been made.

A cube, of course, has eight corners or coigns as well as six faces. Similarly, the crustal plates intersect in so-called *triple junctions*. These play a fundamental role in geotectonics which still remains to be fully elucidated. A few types of triple junctions

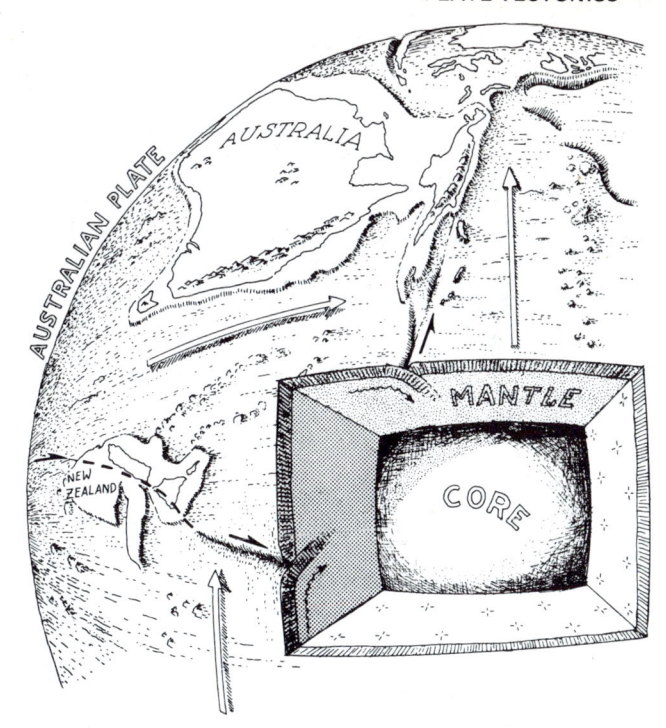

Figure 2. The Australian plate is drifting north about 8 cm/yr. Australia was detached from Antarctica only about 50 million years ago in the Eocene. This plate is moving nearly at right angles to the Pacific plate so that the northern boundary is a subduction zone for the Australian plate and a shear transform zone for the Pacific plate. The reverse holds true for the eastern margin of the Australian plate.

are inherently unstable, that is, they will change their geometry with time, while most are stable. Near the Azores, for example, is the Am/Eur/Af triple junction marking the intersection of the North American, Eurasian, and African plates. It is classified as a ridge/ridge/fault triple junction, as two of its arms are mid-ocean, spreading ridges, while the third is the Azores-Mediterranean transform fault.

Another triple junction near the Galapagos Islands in the east-central Pacific is the Nas/Co/Pac ridge/ridge/ridge nexus marking the intersection of the Nazca, Cocos, and Pacific plates. John Holden and I have inferred that, although this junction has maintained a stable geometry, the juncture itself has migrated 1000 km westward since its inception 20 to 40 million years ago.

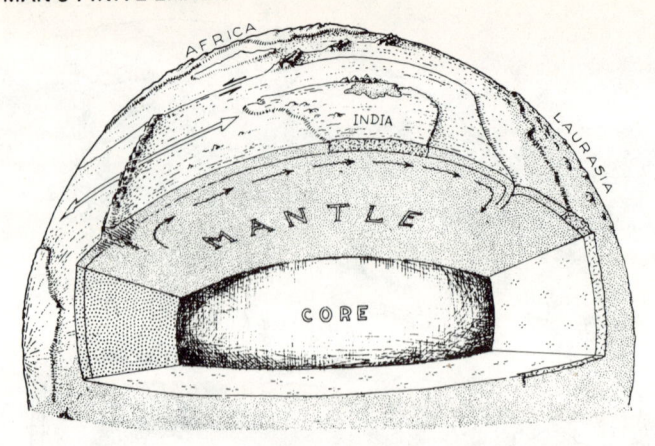

Figure 3. The Indian plate. The Indian subcontinent has undergone a remarkable drift since being detached from Antarctica. New ocean floor has been emplaced between the mid-ocean ridge which has maintained its position equidistant between Antarctica and India, while the northern portion of the plate has been consumed in the Himalayan trench. About 20 million years ago India collided with Asia, throwing in the Himalayan rampart. These mountains are as high as a continental slope (ca. 4500 m), indicating that the underlying sialic slab is two continents thick due to underthrusting and dovetailing.

Shifty Continents

Continental drift is a corollary to plate tectonics since these 35-km-thick, sialic slabs drift passively, embedded in the even larger lithospheric plates. The rotational translations of the continents across the face of the globe quite naturally, however, are of especial interest to man.

According to the continental drift theory, the continents were combined into the universal landmass of Pangaea 200 million years ago in Triassic time. It is probably more correct to say that the continents were sutured together as of this time, for it is likely that the continents were sundered apart and drifted in earlier geologic time, although the details of this "pre-drift drift" remain unclear. As measured to the 2000-m isobath (the true margin of the continents), Pangaea covered 40 percent of the earth while the universal ocean, Panthalassa (the ancestral Pacific) covered the remainder. With Africa as the heartland, the continents were generally farther south and farther east than today. New York and

London were only slightly north of the equator. The ovoid plan of Pangaea was interrupted by a great bight called the Sea of Tethys which separated Eurasia from Africa.

The breakup and dispersal of the continents can best be visualized in a general way by reference to Figure 5. There the drift is shown in absolute coordinates as controlled by "hot spots"—plumes of lava which pierced the lithosphere and laid down strings of islands as the crust moved over a fixed hot spot—or one at least assumed to be fixed. This effect can best be observed in the case of the Hawaiian Chain where a stream of islands and seamounts extends from Hawaii to Milwaukee Seamount, well past Midway Island. There are other methods of tracking the drift of continents (magnetic reversal anomalies on the ocean crust; the strike of fracture zones; and matching conjugate points on opposing continental margins across a rift ocean, e.g., the South Atlantic), but these methods give only relative solutions to drift rotations.

The major aspects of drift can be summarized as follows. The Atlantic Ocean (including the Gulf of Mexico and the Caribbean Sea) is a new ocean basin, as is the Indian Ocean. The Mediterranean Sea is a remnant of Tethys, as it is still undergoing

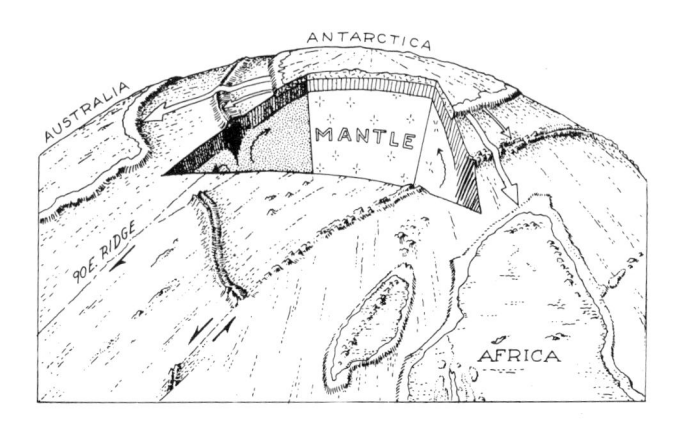

Figure 4. The Antarctic plate, because it is surrounded entirely by a system of ridges and transform faults and without any trench, is the most fixed of all of the plates, but has grown ever larger by peripheral addition of new oceanic crust. The detachment of Australia and Africa from Antarctica and their northward drift has been accommodated by migrating, mid-ocean ridges which have moved one-half the velocity of the drifting continents.

1: Permian, 225 million years ago

2: Triassic, 200 million years ago

3: Jurassic, 135 million years ago

4: Cretaceous, 65 million years ago

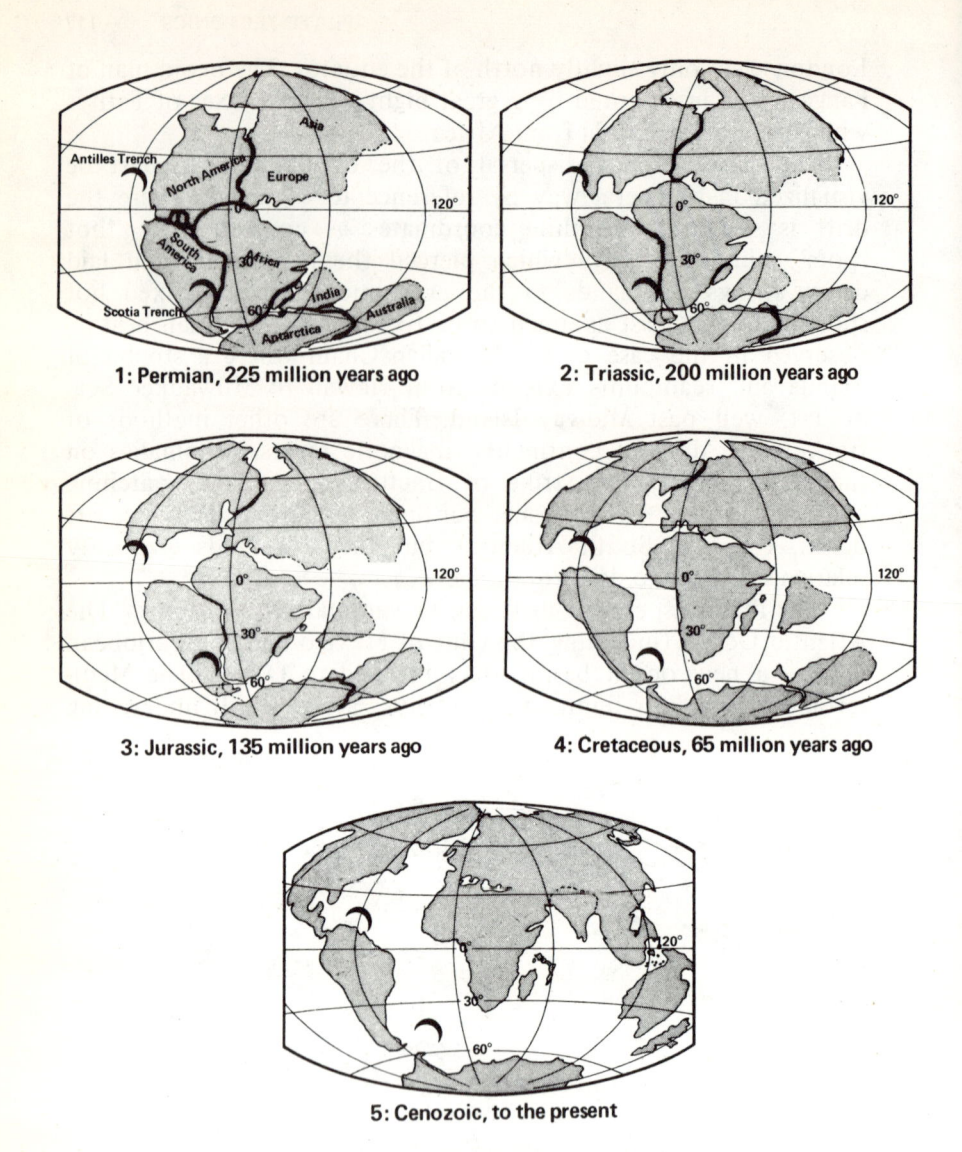

5: Cenozoic, to the present

Figure 5. The breakup and dispersion of Pangaea. See text for explanation. To help clarify the drifting, two well-known landmarks, the Antilles and the Scotia arcs (lunate crescents), are placed in their present positions on all maps. Maps adapted and simplified from those of R. S. Dietz and John Holden (*Journal of Geophysical Research*, Sept. 10, 1970).

closure. The Pacific Ocean is a closing ocean, for the earth as a whole is growing no larger. For each addition of new oceanic crust (ca. 1.5 km^2/yr), an equal amount of crust is lost by subduction in the oceanic trenches. At this rate the entire ocean basin undergoes renewal, on an average, once every 200 million years.

Turning now to the continents, the New World has drifted west while those continents around the Indian Ocean have moved north. Antarctica has remained almost stationary, as has Eurasia, while undergoing clockwise rotation such that Europe has moved north and China southward. The latest major event was the detachment of Australia from Antarctica in the Eocene only about 55 million years ago.

Actualistic Geosynclines, Mountains, and Continent Building

Plate tectonics also provides fresh insight into the geosyncline-mountain building cycle. A drifting continent has a leading and a trailing edge—for North America, the west coast leads while the

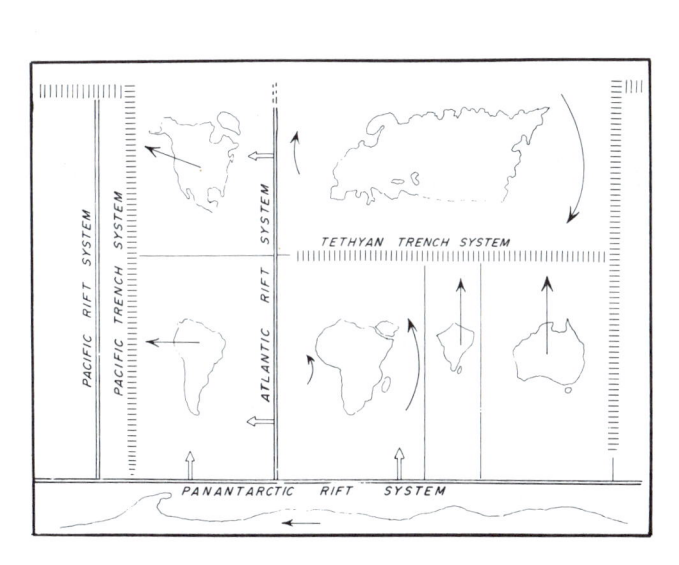

Figure 6. A summary of the drift motion of the continents. The New World has drifted west. Africa has moved north and counterclockwise, India and Australia both have moved north. Eurasia has rotated clockwise so that Great Britain has moved from the tropics to a northern clime, while Japan has become more tropical. Antarctica has remained almost stationary.

Atlantic coast trails. The trailing margin, fully coupled with the crustal plate, is tectonically stable; and, since the continental divide is near the mountainous Pacific perimeter, most sediments are dumped onto the Atlantic floor (including the Gulf of Mexico). So that, along this trailing edge, a thick prism of sediments has been deposited at the base of the continental slope, the continental rise. This prism appears to be a living eugeosyncline, or, as plate tectonicists are inclined to say, eugeocline. The Atlantic Ocean is now opening, but suppose the Atlantic crust eventually becomes heavy and unstable along the Atlantic margin. The lithosphere would then form a new break and dive, creating a trench, and the Atlantic would revert to being a closing ocean. Eventually, the continental rise prism would collide with this convergent juncture, and the continental rise would collapse into a classical eugeoclinal foldbelt.

The probable reality of this story line is enhanced by the fact that this sequence of events apparently has already transpired once before with the creation in the lower Paleozoic of the crystalline Appalachian eugeoclinal foldbelt. The Atlantic Ocean seems to have opened, closed, and reopened again (the last-named event commencing 190 million years ago), and in the future it may once more undergo closure. One of the grand themes of plate tectonics is that ocean basins are transient features, constantly changing their size and shape, while the continents not only persist, but also grow by the accretion to their margins of new eugeoclinal foldbelts.

Paired with the ancient, folded eugeoclines are miogeoclines (or miogeosynclines) of which the loosely folded Appalachians, lying parallel and inside of the crystalline Appalachians, is the type, example for North America. The thick wedge of Cretaceous and Cenozoic sediments underlying the coastal plain and capping the Atlantic shelf provides an actualistic example paired to the modern continental rise. Plate tectonics also provides an explanation for the extensive downwarping needed for the deposition of this wedge entirely laid down in shallow water. With the initial splitting of North America from Africa, a spreading ridge must first be inserted beneath the craton like that under high Africa today. Erosion ensues, thinning the cratonic slab along the new rift. Then, as the new ocean basin opens and the underlying upper mantle detumesces, the continental margin subsides, providing a submerged platform on which a wedge of sediments accumulates. Thus, a living miogeocline is deposited.

A Unifying Model

The debate between the fixists and the mobilists dragged along for a few decades. Then it was suddenly marked by a wholesale conversion to drift in 1967. Now the burden of proof lies with those who prefer to hold the continents still. A revolution has been wrought. The new global tectonics now is having a salubrious and unifying effect on the geosciences. As Sir Edward Bullard has noted, "the acquisition of new data is likely to re-establish the

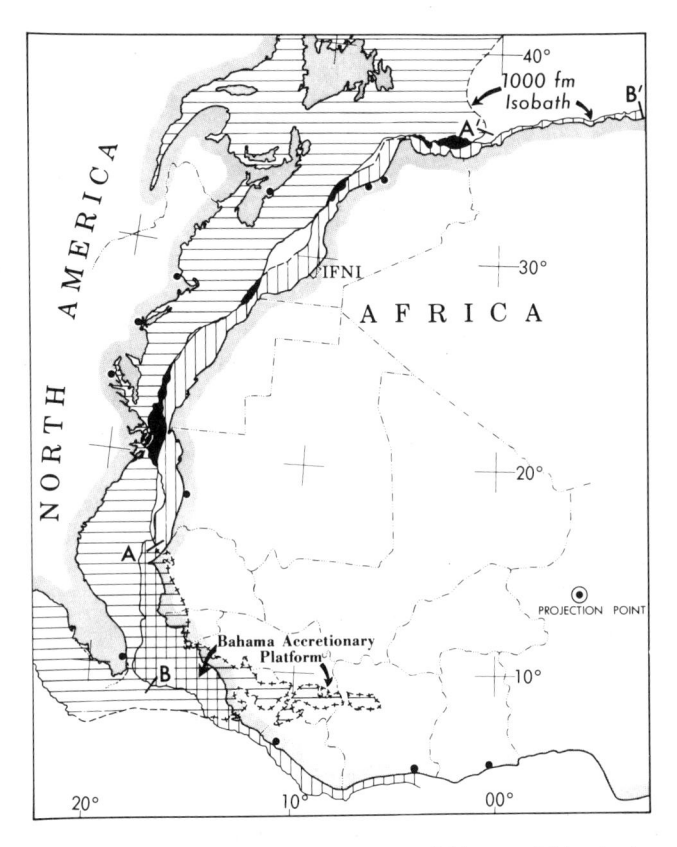

Figure 7. A computerized fit between Africa and North America matched to minimize total overlap (black areas) and underlap (white areas). The margin of North America was matched between points A-A' and Africa between B-B'. The Bahama platform was permitted to overlap on the premise that it is a post-breakup sedimentary deposit lying directly on oceanic crust. The fit achieved seems remarkably good.

usual chaos in earth sciences, but at least we do now have a universal model as a point of departure."

The transformation of geology has been so complete that Kenneth Deffeyes of Princeton has said, "Ninety-nine percent of the profession has had to admit that they were wrong, including a great many who were in print saying continental drift couldn't possibly happen.... Everything has got to be re-written. We've been caught with our textbooks down."

References

Dewey, J. 1972. Plate tectonics. *Scientific American* 226(5):56-68.
Dewey, J., and Bird, J. 1970. Mountain belts and the new global tectonics. *Journal of Geophysical Research* 74(14):2625-47.
Dietz, R. S. 1972. Geosynclines, mountains and continent building. *Scientific American* 232(3):30-38.
Dietz, R. S., and Holden, J. 1970. The breakup of Pangaea. *Scientific American* 223(4):30-41.
Irving, E. 1964. *Paleomagnetism*. New York: John Wiley and Sons.
McKenzie, D. P., and Morgan, W. J. 1969. Evolution of triple junctions. *Nature* 224(5215):125-33.
Runcorn, S. K., ed. 1962. *Continental drift*. New York: Academic Press.
Takeuchi, H., Uyeda, S., and Kanamori, H. 1967. *Debate about the earth*. San Francisco: Freeman, Cooper and Co.
Tarling, D., and Tarling, M. 1971. *Continental drift: a study of the earth's moving surface*. New York: Doubleday and Co.
Vine, F. J. 1966. Spreading of the ocean floor. *Science* 154(3755):1405-15.
Wilson, J. T., ed. 1972. *Continents adrift: readings from Scientific American*. San Francisco: W. H. Freeman and Co.

Additional Readings

Bird, J. M., and Isacks, B. 1972. *Plate tectonics: selected papers from the Journal of Geophysical Research*. Washington, D. C.: American Geophysical Union.
Dewey, J. F. 1972. Plate tectonics. *Scientific American* 226(5): 56-68.
Du Toit, A. L. 1937. *Our wandering continents*. Edinburgh: Oliver and Boyd.
Garland, G. D., ed. 1966. *Continental drift*. Toronto: University of Toronto Press.
Hallam, A. 1972. Continental drift and the fossil record. *Scientific American* 227(5):56-66.
Meyerhoff, A. A. 1970. Continental drift: implications of paleomagnetic studies, meteorology, physical oceanography, and climatology. *Journal of Geology* 78 (1):1-51.
Meyerhoff, A. A., and Meyerhoff, H. A. 1972a. "The new global tectonics":

major inconsistencies. *American Association of Petroleum Geologists Bulletin* 56(2):269-336.

———. 1972b. "The new global tectonics": age of linear magnetic anomalies of ocean basins. *American Association of Petroleum Geologists Bulletin* 56(2):337-59.

Rona, P. A. 1973. Plate tectonics and mineral resources. *Scientific American* 229(1):86-95.

Takeuchi, H., Uyeda, S., and Kanamori, H. 1970. *Debate about the earth.* 2nd ed. San Francisco: Freeman, Cooper and Co.

Tarling, D. H., and Tarling, M. 1971. *Continental drift: a study of the earth's moving surface.* Garden City, N. Y.: Doubleday and Co.

Vacquier, V. 1972. *Geomagnetism in marine geology.* New York: Elsevier.

Wegener, A. L. 1966. *The origin of the continental oceans* (Translated from 4th revised German edition by J. Biram.) New York: Dover.

Wilson, J. T., ed. 1972. *Continents adrift: readings from Scientific American.* San Francisco: W. H. Freeman and Co.

Part Four

Astrogeology

At this instant through this very document are perhaps passing radio waves bearing the conversations of distant creatures, which we could record if we but pointed an existing radio telescope in the right direction and tuned it to the proper frequency.

Committee on Science and Public Policy, National Academy of Sciences

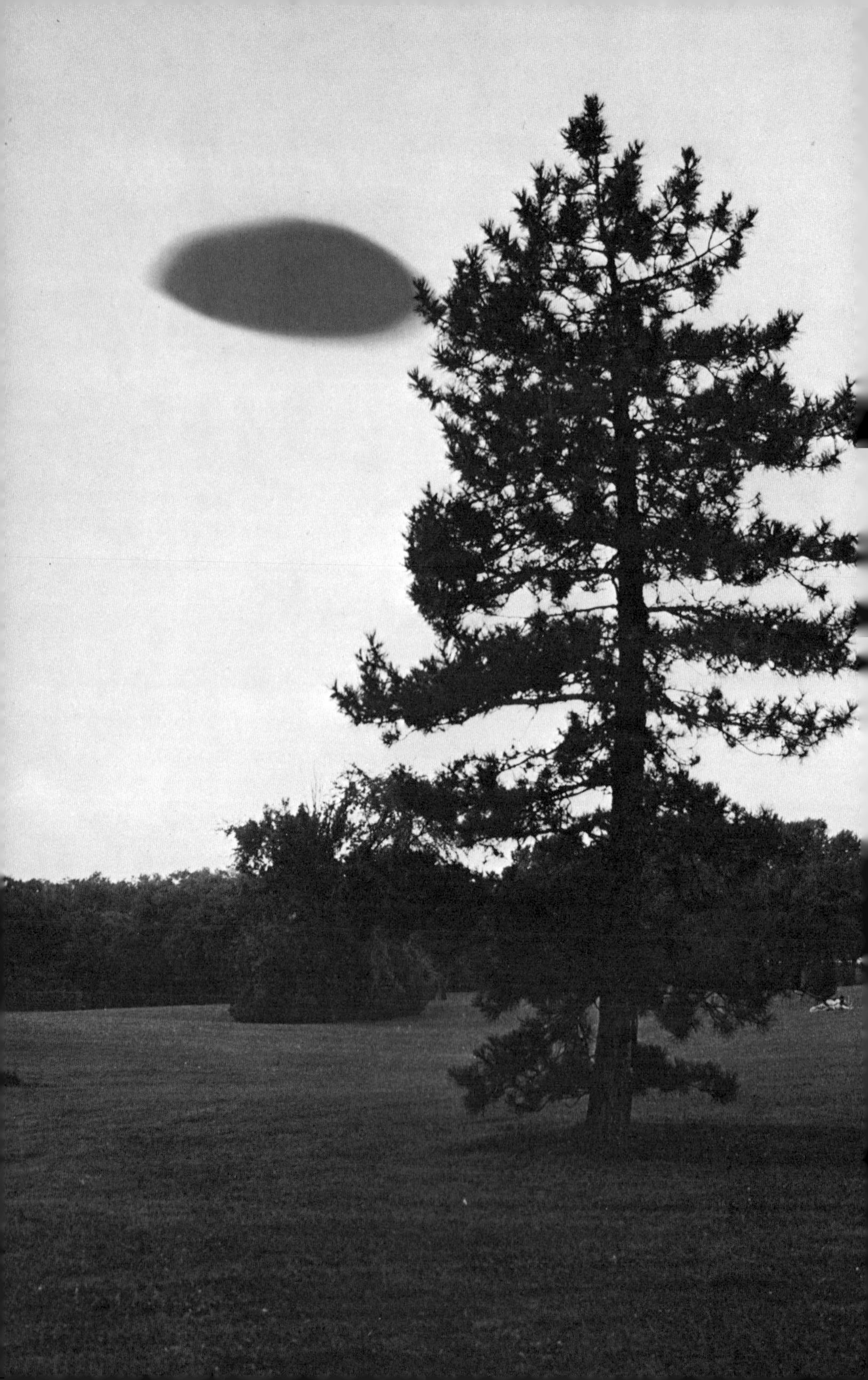

*A*strogeology is a science that applies the principles and techniques of geology, geochemistry, and geophysics to the study of the nature, origin, and history of the condensed matter and gases in the solar system. It includes the terms extraterrestrial geology, lunar geology, space geology, and planetology. The term was first used almost 100 years ago; however, the serious large-scale exploration of extraterrestrial bodies began with the announcement by President Kennedy on May 25, 1961, that "This nation should commit itself to achieving the goal, before this decade is out, of landing a man on the moon and returning him safely to earth." This goal was achieved in July, 1969 with the Apollo 11 mission. The Apollo program thus made a major contribution to planetary science's goal of achieving an understanding of the origin and evolution of the solar system based on observational facts from many different bodies within the system.

Since the program began, almost 400 kilograms of lunar samples have been returned to earth. Although many questions have been answered, as is often the case with science many more have been asked, for which we still do not have satisfactory answers. Data continue to be generated from analysis of these samples and will for several decades as solutions to unresolved puzzles are sought.

In developing a model for planetary evolution, it now appears that the moon differs greatly from the earth in chemistry and internal structure. Many questions remain, such as the nature of the origin of the moon, the differences in crustal thicknesses from one place to another, the distribution of moonquakes, the origin of the stable remanent magnetization in crystalline and breccia samples from the moon (was there a molten core with a dynamo process, or a deep internal magnetization created during formation of the moon by a strong magnetic field of the sun?), the age and origin of the major and minor rock types on the moon, and the early history of bombardment of the moon. There are many suggestions to explain the many problems in the major categories above, and, as the analyses come in, our understanding of the moon, and possibly the sun and other objects in the galaxy as recorded in the lunar surface, will continue to evolve.

The first two articles cover lunar exploration and some of the *many questions about the moon that remain. One of the most publicized discoveries on the Apollo 17 mission was the orange regolith or soil found by Harrison Schmitt, who thought that it came from the nearby volcanic vent and might indicate a geologically active moon about 10 million years ago. Radiometric dating later established the age of the soil at 3.7 billion years in agreement with current hypotheses on lunar activity. The color of the soil was not due to rust but to a red-tinted glassy material with a composition similar to basalt. Its origin might have been meteoritic impact or volcanism.*

The latter two articles in this section are about Mars. Hammond presents an excellent summary of the current interpretations and exploration of this planet, followed by photographs of some of the findings of the Mariner 9 flight from the paper by Masursky and others. These papers are based on observations by the Mariner 9 spacecraft which went into orbit around Mars on November 13, 1971, and, before the end of its mission, had transmitted over 7000 photographs to earth. As photographs were received, new ideas on the history of Mars were developed, only to be discarded or modified later in the light of evidence from subsequent photographs. Erosional and depositional features and the importance of dust in creating light and dark markings viewed from earth became known. Other surprises included the channels on the surface which led to the idea that running water had once and perhaps cyclically played a part in shaping the surface. The structural geologists, in addition to the geomorphologists, had surprises when the trenchlike valleys and the apparent youthful aspect of many features of Mars were seen. And maybe the ice caps, that showed rapid fluctuations during the time of the observation and that had apparently built end moraines, were not made of carbon dioxide.

The possibility of life on the planet brightened with the discovery that liquid water probably existed on the surface. A later probe to check for biological activity will be made. If there is life on Mars, what are the possibilities for life elsewhere in the universe?

The latter question is one that is becoming more respectable to ask, as might be gleaned from the quotation on the introductory page for this section. Calculations on the occurrence of planetary civilizations suggest that within 10,000 light years (60 million billion miles) from earth are four million stars, one million planetary systems, 10,000 civilizations, and one civilization of enough intelligence to be detected from earth. Searches for this civilization will probably include both space vehicles and radio telescopes. Radio telescopes have been employed for a short time in this country and are now in use for this purpose in the Soviet Union. As the number and intensity of such searches increase and as understanding of the evolution of the universe develops, including the black holes which apparently are dense remnants of collapsed stars (from which no light escapes) ten times as massive as the sun, the importance of the Apollo program will then become clear to many more people throughout the world.

11

Apollo 15:
Scientific Journey
to Hadley-Apennine

Joseph P. Allen

Joseph P. Allen has recently served as a staff member on the President's Council on International Economic Policy and is currently working with the NASA Shuttle Project at the L. B. Johnson Space Center, Houston, Texas.

On July 30, 1971, the Apollo 15 lunar module *Falcon*, with commander D. R. Scott and lunar module pilot J. B. Irwin, descended along a steep trajectory over the 5000 meter Apennine mountain front and landed just inside this towering range on the hummocky plain of Mare Imbrium some 800 km north of the lunar equator. The safe landing marked the beginning of the scientific phase of the most ambitious expedition yet undertaken to explore and systematically study the moon. Although the mission itself ended formally when the command module *Endeavour*, piloted by A. M. Worden, splashed into the Pacific eight days later, the scientific adventure continues unabated as the returned samples and photographs are examined, and as the data which continuously stream from the emplaced experiments on the moon and from the subsatellite now orbiting the moon are analyzed.

From the outset the objectives of the Apollo 15 mission were distinctly different from those of the three successful lunar-landing missions that preceded it. The reliability of the launch vehicles, the spacecraft, and the associated support systems, including earlier versions of the spacesuits to be used on the lunar surface, had been proved; and the capability of the mission teams and astronaut crews to achieve safe, pinpoint landings had been extensively tested in previous missions. Consequently, the course of Apollo 15 had been determined almost exclusively by the experimental goals the program scientists wished to attain. These scientific objectives dictated that Apollo 15 be different from its predecessors in four major ways: (1) the *Falcon* carried with it an electrically powered lunar roving vehicle (the Rover) which could increase the traverse distance of the crew up to ten times that possible by walking, (2) both spacecraft, *Falcon* and *Endeavour*, had the capability of remaining on and round the moon considerably longer than the spacecraft of previous missions because of the increased consumable supplies carried on board, (3) the command-service module carried with it a complement of special instruments—the scientific instrument module (the SIM bay)—designed to study the lunar surface from orbit, and (4), perhaps most important, the mission was targeted into a landing site chosen for its potential for new scientific results and insight.

Scientific Objectives

The scientific objectives of the Apollo 15 mission to Hadley-Apennine can be divided into three distinct categories: (1) to carry out orbital measurements; (2) to deploy and activate a series of lunar surface experiments; and (3) to accomplish a comprehensive geological exploration of the complex Hadley Rille, the Apennine mountain front, and the mare surface accessible from the selected landing site. The orbital experiments (Esenwein et al., 1971) carried in the SIM bay of the service module included two large cameras for comprehensive mapping of the sunlit surface, a mass spectrometer, a laser altimeter, a remote-sensing geochemical package containing X-ray, gamma-ray, and alpha-particle spectrometers, and a small subsatellite to be placed in lunar orbit just before the homeward journey began. The subsatellite housed an S-band transponder to allow accurate tracking of its orbit from earth and subsequent determination of the moon's mass distribution, a flux-gate magnetometer to study the induced and permanent magnetic fields of the moon, and a series of charged-particle detectors to map the plasma flux around the moon as it moves in and out of the earth's magnetosphere and magnetopause.

Space limitations preclude discussion of the design, the operation, and the specific objectives of the SIM bay instruments; suffice it to say that their basic objective was to provide synoptic mapping and geochemical data for the large areas of the moon seen from orbit 60 nautical miles above its surface. From these data it will be possible to deduce correspondingly large-scale characteristics of this planet. Detailed knowledge returned from the various Apollo landing sites gives us extremely important calibration points for the interpretation of these remotely made measurements.

The surface instruments emplaced upon and into the lunar surface included a magnetometer, a seismometer, a cold-cathode ion gage, a solar-wind spectrometer, a suprathermal ion detector, a lunar-dust detector, a laser reflector, a solar-wind composition experiment, and a heat-flow experiment. Except for the heat-flow experiment, units similar to each of these instruments were already in operation at one, and in some cases at two of the three previous landing sites. In fact, for the seismometer, cold-cathode ion gage, suprathermal ion detector, lunar-dust detector, and laser reflector, the Apollo 15 package of surface experiments provided the all-important third corner for a triangle of experimental stations operating on the moon, a fact that had been weighted heavily in the selection of the landing site so far from the lunar equator.

The geologic investigation of the landing site was designed to collect data needed for a comprehensive understanding of the geological history of the Hadley-Apennine region, important in itself, but of further importance to the other surface experiments because much of their data must necessarily be interpreted within the framework provided by the regional geology. Earth-based photographs and lunar orbiter photographs available to the planning teams prior to the flight show the Apennine Mountains rising some 5 km above the mare surface which they partially encompass. As shown in Figure 1, these mountains, which form the southeastern boundary of Mare Imbrium, the largest and second-youngest circular basin on the moon, are thought to be large fault blocks uplifted as part of the system of concentric rings resulting from the Imbrium impact. These massive mountains are thus assumed to be composed of pre-Imbrium crustal rocks, perhaps mantled by ejecta from the impact event which produced the Imbrium basin (these ejecta were sampled by the Apollo 14 crew at the Fra Mauro site) and perhaps by ejecta from the older Serenitatis basin.

Hundreds of canyon-like formations, known as rilles, are found on the moon, many of them concentrated around the edges of the circular basins. They are particularly enigmatic features

Figure 1. Lunar orbiter photograph of the regional features surrounding the Hadley-Apennine landing site.

(Baldwin, 1963). Although some meander like rivers and some show straight fault-like walls, in most cases their sheer size sets them apart from any possible terrestrial equivalents. Of these, Hadley Rille is thought to be one of the youngest and freshest sinuous rilles on the moon. It meanders from its probable source, an elongate depression south of the landing site, through the mare surface, and occasionally touches the mountain massifs.

The mare region between the rille and the mountains is a

gently rolling, dark plain at the margin of Mare Imbrium. Lying on the mare plain just north of the landing site is a series of low, irregularly shaped hills (the North Complex) that may consist of ejecta from nearby large craters, or may be landforms of volcanic construction that occurred during the last stages of mare-basin filling. The secondary crater cluster located just south of the landing site lies on an ejecta ray that is traced back to either of the two Copernican-age craters Autolycus or Aristillus found in Mare Imbrium 150 and 240 km to the northwest, respectively.

The major objectives of the geological exploration were to visit, photograph, and sample as thoroughly as possible (in order of decreasing priority) the Apennine mountain front, Hadley Rille, the mare plain, the North Complex, and the secondary crater cluster. These objectives were recognized before the mission to be extremely ambitious, overly optimistic even; yet they were very nearly met during the three geology traverses of the mission in spite of several surprises encountered along the way.

* * * * *

Summary of Surface Geology

I shall discuss initial results from just a few selected experiments and attempt to point out possible interpretations that the new data may suggest. Because many of the results are still preliminary, these interpretations may change as analyses continue.

The mobility provided by the lunar Rover permitted Scott and Irwin to explore an extensive area around Hadley Base. The crew was able to accomplish four of the major objectives established prior to the mission. During the three traverses they investigated the Apennine front along the mountain Hadley Delta, south of the landing site, Hadley Rille at locations west and southwest of the landing site, and the mare plain at various locations. From the stop at Dune crater information was also obtained about the secondary crater cluster. Finally, because of the unanticipated loss of time in recovering and stowing the deep corestem, the North Complex was not visited; even so, some information was collected by photographing, with a telescopic lens, the south-facing exposures of this intriguing feature.

An extensive geological description of the major features in the Hadley area has been constructed (Swann et al., 1971) from the data returned. In general, the Apennine Mountains show gentle-to-moderate slopes, with very subdued, rounded outlines, in sharp contrast to many of the fresh and rugged mountain ranges found on earth. Large blocks are scarce on the mountainsides, even near the few fresh craters, suggesting that most of the debris thrown out by the cratering process has been subsequently transported downhill, leaving only a thin debris cover on the upper slopes. Quite surprisingly, sets of stark, clearly etched, parallel linear patterns appear on most of the mountain faces, an observation totally unexpected before the mission. These major lineaments may represent the expressions of sets of compositional layers, or sets of regional fractures showing through the regolith cover of the slopes; or, in the cases of the very finely spaced lineaments, they may be an optical illusion caused by the unscattered sunlight falling at low angle on the randomly irregular surfaces of these slopes. Although an unequivocal interpretation of these striking patterns has not been possible in the initial analyses of all the photographs, it seems clear that some of the parallel patterns contain exciting clues to the processes which formed the premare lunar surface.

Hadley Rille, shown in Figure 2, was visited during the first and third traverses. Several massive outcrops of bedrock on the near side of the rille rim were sampled. They represent what are perhaps the first genuine bedrock samples returned from the moon, a particularly important identification if the fundamental stratigraphy of the local regolith is to be understood (Goles, 1971). In addition, the exposed rille walls on both the near and the far sides were photographed in detail. The bedrock strata visible in these many photographs have thicknesses as great as 60 meters, are distinctly layered along the horizontal or near-horizontal, show evidence of columnar jointing, and exhibit varying surface textures and albedos. All these features are typical of individual lava flow units.

Thus, we now have clear evidence that the Hadley plain, and perhaps all mare plains, are underlain by a series of lava flows similar to those found in many lava fields here on earth. An unequivocal explanation of the origin of the rille itself is not so easily made. However, initial studies (Howard et al., 1971) of the depth-to-width ratios derived from the Apollo 15 surface and orbital photography of this mysterious feature give tentative support to the theory that it may be the result of incomplete collapse of a buried lava tube of size that is enormous by terrestrial standards.

Figure 2. A view to the north along Hadley Rille, taken by Irwin during the first geology traverse. Scott is unloading equipment from the Rover, parked south of Elbow crater.

The dark plain of the mare surface is generally smooth to gently undulating, heavily cratered, and hummocky. Except for the rougher ejecta blankets around the numerous craters, rocks cover only some one percent of the surface. (This fact, determined only after the landing, was crucial to the planned traverse distances, because the average driving speed attainable by the Rover is critically dependent on surface roughness.) Initial study of the photographs of the landing site indicates a probable subdivision of the mare into four geological units characterized by differences in crater populations and surface textures. The morphology of the craters in the mare suggests its age to be late

Imbrian to early Eratosthenian (Mutch, 1970), and the specific sampling sites visited by the crew span this age range. For example, a particular 15 meter-diameter crater with a widespread glassy ejecta blanket probably represents the youngest surface feature yet sampled on the moon. Age dating of these samples is awaited with high anticipation.

A total of 77 kg of lunar samples was returned by the Apollo 15 crew. These samples comprise rocks that weigh from 1 g to 9.5 kg (the largest sample was scooped up by Scott near the rille rim and is now called the "Great Scott Rock," much to his consternation), three core tubes, a deep corestem, and a variety of soil samples taken from numerous sampling locations including the hillside of Hadley Delta, the rim of Hadley Rille, and across the mare plain. In general, the samples reveal (Lunar Sample Preliminary Examination Team 1971) the immense variety of materials that were accurately recognized and described by Scott and Irwin while they selected the materials. In spite of the awkward sampling procedures made necessary by the bulky spacesuits, the astronauts located, described, documented, and returned samples representative of the area, including anorthosite; vesicular basalts with phenocrysts of plagioclase, pyroxene, and olivine; complex breccias with an assortment of well-defined clasts; rocks in various stages of shock metamorphism; rounded and angular glass fragments; soils of different textures and granularities; and samples displaying color differences, hints of surface coatings on fractures, slickensides, glass coatings, and a variety of fracture patterns.

Even a brief summary of the some 350 individual rock samples is clearly beyond the scope of this article. However, a twofold classification of the returned materials shows a simple relationship between the major rock types and the major geological units. Many of the samples from the mare are basalts roughly similar in composition to the basalts returned from the Apollo 11 (Mason et al., 1970), the Apollo 12, and the Luna 16 mare sites. In particular, these basalts are high in iron, with a correspondingly high iron-to-magnesium ratio, and low in sodium, in notable contrast to typical terrestrial basalts. On the other hand, the majority of samples from the Hadley-Apennine mountain front are breccias that contain clasts of basalts, but basalts which are distinctly different in both appearance and chemical content from the mare basalts. Specifically, these basalt clasts are considerably richer (by up to a factor of 2) in aluminum than the typical mare-type basalt. The possibility of finding aluminum-rich material in the lunar highlands had been pointed out (Wood et al., 1970) prior to the Apollo 15 landing, but it was generally felt that this material would be in the form of anorthosite, a rock of high aluminum content.

Scott and Irwin did indeed return a beautiful specimen of anorthosite (the 269-gram sample #15415, also dubbed the "Genesis Rock"), but we now suspect that this sample is not characteristic of the material from the upper sections of the Apennine Mountains around Hadley; rather it is probably an interloper thrown from some distance onto the mountain slope by some traumatic event in its history. In any case, the discovery of an aluminum-rich igneous rock as an abundant constituent of lunar highlands raises a perplexing question. Why do the highland rocks, formed perhaps around 4½ billion years ago, differ in make-up so strongly from the mare rocks, formed around 3½ billion years ago? Obviously these two types of igneous rock, so fundamental to the moon, were derived from very different magma sources. Could the moon have accreted in inhomogeneous zones that were never thoroughly mixed?

The soils returned from Hadley Base are similar in most respects to soil samples returned from previous missions (Wood, 1970) except for a peculiar component in these samples of green glass spheres never before reported in lunar soils. Interestingly enough, the chemical composition of the soil samples, particularly from the mare regions, is distinctly different from the composition of the rocks from the same locales. However, a linear correlation involving the iron and the aluminum constituents of the soils suggests very convincingly that lunar soils may be derived from a range of rock materials, with the two end-members of this range being the iron-rich basalt so typical of the mare and the aluminum-rich basalt found so abundantly at the Apennine front.

A total of 4.6 kg of material was returned from the Hadley site in the form of core samples. Three of the core tubes were hammered into the lunar surface during the traverses. Initial X-radiographs of the lunar material within these core tubes reveal distinct layering delineated by discontinuities in the spectrum of soil textures, fragment sizes, and material densities. The fourth core was drilled 2.4 meters into the surface with the percussive drill that was also used to emplace the heat-flow probes. The upper layers of the surface regolith were far more resistant to this drilling operation than had been anticipated before the mission. The resulting mechanical problems with the drilling, the extraction, and the subsequent separation of the sections of this deep corestem into lengths which would fit into the spacecraft were among the most exhausting faced by the crew during their surface operations.

These problems also posed difficult questions, requiring immediate answers, to the science team guiding the surface activities from Mission Control. In spite of the difficulties, the deep corestem was ultimately recovered and returned; and the X-radiographs of it show more than 50 individual layers with thicknesses from 0.5 to 21 cm, a truly remarkable stratigraphic record. In fact, the episodes represented by these layers may be a significant fraction of the post-volcanic history of Mare Imbrium.

A study of the layers should provide valuable insight into the only erosional process of apparent current importance on the moon, that of micrometeorite "gardening." Additionally, a study of the various isotopes within these strata could provide a history of solar activity that stretches back hundreds of millions of years into the past. The potentially rich scientific return from this corestem is gratifying in view of the agonizing moments spent by the mission team and the astronaut crew in deciding to give up a portion of the third geology traverse in favor of a last attempt to recover the corestem, even when it seemed at the time to be irretrievably stuck in the lunar surface.

Relative concentrations of noble-gas isotopes measured in the samples from Hadley Base are similar to the abundances and abundance ratios previously found in lunar materials. Amounts of the spallation-produced isotopes, resulting from cosmic-ray bombardment, such as neon-21, krypton-80, and xenon-126, can be used to calculate exposure ages (not to be confused with crystallization ages) of the rocks. The measured exposure ages lie in the range of 50×10^6 to 500×10^6 years. Thus some of the samples picked up at Hadley Base had been lying on the surface, more or less undisturbed, since the beginning of the Cambrian period on earth.

Summary of Surface Experiments

The purpose of the seismic experiment (Latham et al., 1971b) is to study the lunar-surface vibrations, from which interpretations of the internal structure and the physical state of the moon can be determined. Sources of seismic energy may be internal (from moonquakes) or external (from impacts of either meteoroids or spent space hardware). Regardless of whether the source is internal or external, a straightforward determination of source location requires at least three instruments, suitably spaced, monitoring the event of interest. The Apollo 15 seismometer station represents the third in the network of seismometers now operating on the lunar surface; thus the deployment of this unit marked a vitally important step in the investigation of this unique planet.

It was well established from earlier data that seismic energy deposited at the surface by an impacting object is confined by efficient surface scattering for a surprisingly long time within a shallow surface layer. Nevertheless, the energy slowly dissipates through interior propagation to more distant parts of the moon, and probably all but the smallest of impact signals from all parts of the moon are detected at the operating stations. As expected, the impact of the third stage of the Apollo 15 launch vehicle was recorded by both the Apollo 12 and Apollo 14 seismometers, and the much smaller impact of the abandoned ascent stage of *Falcon* was recorded by all three instruments. The seismic refractions from these two impacts provided the first clear-cut evidence for pronounced sound-velocity differences in the lunar subsurface, suggesting the presence of a lunar crust and mantle analogous to what is found on earth.

The thickness of the lunar crust is between 25 and 70 km in the region of the Apollo 12 and 14 landing sites, and the velocity of the compressional waves in this crustal material is between 6.0 and 7.5 km/sec, which is a range that spans the velocities expected for the rocks found on the lunar surface. The transition from the crustal material to the mantle material may be gradual, starting at about 25 km depth, or rapid, with a sharp discontinuity at a depth near 70 km. In either case, the compressional-wave velocity reaches about 9 km/sec in the mantle material, and the contrast in elastic properties of the materials making up these two major layers is at least as great as the contrast that exists between the rocks which make up the crust and mantle of the earth.

The major part of the natural lunar seismic energy detected by the network is in the form of periodic moonquakes that occur near times of perigee and that originate from ten or more separate locations. However, a single source located approximately 600 km south-southwest of the Apollo 12 and 14 sites, at a depth of approximately 800 km, accounts for 80 percent of the seismic energy detected. From the nearly identical signature of the events originating at this focal zone, it is concluded (Latham et al.,

1971a) that the source region is confined to just a few kilometers in extent. Furthermore, the release of seismic energy from this great depth (slightly greater than for any known earthquake sources) indicates that the lunar interior, almost halfway to the center of the planet, must still be rigid enough to support appreciable stress. This fact places strong constraints on thermal models of the moon.

Besides the periodic moonquakes, episodes of frequent small moonquakes (Latham et al., 1971b) have been discovered. Individual events may occur as frequently as every two hours for periods lasting up to several days. The source of the moonquake swarms is at present unknown, but they may result from continuing minor adjustments to stresses in the outer shell of the moon. Even with the periodic moonquakes and the moonquake swarms, the average rate of seismic energy released within the moon is still far below that of the earth; it follows that the outer crust and mantle of the moon must be relatively cold and stable as compared to the crust and mantle of the earth.

The Apollo 15 magnetometer (Dyal et al., 1971) was deployed to study intrinsic remanent magnetic fields at the lunar surface and to observe the global response of the moon to large-scale magnetic fields imposed upon it. Such fundamental properties of the lunar interior as electrical conductivity, magnetic permeability, and internal temperatures can be derived from these magnetic measurements. The three flux-gate sensors of the Apollo 15 instrument show a steady magnetic field of approximately 5 γ ($1\gamma = 10^{-5}$ gauss) at the Hadley site, which is considerably smaller than the 38γ field measured at the Apollo 12 site, and the $50\text{-to-}100\gamma$ fields measured at the Apollo 14 site. Far more surprising, however, is the fact that the majority of the samples returned from all the sites show remanent magnetic fields so large that they must have formed in the presence of magnetic fields several orders of magnitude greater than we now measure at the surface. What was the original source of this magnetism—the earth in close proximity to the moon, or the interior of the moon itself?

The response of the moon to the externally imposed variable magnetic fields of the sun and the earth yields data that can be interpreted in terms of a spherically symmetric, three-layer model of the moon with a thin outer crust (of approximately 80 km thickness) of temperature about 440°K, and intermediate layer of temperature about 800°K, and an inner core (of approximately 1000 km radius) of temperature around 1100°K, where one assumes that the chief constituent of the interior is olivine, a mineral of possible abundance within this planet.

The third and largest U. S. retroreflector for laser ranging from earth (Faller et al., 1971) was deployed at the Hadley site during the first traverse and now provides the crucial north-south baseline separation with the Apollo 11 and the Apollo 14 reflectors. Successful range measurements to this 300-corner-cube array were made shortly after the *Falcon* lifted off several days later. The better signal-to-noise ratio available with this larger reflector enables more frequent ranging measurements to be made, and it will enable measurements to be carried out by telescopes of smaller aperture than heretofore possible. It is encouraging to note that none of the reflectors on the lunar surface seem to be suffering any deterioration of their optical properties as they are exposed to the space environment. This durability is important since ranging measurements to these benchmarks in space over many years will be required to accumulate the data for the planned astronomical, geophysical, and general-relativity investigations.

Emplacement of the first heat-flow experiment (Langseth et al., 1971) into the lunar surface was begun during the first surface extravehicular activity period and, because of difficulties encountered with the emplacement operation, was finally finished at the end of the second extravehicular activity. The drilling of the 3-meter-deep holes into which to drop the two slender temperature-sensing probes proved to be far more demanding than had been anticipated. In spite of the problems encountered, one of the two probes is now placed at about 1.5 meters and the other at about 1.0 meters into the lunar soil, and the subsequent measurements of the temperature gradients and the thermal conductivity of the lunar regolith have provided a remarkably accurate value of the heat flux from the interior of the moon.

The heat loss of a planet is, of course, related directly to its rate of internal heat production and to its internal temperature profile; hence, a heat-flow measurement gives information about the abundances of long-lived radioisotopes within a planet and, in turn, leads to an increased understanding of its thermal evolution. In the case of the moon, we now know that the lowest temperature anywhere within it exists just 70 cm below the surface because of the impressive insulating properties of the upper few centimeters of soil. The temperature increases from this depth at the rate of $1.75 \pm 2\%$ °K/m, and the thermal conductivity of the regolith increases with depth as well. These results yield a heat-flow from below Hadley Base of $3.3 \times 10^{-6} \pm 15\%$ watts/cm^2, a value very close to one-half the average heat-flow from the planet earth. The most fashionable models of the moon to date predict a heat-flow value at least a factor of two below what is now being measured. Consequently, this surprising number, if it is representative of the whole moon and not just a local anomaly, implies a radioactive content for the moon considerably higher than previously assumed and, more important, considerably higher than the radioactive content usually accepted for earth.

In spite of the severe demands placed on the crew in the final minutes of the last traverse period, a quick demonstration experiment was conducted just before they left the lunar surface. Making good use of the superb vacuum and the weak gravity unique to the lunar environment, Scott released simultaneously, from shoulder height, a suitably heavy object (a metal geology hammer) and a suitably light object (the tail feather of a falcon, donated out of molting season under strong objection) while in clear view of the TV camera. These two objects were observed to undergo the same acceleration and to strike the lunar surface simultaneously, a result which has been verified in the instant-replay TV tapes to within the accuracy of the simultaneous release of the objects. The result had been long predicted by well-established theory, but it was nonetheless gratifying, considering the number of viewers that witnessed the experiment, and particularly reassuring, considering the fact that the homeward journey was based critically on the validity of the gravitational principle being demonstrated.

Summary of Orbital Experiments

The purpose of the X-ray experiment (Adler et al., 1971) carried in the SIM bay was to map the elemental constituents (specifically aluminum, silicon, and magnesium) of the upper layer of the lunar surface by measuring the fluorescent X rays produced by the interaction of incident solar X rays with the lunar surface material. Starting with the first hours of data return, this experiment has given very exciting information. In general, the long-suspected major compositional differences between the two fundamental lunar features, the maria and the highlands, are confirmed; and more subtle compositional differences within both the maria and the highlands are strongly indicated.

144   MAN'S FINITE EARTH

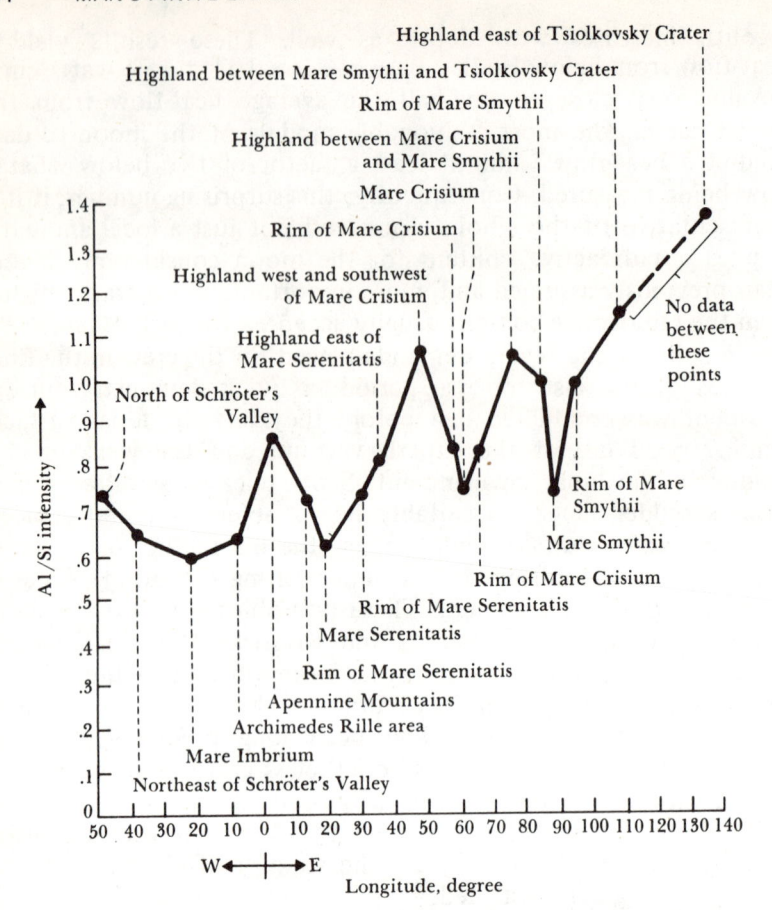

Figure 3. Plot of Al/Si intensity ratios along a northerly ground track (from Adler et al., 1971), derived from data from the X-ray spectrometer experiment.

For example, as shown in Figure 3, the aluminum-to-silicon intensity ratio is highest over the highlands, lowest over the maria, and, as would be expected, intermediate over the boundary regions between these major units of the moon. The extremes for this ratio vary from about 0.6 to 1.4, with a tendency for the value to increase from the western mare to the highlands of the eastern limb. Furthermore, a striking correlation exists between the aluminum-to-silicon intensity ratio and the values of surface albedo along the ground track surveyed by the X-ray experiment. In essence, the data confirm the idea that the maria and the highlands are indeed chemically different. The distinguishing albedo differences between these major lunar features, easily seen from earth by anyone who has examined the face of the "man in the moon," must be, in part at least, the signature of this chemical difference.

The high aluminum component of the samples returned from the Hadley-Apennine front is certainly related to the high aluminum content measured across all the highland regions, and the correspondingly low aluminum content of the mare basalts returned from all the mare sites is consistent with the low aluminum values measured across all the mare regions. It is interesting to note that the sharp change in the aluminum-to-silicon ratio between the highland and the mare areas places stringent limitations on the amount of horizontal displacement the highland materials could have undergone after the aluminum-poor lavas flooded into the enormous mare basins. This strongly rules out certain theories which invoke large-scale horizontal transportation mechanisms to explain mare filling. Finally, the X-ray data support very convincingly the theory that the moon developed a well-differentiated, aluminum-rich crust shortly after its formation.

The gamma-ray spectrometer (Arnold et al., 1971) was designed to measure from lunar orbit the gamma-ray activity of the lunar-surface materials over which the spacecraft flies. It is well known that the near-surface abundance of naturally occurring radionuclides is a sensitive function of the degree of chemical differentiation undergone by the moon, and thus their measured abundances relate directly to the origin and evolution of this planet. The gamma-ray data show notable regional differences in the amounts of radioactive elements across the surface of the moon. Specifically, the typical mare regions have a higher radioactive level than the average highland regions; and, although there seems to be a wide range of radionuclide content across the highlands, the average highland radioactivity content is considerably lower than that inferred from the few samples returned on previous missions, which were thought to have originated in highland areas. This is particularly true in the case of the Apollo 14 samples. Their high radioactivity, with respect to the level that seems best to represent highland material in general, throws open to question the interpretation of these samples as ancient highland rocks that were excavated and thrown onto the Fra Mauro site by the Imbrium impact.

Local mass concentrations (Sjogren et al., 1971) and local magnetic field anomalies (Coleman et al., 1971) are being systematically mapped by the small subsatellite that was launched from the orbiting command-service module just before the homeward journey began. The strongest of the measured magnetic anomalies, which are thought to be due to the remanent magnetization of surface rocks, ironically is found over the crater Van de Graaff, suggesting to electrostatic accelerator physicists at least that other electromagnetic anomalies should be found over the large craters Gauss and Weber!

* * * * *

Conclusions

It is certainly premature to offer, at this stage, any far-reaching theories for the origin, evolution, and present condition of the moon, much less to indicate major revisions required in these theories based on the newly returned Apollo 15 data. Nevertheless, over the past few years an enormous increase in our understanding of this fascinating planet has occurred (for an excellent review of current lunar knowledge and theories, see Hinners, 1971).

* * * * *

We now know that the moon was formed about 4.6 billion years ago during the beginning of the solar system. It is a heterogeneous planet, as reflected both in the wide spectrum of materials it comprises and in the enormous variety of landforms it exhibits. Furthermore, we have a general idea of the fundamental chapters that make up its early history, including its accretion, the formation of its initial crust showing pronounced chemical differentiation, the construction of its vast circular basins by massive impact events, the volcanic flooding of these basins to form the maria, and the subsequent modification of its surface by continued volcanic activity and impact events.

The moon harbors no life and apparently none of the complex organic molecules that serve as the building blocks of the forms of life familiar to us. The rocks we find on the surface of the moon are of igneous origin. Perhaps more important, they are extremely old, with the oldest going back to the very first days in the birth of the moon and with the youngest corresponding in age (as sheer good fortune would have it) to some of the oldest rocks found on the surface of the earth; thus, an essentially unbroken record of the history of the solar system has been left within the materials of these two planets together. The birth and early adolescence of the solar system is written in the lunar samples; the late adolescence and adulthood of the solar system is written in the terrestrial and the lunar samples. How completely we will be able to decipher this record remains to be seen. In any case, the knowledge that the record exists, and the hope that much of it can be unraveled, provides an intriguing challenge to the ingenuity of man.

The moon is only the second planet that man has been able to study in situ and later at great lengths in earthbound laboratories. This is an obvious, yet vitally important, fact because, just as a nuclear physicist studies the widest possible spectrum of nuclear isotopes in order to understand any one particular nucleus more clearly, so must a planetologist study in detail as many examples of planets as possible in order to understand the planet earth more clearly.

In these days when environmental science has finally found a new and much needed relevance, it is becoming increasingly obvious that nothing will contribute more to the continued life of man on earth, and to the continued quality of this life, than a comprehensive understanding of the vulnerable planet on which he lives. Our study of the primitive moon, in relation to the complex earth, is contributing significantly to this comprehensive understanding. Ironically, despite this, there is no doubt that the initial sense of excitement and adventure involved in the space journey to the moon has diminished greatly in many quarters. (See Mendelssohn, 1971, for an interesting discussion of another major national effort with a far less obvious final goal.)

Nevertheless, the scientific adventure of these missions continues to increase markedly as the present surface explorations unfold, and as the resulting samples and returning data are received and studied on earth. The study of the moon is a great adventure in thought. A comprehensive understanding of the origin and evolution of the earth's only natural companion in space will be a milestone in the history of man. It is particularly satisfying to realize that the scientific contributions of the Apollo 15 expedition have provided an enormous quantum jump in the information available to this continuing adventure.

References

Adler, I., Trombka, J., Gerard, J., Schmadebeck, R., Lowman, P., Blodgett, H., Yin, L., Eller, E., Lamothe, R., Gorenstein, P., Bjorkholm, P., Harris, B., and Gursky, H. 1971. X-ray fluorescence experiment. *Apollo 15 Preliminary Science Report*, NASA SP-289, Chap. 17.

Arnold, J. R., Peterson, L. E., Metzger, A. E., and Trombka, J. I. 1971. Gamma-ray spectrometer experiment. *Apollo 15 Preliminary Science Report*, NASA SP-289, Chap. 16.

Baldwin, R. B. 1963. *The measure of the moon.* Chicago: University of Chicago Press.

Coleman, P. J., Schubert, G., Russell, C. T., and Sharp, L. R. 1971. The particles and fields subsatellite magnetometer experiment. *Apollo 15 Preliminary Science Report*, NASA SP-289, Chap. 22.

Dyal, P., Parkin, C. W., and Sonett, C. P. 1971. Lunar-surface magnetometer experiment. *Apollo 15 Preliminary Science Report*, NASA SP-289, Chap. 9.

Esenwein, G. F., Roberson, F. I., and Winterhalter, D. L. 1971. Apollo in lunar orbit. *Astronautics and Aeronautics* 9(4):52-63.

Faller, J. E., Alley, C. O., Bender, P. L., Currie, D. G., Dicke, R. H., Kaula, W. M., MacDonald, G. J. F., Mulholland, J. D., Plotikin, H. H., Silverberg, E. C., and Wilkinson, D. T. 1971. Laser ranging retroreflector. *Apollo 15 Preliminary Science Report*, NASA SP-289, Chap. 14.

Goles, G. G. 1971. A review of the Apollo project. *American Scientist* 59:326-31.

Hinners, N. W. 1971. The new moon: a view. *Review of Geophysics and Space Physics* 9(3):447-522.

Howard, K. A., and Head, J. W. 1971. Regional geology of Hadley Rille. *Apollo 15 Preliminary Science Report*, NASA SP-289, Chap. 25, Part F.

Langseth, M. G., Clark, S. P., Chute, J. L., Keihm, S. J., and Wechsler, A. E. 1971. Heat-flow experiment. *Apollo 15 Preliminary Science Report*, NASA SP-289, Chap. 11.

Latham, G., Ewing, M., Press, F., Sutton, G., Dorman, J., Nakamura, Y., Toksoz, N., Lammlein, D., and Duennebier, F. 1971a. Passive seismic experiment. *Apollo 15 Preliminary Science Report*, NASA SP-289, Chap. 8.

———. 1971b. Moonquakes. *Science* 174:687-92.

Lunar Sample Preliminary Examination Team. 1971. Preliminary examination of lunar samples. *Apollo 15 Preliminary Science Report,* NASA SP-289, Chap. 6.

Mason, B., and Melson, W. G. 1970. *The lunar rocks.* Washington, D. C.: Wiley-Interscience.

Mendelssohn, K. 1971. A scientist looks at the pyramids. *American Scientist* 59:210-29.

Mutch, T. A. 1970. *Geology of the moon*. Princeton, N. J.: Princeton University Press.

Sjogren, W. L., Gottlieb, P., Muller, P. M., and Wollenhaupt, W. R. 1971. S-band transponder experiment. *Apollo 15 Preliminary Science Report*, NASA SP-289, Chap. 20.

Swann, G. A., Bailey, N. G., Batson, R. M., Freeman, V. L., Hait, M. H., Head, J. W., Holt, H. E., Howard, K. A., Irwin, J. B., Larson, K. B., Muehlberger, W. R., Reed, V. S., Rennilson, J. J., Schaber, G. G., Scott, D. R., Silver, L. T., Sutton, R. L., Ulrich, G. E., Wilshire, H. G., and Wolfe, E. W. 1971. Preliminary geologic investigation of the Apollo 15 landing site. *Apollo 15 Preliminary Science Report*, NASA SP-289, Chap. 5.

Wood, J. A. 1970. The lunar soil. *Scientific American* 223:14-23.

Wood, J. A., Dickey, J. S., Marvin, U. B., and Powell, B. N. 1970. Lunar anorthosites and a geophysical model of the moon. *Geochim. Cosmochim. Acta. 34*, Suppl. 1:965-88.

12 Puzzling Facts from Lunar Exploration

Thornton Page

Thornton Page is a research astrophysicist in the Naval Research Laboratory working at the L. B. Johnson Space Center, Houston, Texas.

Ever since Galileo was the first to look at the moon through a telescope in 1609, men have studied our large satellite with more and more complex instruments—large telescopes, spectrographs, radar, radio telescopes, and infrared sensors. By 1968, astronomers thought that they understood the moon accurately (1)—its maria (plains), highlands (mountains), thermally insulating surface layer and temperatures (−170 to +100°C), lack of an atmosphere, its craters, volcanic domes, shape (prolate spheroid), size (3476 km diameter), mass (1/80 of the earth's), surface gravity (1/6 of the earth's), average density (3.34 g/cm^3), orbit (384,405 km, inclined 5° to the ecliptic, eccentricity 0.055), librations (17° shift in the point facing the earth), and history (increasing distance from the earth due to tidal effects). There was some uncertainty about the origin of the moon, but it was generally agreed (2) that both moon and earth were formed from the same material in the "solar nebula," a vast cloud of gas and dust that collapsed to form the sun about 5 billion years ago.

Starting with NASA's unmanned Surveyor-V, -VI, and -VII landings on the moon in 1967 and 1968, doubts were cast on several of these well entrenched concepts. Chemical analysis of the lunar surface materials shows small but important differences from the composition of earthly materials (higher radioactive content, no water in minerals, higher average content of Ca, Al, Ti, and rare

Source: NASA.

earths, and lower content of Na, Mg and Fe). The trackings of spacecraft near the moon have determined its mass with high accuracy (±0.05%) and indicated large mass concentrations ("mascons") of higher than average density near the surface. The six Apollo manned landings (Figure 1) have produced over 350 kg of samples carefully stored at the Johnson Space Center (JSC) near Houston, Texas; collected a remarkable amount of data on surface formations, the interior, and the space between earth and moon; and left dozens of instruments on the surface that will telemeter more data to JSC for several years. Preliminary reports have been discussed at several Lunar Science Conferences (3) held annually at JSC. I will try to summarize the findings that puzzle the several hundred scientists involved. These inconsistencies center on the moon's structure, temperature, magnetization, composition and origin.

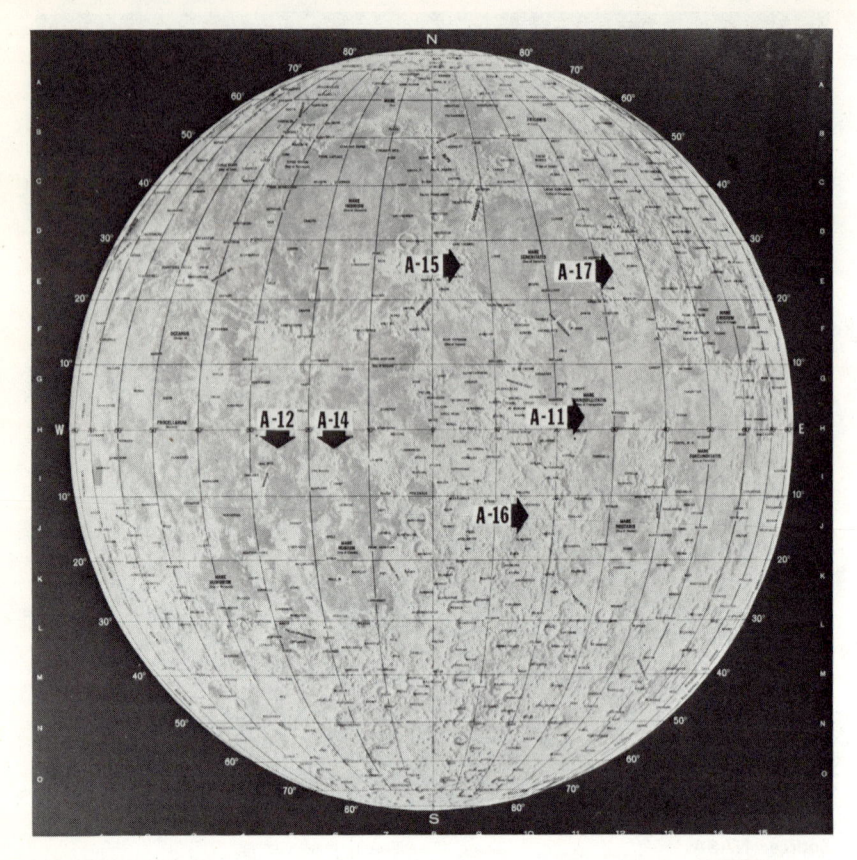

Figure 1. Locations of the six Apollo landings on the lunar surface.

Surface Features

There is no doubt but that the lunar craters were formed by meteroid impacts of various sizes and at various times in the past. "Young" craters are whiter than "old" ones, in general. A small crater, 10 to 20 meters in diameter, requires a ton-sized meteoroid impact—an infrequent event. However, one gram-size meteoroid hits each square meter of the lunar surface every million years, on the average, and these missiles "plow up" the lunar soil into a rubble of rock fragments, dust, and glass beads that result from the melting in high-energy impacts. It is generally agreed that the large "seas" or maria were formed by early impacts of very large meteoroids up to 100 km in diameter which pushed up the mountains or highlands around the rims of the maria. One type of feature still unexplained is the canyon-like rille (one of which is shown in Figure 1 on p. 133 and Figure 2 on p. 136). There is no water on the moon to dig canyons; dust flows might do it, but dust is sticky—not slippery—in the vacuum conditions on the moon. If a lava tube were emptied and then collapsed, it would leave a rille-like trough, but Figure 1 (p. 133) shows no obvious lava flow at the end of Hadley Rille.

Mascons, Layering, and Rigidity

Very accurate tracking of lunar orbiters, using the Doppler shift in their 2.3-megacycle radio signals to measure minute changes (1 mm/sec) in orbital velocity, showed 200-milligal gravity anomalies near the centers of seven circular maria. These local maxima in lunar gravity show that each mascon is roughly a disk of higher density material about 200 km diameter and 15 km thick— presumably the flattened remains of a large meteoroid whose impact formed the mare. Some, but not all of the maria were later filled by liquid lava that froze in a smooth sheet, giving the resemblance to the level water surface of a sea (mare). The impact energy undoubtedly melted parts of the original rock surface and meteoroid, but geophysicists are not sure that this melting would provide enough lava to fill each mare.

By analogy to the earth, you might think that a major impact would tap pools of molten lava under the surface rocks, but there is good reason to doubt high-temperature lava beneath the moon's crust. In fact, the heavy mascons have been prevented from sinking to a depth where the moon's internal density is equal to their density (about 4 g/cm^3). Moreover, the moon's shape and measured moments of inertia definitely show that it is not in hydrostatic equilibrium. The rigidity of cold material seems to be required also for the moonquakes at 800 km depth, detected and located by the three highly sensitive seismographs telemetering records back from the Apollo-12, -14, and -15 sites. These seismic records show three major layers below the rubble and rocks exposed on the surface: The crust material is solid basalt, with seismic-wave velocity less than 5 km/sec down to about 25 km depth where the velocity jumps to 7 km/sec. From 25 to 65 km depth the density and elastic properties remain the same and the material is thought to be rigid eclogite or anorthosite rock.

At 60 to 65 km depth the velocity jumps again, and the 9 km/sec indicates rigid olivine rock. The topmost strata can be seen as layers in various places on the walls of Hadley Rille in Figure 2 (p. 136): the "regolith" (rubble) about 1 m deep on top, then a dark layer 2 or 3 m thick, then a light-colored cliff 25 m high, then some horizontally striped rocks about 10 m thick, then talus (rubble) for the rest of the 350-m depth.

The very existence of these layers is an enigma; the upper ones seem to indicate several lava flows, possibly interspersed with dust and rubble thrown out of large impact craters. Other layers can also be seen on the side of Mt. Hadley, the ridge seen in the upper right of Figure 1 (p. 133), at an elevation of 4000 m above the Mare Imbrium plain. On earth, horizontal strata in sedimentary rocks (formed on sea bottoms) are common, but no one had expected similar structures on the waterless, airless moon.

Temperature and Shape

The puzzle of internal temperature was further compounded by measurements of heat flow on the Apollo-15 mission.* Very accurate thermometers in two long tubes were placed in two holes drilled 100 and 140 cm deep, and their readings were radioed to Houston, giving a continuous record of the temperature. After things had settled down, the temperatures (accurate to $0°.001$ C) showed a thermal gradient of $1.°75$ C/m downward, indicating 3.3 microwatts/cm^2 (heat flow 1/3 of the earth's). If this gradient continues downward, the temperature at 1 km depth would be 1950° K, and the rock material would not be rigid. Although the higher conductivity of solid rock (compared to the regolith) would reduce the deeper thermal gradient, this heat flow measurement indicates high internal temperature—at odds with mascon support, layering, deep moonquakes, and a rigid shell to hold the non-hydrostatic shape of the moon.

The figure of the moon, previously thought to be prolate—a "frozen tide" from an earlier time when the moon was closer to earth—has been measured with high accuracy from the orbiting Command Module (CM) during the Apollo-15, -16, and -17 missions. Three cameras and a pulsed-laser altimeter were mounted on the CM, one camera pointed 6° above the horizon to photograph stars, one straight down, and one rocking about 80° either

---

*Confirmed by measurements on Apollo 17, the last Apollo mission.

side of downward to take panoramic photos from horizon to horizon. The "Pan Camera" gives a map of the sunlit surface; the vertical ("Metric") camera shows where the CM was on that map; the Star Camera shows the accurate downward direction in space, relative to the stars, and the pulsed ruby laser gives the CM height above the ground, accurate to ±2 m. Radio tracking from the earth helps to determine the CM's exact orbit in space around the moon's center of mass, and the laser gives the distance (about 95 km) from that orbit down to the surface. The results for several bands around the moon show that it is oblate—tomato-shaped— with a 6-km "dimple" on the back side (no frozen tidal bulge). The center of mass is about 2 km closer to the earth than the center of the best-fitting sphere, which has 3476 km diameter.

Magnetic Field

Accurate magnetic measurements have been made from lunar orbiters and on the surface. As expected, these show that the general dipole magnetic field of the moon (5 gammas) is only 10^{-4} of the earth's (500 mG) at the surface. However, there are many local anomalies of 40 to 50 gammas, some associated with craters that deflect the solar wind plasma as it passes them near the sunset-sunrise line. It is likely that if the interior were molten the moon would have a stronger magnetic field, generated like the earth's by dynamo action in the rotating liquid core. Lab measurements of lunar rock samples show that they are magnetized, as many terrestrial rocks are, indicating that they were formed in a magnetic field, the direction and strength of which can be estimated from the rock's magnetization. This accounts for the local magnetic anomalies on the lunar surface, but it raises the question of why the lunar rocks got magnetized when they formed 3.5 to 4 billion years ago. The magnetic field then must have been several thousand gammas to account for the rocks' magnetization (which is caused by lining up the small magnetic elements in liquid rock as it freezes).

Composition

Lunar surface materials are classed as mare-type and highland-type, of which the latter is probably the original crust. Mare-type rocks, breccia, soil, and dust consist of hardened lava from impact meltings or volcanoes, and probably lie on top of highland-type bedrock. In any one place there is a mix of meteoroid material and fragmentary ejecta of both mare and highland types thrown out of distant craters. The average compositions of all three materials differ from each other and from the earth's and the question is: What caused these differences?

The meteoroids, mostly coming from a different part of the solar system, are expected to differ because the primordial solar nebula (from which the solar system was formed) had a different composition where the meteoroids were formed than where the earth and moon were formed. The differences between lunar and terrestrial materials were at first explained by geochemists as due to different mechanisms of differentiation (separating different minerals) on moon and on earth. The following simplified table of compositional differences shows what must be explained:

Mare-type material (MT)	Highland-type material (HT)
Twice as much Al_2O_3 as in HT	Twice as much FeO as in MT
10 times as much KREEP as in HT	20 times as much KREEP as in meteoroids and earth
5 times as much Ca, Al, U, and Th as in meteoroids and earth	10 times as much U and Th as in meteoroids and earth
More Si than in HT	Higher albedo than MT
Age 3.2 to 3.9 billion years	

(KREEP stands for potassium [K], Rare Earth Elements, and Phosphorus, which seem to form an excellent indicator of HT material, even though the rare-earth content is measured in parts per million. Albedo is the fraction of sunlight reflected by the surface.)

These strange differences between MT and HT materials cannot all be due to differentiation during several meltings and refreezings or to added meteoroid material. So the geochemists conclude that the moon must have been formed with a difference between the chemical composition of the original crust (HT) and the composition of a layer 100 to 200 km down, from which the MT material later came to flow out over the HT and meteoroid floors of the maria.

In order to explain the moon's oblate shape and the support of the mascons despite high thermal gradient near the surface, it is suggested that the moon's interior is in *convective* equilibrium rather than hydrostatic equilibrium. That is, large-scale motion of the internal material may carry heat outward; the existing mascons may be over "updrafts"; and the shape of the thin crust may be set by some stable vortex motion in the interior. At 700° K or higher temperature, there is a solid-state creep of about 10^{-7} cm/sec to relieve stresses in such rocks as basalt and olivine. This slow motion, allowing hotter rock from the interior to move outward, cooler rock near the crust to move down, would keep the interior temperature below 1000° K and transport the expected radioactive heat outward by convection to the more rigid crust, through which it moves by conduction at the measured rate of about 3 microwatts/cm^2. The shell just under the crust may have been hot enough, and at low enough pressure, to provide liquid lava for volcanoes, lava flows, and mare-type material in the past. Similar convection in the earth's mantle has been postulated; both theories need a good deal more analysis.*

Origin

The major questions left unanswered in this discussion of the facts obtained from NASA's lunar exploration are concerned with the formation of the moon as solid body some 4.5 billion years ago. If the moon was formed in the solar nebula near the earth, why should its composition differ from the earth's? If it was formed at a different distance from the sun, where the composition of the solar nebula differed from that where the earth was formed, how did the earth capture the moon?

To escape this difficulty, we might suggest that the moon was formed from the proto-earth, stripping off the surface materials of

---

*Just after this was written (November, 1972), there was a large meteoroid impact on the far side of the moon, detected by the seismometers on the near side. They showed the normal compressional waves from the impact, but not the transverse (shear) waves. The same thing has long been observed by seismologists on earth; when an earthquake is on the opposite side of the earth's core from a sensitive seismograph, only the compressional (P) waves are recorded. The shear (S) waves cannot pass through the liquid core.

So it seems that the moon *does* have a liquid, or partly molten core about 900 km in radius. It cannot be dense material (Iron), as in the earth, because the moon's average density is too low. Heat released by radioactive minerals has slowly melted the rocks in the lunar core, and convection in the outer 900 km (if there is any) was not sufficient to keep the core cool. This newest fact, though unexpected, does not alter the puzzle of the moon's origin.

lower density after the denser materials had sunk toward the center of the rotating, evolving earth. But how could the ring of material spun off the proto-earth condense into one solid object?

If the moon were formed close to the earth, it may be that the earth's magnetic field accounts for magnetization of lunar rocks. But a dipole field decreases as $1/r^3$; in order to have 2000 gammas magnetic field during solidification, the moon must have been within $r = 3$ earth radii (12,000 miles). At such a close distance, tidal forces would have disrupted the forming moon. At $r = 5$ earth radii (20,000 mi), the tidal forces are acceptable, but the earth must then have had five times its present magnetic field (possible, since the earth would have been rotating about 4.3 times faster and might have had a larger liquid core than at present.)**

Starting with an HT crust formed from hot liquid rock of different composition than deeper (MT) material, the moon must have cooled rapidly so that no mixing of HT and MT took place in the half-billion years before large meteoroid impacts (forming mascons and maria) could bring up MT material. In this 500-million year period, the moon would spiral out about 30 earth radii (depending on the large tidal friction in the two cooling bodies). But if seven or more 100-km meteoroids hit the moon then, why did none hit the earth, only 120,000 miles away?

It will take a good deal more study, and further lunar exploration, to understand the origin and history of earth's closest neighbor.

**A more recent article ("Notes on the Fourth Lunar Science Conference-1" [*Sky and Telescope*, June, 1973]) presented a different approach to planet formation. "The first step in planet formation from the primordial solar nebula was the condensation of *moon-sized objects*, rotating balls of hot gas having about the moon's mass. It is likely that there were magnetic fields in the solar nebula, and when the spining gases condensed to form the primordial moon, the magnetic field trapped in the hot plasma increased to over 100,000 gammas (several gauss) in the hot core. Then the sun began to shine, blowing away the outer gases from the moon, in which most of the solid materials had already condensed. (More matter was collected later by impacts.) Cooling of the lunar core proceeded past the Curie point of iron (780° centigrade) when a field of some 20 gauss was "frozen" in the core material. Strangway notes that this provided a surface field of 1,000 to 2,000 gammas.

The outermost 200 kilometers of crust was still hot from impacts and differentiation (lighter materials rising from the interior), so it was not magnetized. In the second stage, the crust cooled and became magnetized in the 1,000-gamma field, but because of its low iron content (one percent) its magnetization was less. During this stage "breccia flows" took place, where loose rocks and soils were compacted in the 1,000-gamma field, and basaltic lava flows crystallized in the field, too.

Finally, in the third stage, the moon's core heated up due to the radioactive decay of uranium, thorium, and plutonium, becoming hotter than the Curie point of iron, causing the "frozen" magnetic field to disappear. This lowered the surface field to a few gammas, as observed today."

References

1. Page, T., and Page, L. W., eds. 1965. *Wanderers in the sky*. New York: Macmillan Company.
2. ———, eds. 1966. *Origin of the solar system*. New York: Macmillan Company.
3. Page, Thornton. 1972. The third lunar science conference. *Sky and Telescope* 43:145, 219.

13 The New Mars: Volcanism, Water, and a Debate over Its History

Allen L. Hammond

Allen L. Hammond has been the editor of the Research News Department of Science *at the American Association for the Advancement of Science in Washington, D. C., since 1970.*

Mars has been studied for centuries with optical telescopes, and more recently with radar and with instrumented spacecraft on "flyby" trajectories. But the wealth of data obtained from the Mariner 9 spacecraft during nearly 700 orbits around the red planet has vastly improved knowledge of that body and has provided a third reference point—in addition to the moon and the earth—for understanding planetary physics. In the 14 months since the Mariner spacecraft first reached Mars, it has become clear as never before that the sun's fourth planet is a geologically active body with volcanic mountains and calderas larger than any on the earth. Interactions of wind, dust, and surface materials are now considered to account for the changes in the planet's appearance which have puzzled observers for so long. And there are strong indications that water, probably trapped in the polar caps, at one time flowed freely over part of the planet's surface, a possibility that has raised anew hopes that some form of life may exist on Mars.

Between 13 November 1971, when the Mariner spacecraft became the first man-made object to orbit another planet, and 27 October 1972, when depletion of the supply of gas used to keep the spacecraft positioned with its radio transmitter directed toward the earth caused it to be shut down, Mariner 9 took more than 7000 photographs in mapping the entire surface of the planet with its television cameras. Thermal and chemical maps of the surface, as well as studies of surface pressure, atmospheric composition, and the martian gravity field, were made with ultraviolet and infrared spectrometers, an infrared radiometer, and the telemetry signals from the spacecraft to earth. Officials at the National Aeronautics and Space Administration count the mission a tremendous success, as do the scientists who are now digging into the vast amount of new information and offering the first tentative accounts of the planet's history.

Both atmospheric and surface features of the planet were studied. The planet-wide dust storm that obscured the surface in the early months of the mission proved a boon to those studying the martian atmosphere. Carl Sagan of Cornell University mapped streaks of dust extending from the leeward side of many craters and found that they formed a distinct global pattern. The streaks appear to have been laid down during the period of strongest winds and hence to give an indication of the surface winds during the dust storm. The circulation pattern, according to Conway Leovy of the University of Washington, agrees with theoretical models of meteorology at low martian latitudes and seems to be peculiar to its southern summer solstice—during which energy input from the sun is concentrated in a belt near 20° to 30° south of the equator. It appears likely that the coincidence of the summer solstice and the closest approach of Mars to the sun (perihelion) gave rise to the conditions that resulted in the dust storm. An initiating factor in the dust storm, according to Leovy, may have been the increased absorption of the sun's radiation due to dust particles raised by the strong winds that blow along the edges of the south polar cap during its yearly retreat.

Water in the Polar Caps

The polar caps had been thought to be carbon dioxide, which is the main component of the martian atmosphere. Measurements with the infrared spectrometer, however, showed the presence of water vapor in the atmosphere. At times, the atmosphere near the north pole was saturated with water vapor, although the total amount, because of the low temperatures, was small. Variations in the amount detected were correlated with the retreat of the northern polar cap, suggesting that it is the source of the water vapor. Other evidence is consistent with the idea that the residual polar caps are largely water.

Ozone was detected in the martian atmosphere with the ultraviolet spectrometer. The observed variations in ozone may add to the understanding of the stability of the protective ozone layer in the earth's stratosphere. The Mariner results show that ozone concentrations decreased in the summer when more water vapor was present in the atmosphere, presumably because of photochemical reactions involving water and ozone.

Several cloud systems were observed in the martian atmosphere. Among the most spectacular, according to Leovy, were those observed during summer on the slopes of the large volcanos. The clouds were composed of water-ice crystals and had a cellular structure indicating convective activity, possibly the result of being heated from the surface, although the possibility that the clouds arise from vapors emitted by the volcanos is also being studied.

Wind speeds during the dust storm are believed to have averaged at least 30 meters per second in the equatorial region, and some investigators believe that velocities twice that high may have prevailed at the beginning of the storm. These high winds seem consistent with a martian surface marked by widespread erosion. Indeed, according to Hal Mazursky of the U.S. Geological Survey laboratory in Flagstaff, Arizona, eolian erosion is a dominant feature of Mars and is apparently so intense in some areas as to have completely eroded away preexisting volcanos. The edge of the largest existing volcano on Mars, Nix Olympica, is apparently being rapidly eaten away, exposing a 1- to 2-km cliff around its base.

Nix Olympica and the three volcanos along the Tharsis ridge area are huge by any scale. One of the latter rises some 26 km above the surrounding floor. As indicated by the relative scarcity of craters, these volcanic structures are also geologically recent. One estimate by William Hartmann of Science Applications Incorporated in Tucson, Arizona, puts the age of Nix Olympica at 100 million years and that of the Tharsis area at 300 million. But volcanic activity on Mars is not just a recent phenomenon. The Mariner photographs show many older volcanic structures, ranging from small calderas and volcanic vents to volcanic flows reminiscent of the lunar maria, and including at least one large primitive volcano. Many of these older structures are heavily cratered and severely eroded, and they appear to date back as far as 3 billion years, if crater counts are any indication.

East of the Tharsis volcanos is found a high plateau, and still further to the east, a large rift system that stretches nearly 5000 km along the equator. The plateau is broken by faults that appear to indicate vertical movement of the crust. The faults coalesce to form the rift valley, a feature that presumably resulted from tensional forces in the crust whose cause is still unknown. Whatever its origin, the rift valley is the most dramatic evidence of tectonic activity on Mars. On a smaller scale, however, there appears to be evidence of faulting and other tectonic activity over much of the planet.

Perhaps the most startling finding of the Mariner effort has been the discovery of channels in the martian surface which appear to have been cut by running water. Three types of channels have been observed, all of them quite distinct from the lava channels seen on the moon and also on Mars. The largest of the putative stream beds emerge from the foot of landslide areas north of the rift valley and many have originated in the melting of permafrost uncovered in the slide debris. Smaller sinuous channels that originate nowhere in particular and run downhill, coalescing with other channels and becoming broader, have also been found. A third type is composed of a complex, interwoven pattern of channels that, according to Mazursky, is characteristic of intermittent stream flow. All except the first type of channel originate in flat terrain and seem to imply rainfall runoff as the source of the water.

The channels appear to be geologically very recent but not all of the same age, so that their formation could not have been due to a one-time event, and they occur largely in the warmer equatorial regions of Mars. That they exist at all is quite remarkable, because under present conditions water could not exist on the martian surface in liquid form—the liquid phase is unstable at the prevailing temperatures and pressures and would immediately freeze or evaporate. The consensus of those who have examined the evidence, however, is that other explanations for the channels are even less likely; liquid carbon dioxide, for example, would require nearly 1000 times the existing atmospheric pressure on Mars. Hence the evidence is suggestive that the past environment of Mars must at some time have differed considerably from present conditons.

Another peculiar feature of the martian terrain is the laminated structures exposed by the retreating polar caps. What appears to be a series of overlapping plates, each composed of dozens of continuous layers of alternately dark and light material, presents a banded appearance. A convincing hypothesis to many investigators is that the alternating layers represent deposits of dust and ice or frozen carbon dioxide, although what formed the intricate pattern of the plates is a subject of considerable debate. The number and uniformity of the layers, however, suggests a regular pattern of dusty and dust-free epochs in the planet's recent history.

There is a great deal of dust on Mars. Shifting dust seems to have been confirmed as the agent responsible for changes in the dark spots visible on the martian surface as well as for other variable surface features. The source of the dust, which seems to be composed of grains comparable in size to sand, is thought to be the continuing erosion of rocks. Erosion may be most intense in the equatorial regions. From there, dust may be carried poleward, deposited, and gradually redistributed over the planet.

The largest portion of the planet's surface is old, heavily cratered rock that some investigators believe to be the ancient crust. The incidence of craters that can be discerned is considerably less, however, than on the martian moon Phobus. Exactly how far back martian geologic history can be traced, if intense erosive processes have always been present, is uncertain.

Bruce Murray of the California Institute of Technology has proposed that the planet was without any significant atmosphere until recently. He believes that the planet is just beginning to be geologically active and that it created its present atmosphere volcanically during the formation of the Tharsis and Nix Olympica structures. Murray is correspondingly pessimistic that life could have evolved under the harsh, moonlike conditions that in his view prevailed over most of martian history.

Murray offers no explanation for the formation of the channels, but he does propose that the regular pattern of the polar laminated terrain is associated with periodic alteration in the martian climate. He finds that the planet's orbit changes shape from nearly circular to more elliptical with a period of about every 2 million years, perturbations caused by the influence of other planets, primarily Jupiter and Saturn. The resultant variation in the sun's radiant energy reaching the poles could under some circumstances, Murray speculates, lead to changes in the growth and sublimation of the polar caps. If dust is laid down alternately with layers of frost, thin laminae of the type observed could be formed.

On the question of the planet's ancient history, other investigators are less inclined to agree with Murray and cite evidence of volcanism and erosive processes stretching back over much of the planet's lifetime. Hartmann, for example, believes that in the past the planet must have had more rather than less atmosphere than it now has. Hartmann agrees, however, that tectonic activity of the type well known on the earth may just be getting started. Mars thus contrasts strongly with the moon, whose evolution shows no evidence of horizontal movement by crustal plates, and the earth, where convection currents in the mantle have been vigorous enough to completely remodel the surface over much of its history. Hartmann believes that the martian crust in the Tharsis region has been pushed upward in the last few hundred million years, possibly by a mantle convection current.

Others have suggested more explicit analogies with tectonic patterns on earth. Mazursky, for example, believes that the high plateau adjacent to the Tharsis ridge represents light, continental-type rock that overlies heavier materials found in nearby low-lying areas. The heavier material, Mazursky suggests, may be comparable to the basaltic rock of the earth's ocean basin floors. In this admittedly speculative view, the line of volcanos along the Tharsis ridge may represent the incipient stages of a tectonic process in which a plate of the heavier basaltic material is being thrust under the lighter continental plate. Others, such as John McCauley of the U.S. Geological Survey laboratory in Flagstaff, Arizona, do not see any evidence in the martian photographs of the compressional movements that such a process would entail. He points out that the density of Mars is less than that of the earth, that the tectonic activity on Mars happened late in its history, and that its apparent lack of vigor may indicate that no further crustal movements are in prospect.

In regard to the more recent evolution of the planet, the most puzzling problem appears to be the climatic changes that could have led to the formation of the channels by liquid water. Sagan has proposed that the climatic instability responsible for this is not inconsistent with that suggested by Murray for the formation of the polar laminae. If the polar caps were to entirely evaporate, Sagan calculates, the atmospheric pressure on Mars would be about 1 bar, the same as on the earth. Only about one-tenth this much atmosphere, assuming it was 1 percent water vapor, would be sufficient to permit liquid water as a stable phase at the daytime temperatures characteristic of the martian equatorial region. Hence Sagan proposes that rainfall and rivers may be recurring phenomena on Mars during each successive "interglacial" period.

Sagan believes that an advective instability in the martian atmosphere is the most probable cause of the drastic climatic changes he envisions. Changes in the absorption of sunlight at the poles—for example, by a layer of dust deposited in a major storm—would cause the pole to heat up and the total atmospheric pressure to increase slightly. With a more dense atmosphere, the normal circulation patterns that carry heat poleward from the equator would operate more efficiently, heating the poles still further. Continuation of the process would eventually lead to conversion of the polar caps to atmosphere and rivers. The process is also reversible, Sagan believes, because liquid water could serve as a trap for dust, cleaning the atmosphere and allowing it to recondense at the poles.

If indeed the environment on Mars alternates between wet and dry, at least in the tropics, then speculations about life on the planet become more interesting. Life forms that could go into extended repose while awaiting the availability of liquid water, for example, would seem well within the realm of possibility, if such forms exist. Experiments that expose martian soil to liquid water and test for biological activity might therefore be of particular interest. As it happens, two such experiments are planned for the Viking spacecraft that will attempt to land on Mars, in 1976.

14 Mariner 9 Television Reconnaissance of Mars and Its Satellites: Preliminary Results

Harold Masursky, et al

Harold Masursky is chief scientist at the U. S. Geological Survey's Center of Astrogeology in Flagstaff, Arizona. (See the editor's note on p. 171.)

Figure 1 shows the south polar cap as viewed by Mariner 7 late in the Martian southern winter and by Mariner 9 during the Martian southern summer. The planimetric similarity of the curvilinear patterns in both views is evident. However, in the 1969 picture the streaks are bright, being highlighted by frost, whereas they are

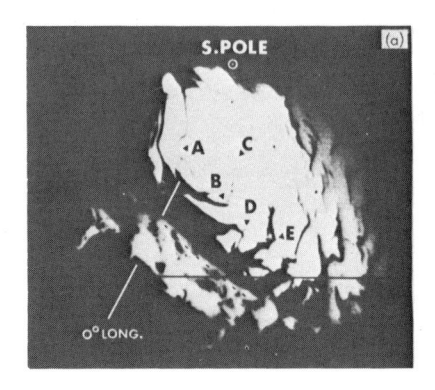

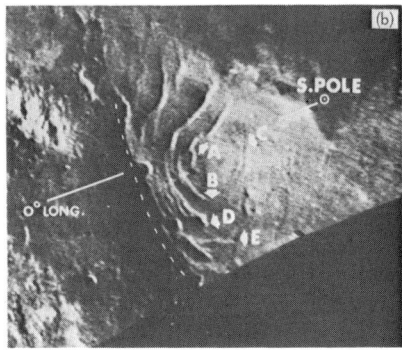

Figure 1. (a) Mariner 9 (JPL 4019-25) and (b) Mariner 7 (7N17) views of the south polar cap. The dark markings in the Mariner 9 wide-angle view of cap are correlated with the light markings in the Mariner 7 narrow-angle frame. The Mariner 7 image has been rectified and then enhanced. Although (b) shows strong contrast, the entire area actually was covered by frost.

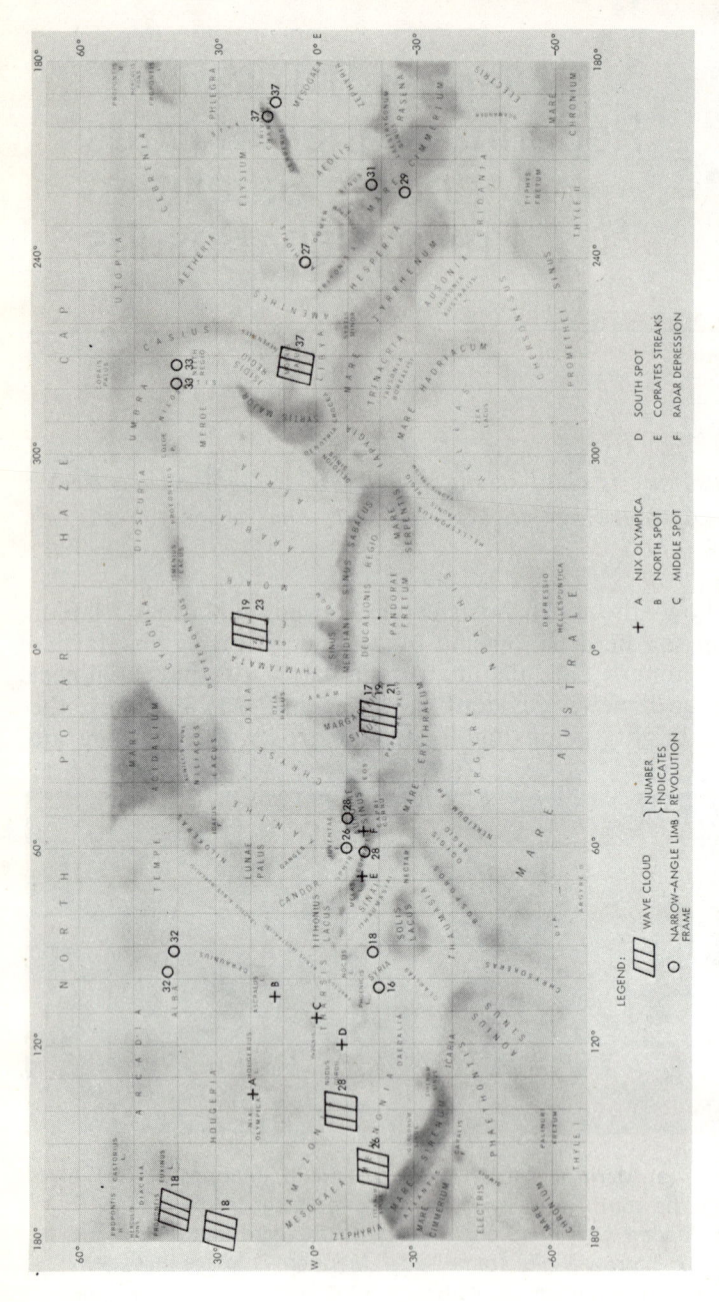

Figure 2. Mercator map showing the approximate locations of wave clouds visible in violet and blue frames taken with the wide-angle camera. The lines show the orientation of the wave bands. The circles show the positions of narrow-angle frames of the limb. The numbers near the lines and circles designate the revolutions on which the observations were made. The aerographic features, Nix Olympica (A), the three dark spots (B, C, D), the bright streaks in the Eos-Ophir region (E), and the radar depression near Eos (F), were observed in Mariner 9 photographs.

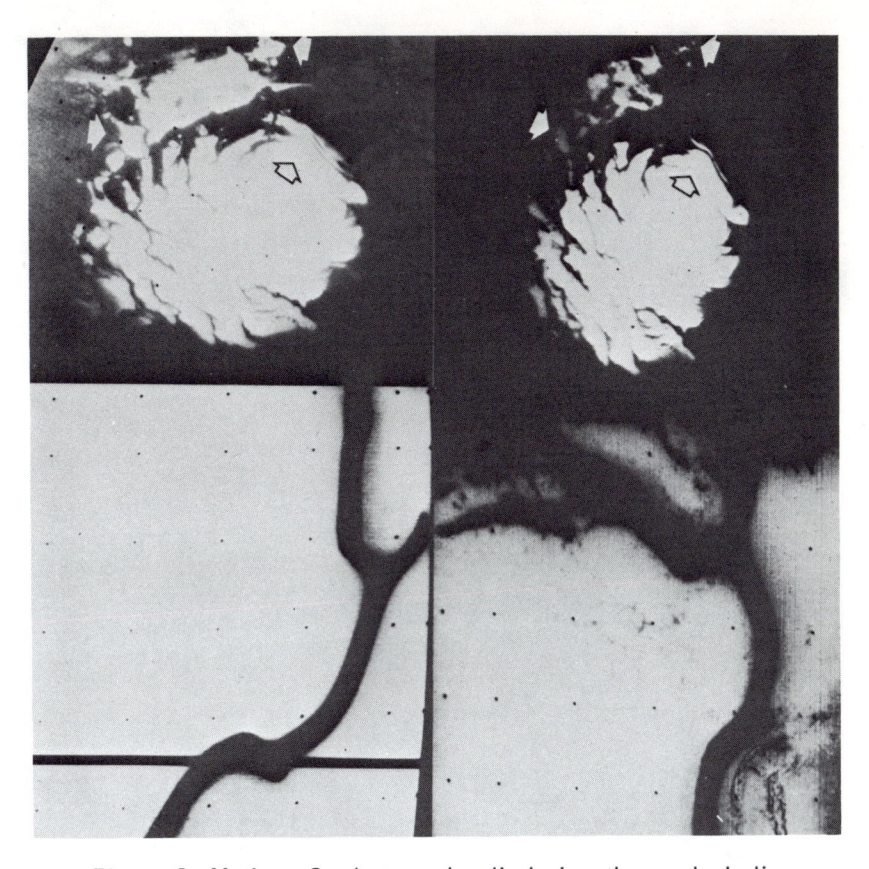

Figure 3. Mariner 9 photographs displaying the gradual disappearance of the residual polar cap. At the top are two Mariner 9 wide-angle pictures of the polar cap acquired about 9 days apart on Revs 11 (left) and 29 (right). These two images have been rectified to a polar stereographic projection. The relative distortion between the two projections is due to uncertainties in the preliminary pointing data. The cap is approximately 450 km in diameter in the earlier image. The white arrows indicate common areas that have suffered significant defrosting over the 9-day period. A residual image of the cap can be seen to its right in the left-hand picture. At the bottom are two Mariner 9 narrow-angle frames from Revs 11 (left) and 34 (right); these show changes in the cap over 11½ days. The area of the narrow-angle frames is indicated by black arrows in wide-angle frames. Since certain fine details can be seen in both images, the difference in the appearance is real and not solely the result of varying obscuration. Detailed patterns parallel to the main frost-free corridor are emerging in the later version.

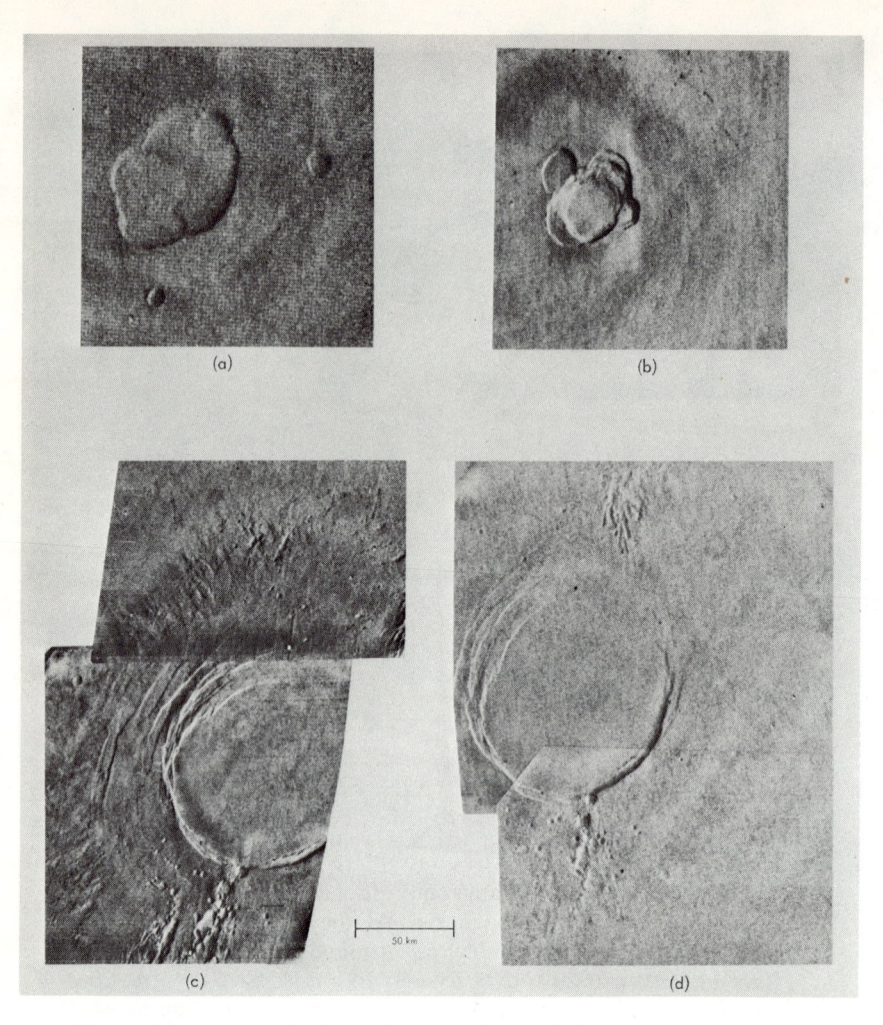

Figure 4. Narrow-angle views of craters in the dark spots, rectified on the basis of preliminary orbital data. (a) Crater complex in Nix Olympica. Arcurate scarps continuous with the scallops on the wall indicate that the floor is not fully obscured. Smaller craters to the east and south may be superposed, unrelated features (IPL roll 882,025127). (b) Crater complex of North Spot. The central crater with terraced walls apparently truncates several subsidiary craters (IPL roll 882, 030159). (c) The South Spot crater photographed on Rev 28 (IPL roll 882, 022534 and 023951). A concentrically fractured zone extends about one crater radius out from the crater edge; the surface beyond has a complex subradial texture. Several rimless craters in the smooth zone appear to be aligned along the arcuate fractures. Troughs composed of numerous coalescing cra-

dark in the Mariner 9 pictures. It is now evident that these curvilinear patterns are permanent topographic forms that control the outline of the residual cap as it retreats. It was observed in 1969 that local topography, principally craters, strongly controlled the detailed form of the margin of the retreating cap near 60°S latitude. Parts of the boundary and interior of the retreating cap, as seen in the Mariner 9 pictures, are remarkably regular and uniform and, therefore, appear to contain few craters. The topography seems to be limited primarily to curvilinear ridges or troughs, or both, perhaps of a highly subdued character.

* * * * *

Editor's note: The following authors (listed with their affiliations current at the time the article was originally published) contributed to the article. R. M. Batson, J. F. McCauley, L. A. Soderblom, and R. L. Wildey (U. S. Geological Survey, Flagstaff, Arizona); M. H. Carr, D. J. Milton, and D. E. Wilhelms (U. S. Geological Survey, Menlo Park, California); B. A. Smith, T. B. Kirby, and J. C. Robinson (Department of Astronomy, New Mexico State University); C. B. Leovy (Department of Atmospheric Sciences, University of Washington); G. A. Briggs, T. C. Duxbury, and C. H. Acton, Jr. (Jet Propulsion Laboratory, California Institute of Technology); B. C. Murray, J. A. Cutts, R. P. Sharp, and S. Smith (Division of Geological and Planetary Sciences, California Institute of Technology); R. B. Leighton (Division of Physics, Mathematics, and Astronomy, California Institute of Technology); C. Sagan, J. Veverka, and M. Noland (Laboratory for Planetary Studies, Cornell University); J. Lederberg and E. Levinthal (Department of Genetics, Stanford University School of Medicine); J. B. Pollack and J. T. Moore, Jr. (Space Sciences Division, Ames Research Center); W. K. Hartmann (IIT Research Institute); E. N. Shipley (Bellcomm, Inc.); G. de Vaucouleurs (Department of Astronomy, University of Texas); and M. E. Davies (Rand Corporation).

terlets and grooves extend to the north-northeast and south-southwest of the main crater. Small rimmed craters to the west may be impact craters unrelated to the regional structure. (d) South Spot viewed on Rev. 32, 48 hours after the previous view (IPL roll 882, 033019 and 033901). The differences are due to real changes in the dust cloud or in the lighting and view paths through the dust. The disappearance of the radial facies on the northwestern rim suggests that other apparently smooth areas, such as the crater floor, actually may be concealed by atmospheric dust.

Additional Readings

Davies, M. E., and Murray, B. C. 1971. *The view from space: photographic exploration of the planets*. New York: Columbia University Press.

Ebbighausen, E. G. 1971. *Astronomy*. 2nd ed. Columbus, Ohio: Charles E. Merrill.

Fairbridge, R. W., ed. 1967. *Encyclopedia of the atmospheric sciences and astrogeology*. Encyclopedia of Earth Sciences Series, vol. 2. New York: Reinhold.

Fielder, G. 1971. *Geology and physics of the moon*. New York: Elsevier.

Hartmann, W. K. 1972. *Moons and planets: an introduction to planetary science*. Belmont, Calif.: Bogden & Quigley.

Hess, W. N., Menzel, D. H., and O'Keefe, J. A. 1966. *The nature of the lunar surface*. Baltimore: Johns Hopkins Press.

King, E. A., Jr. (vol. 1), Heymann, D. (vol. 2), and Criswell, D. R. (vol. 3), eds. 1972. *Proceedings of 3rd lunar science conference, Houston, Texas, 1972*. 3 vols. Cambridge, Mass.: MIT Press.

Koenig, L. R., Murray, F. W., Michaux, C. M., and Hyatt, H. A. 1967. *Handbook of the physical properties of Venus*. NASA SP-3029. Washington, D. C.: U. S. Government Printing Office.

Levinson, A. A., ed. 1971. *Proceedings of 2nd lunar science conference, Houston, Texas, 1971*. 3 vols. Cambridge, Mass.: MIT Press.

Lowman, P. D., Jr. 1969. *Lunar panorama: a photographic guide to the geology of the moon*. Feldmeilen/Zurich, Welfflugbild: Reinhold A. Müller.

Lucas, J. W., ed. 1972. *Thermal characteristics of the moon*. Progress in Astronautics and Aeronautics, vol. 28. Cambridge, Mass.: MIT Press.

Lunar Science Institute. 1972. *Post-Apollo lunar science*. Houston, Texas: Lunar Science Institute.

Mason, B. H., and Melson, W. G. 1970. *The lunar rocks*. New York: Wiley-Interscience.

McDonald, R. L., and Hesse, W. H. 1970. *Space science*. Columbus, Ohio: Charles E. Merrill.

Mehlin, T. G. 1973. *Astronomy and the origin of the earth*. 2nd ed. Dubuque, Iowa: Wm. C. Brown.

Mutch, T. A. 1970. *Geology of the moon: a stratigraphic view*. Princeton, N.J.: Princeton University Press.

Penrose, R. 1972. Black holes. *Scientific American* 226 (5):38-46.

Sagan, C. 1970. *Planetary exploration*. Eugene, Oregon: Condon Lectures, Oregon State System of Higher Education.

Waters, A. C. 1967. *Moon craters and Oregon volcanoes*. Eugene, Oregon: Condon Lectures, Oregon State System of Higher Education.

Part Five

Environmental Geology

"Environmental geology" is a
ridiculous term.
Gordon B. Oakeshott, 1970

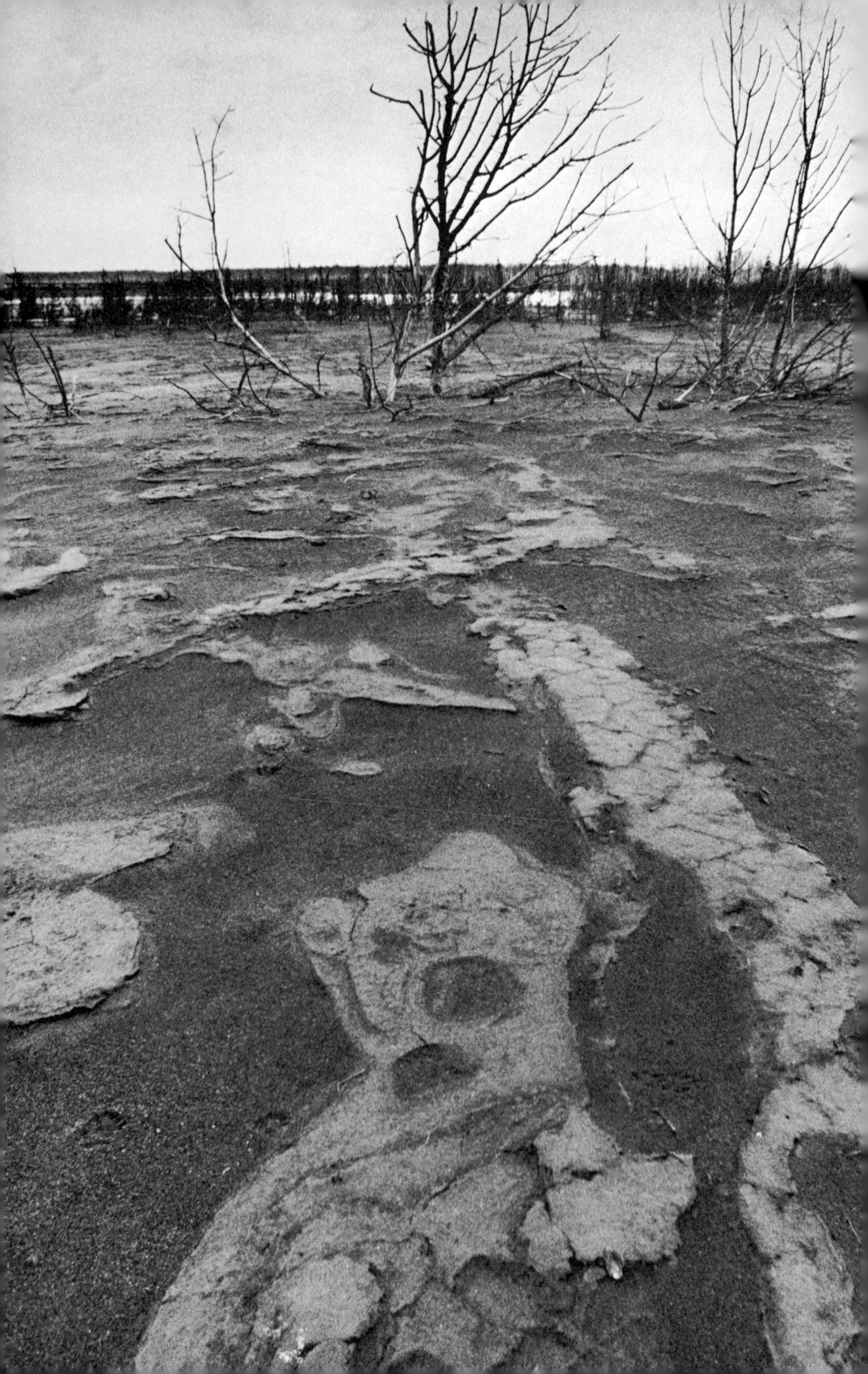

Environmental geology is a relatively new term initiated by John Frye and Jim Hackett to describe some applied geology activities carried out by the Illinois Geological Survey in the mid 1960s. According to the American Geological Institute Glossary, environmental geology is the collection, analysis, and application of geologic data and principles to problems created by human occupancy and use of the physical environment, including the maximization of a rapidly shrinking living space and resource base to the needs of man, the minimization of the deleterious effects of man's interaction with the earth, and the accommodation of the exponentially increasing human population to the finite resources and terrain of the earth. The term was considered ridiculous, not only by Oakeshott, but by many others, since geology is the study of physical environment and thus all geology is environmental. In fact, geologists had been using the term environment for years in their studies of paleoenvironments. Even though all geology is environmental, not all geologists are environmentalists. The term environmental geology is now with us no matter how incorrect semantically. In the first article in this section, John Frye presents an overview of environmental geology. The overview is followed by readings that concentrate on geologic hazards and medical geology. Readings on mineral resources, actually a part of environmental geology, are included in a separate section.

Geologic hazards occur naturally or are man-made geologic conditions or phenomena that are a risk or potential danger to life and property. They include earthquakes, volcanoes, landslides, subsidence, and floods, and they could also include beach erosion, waste disposal, salt-water intrusion of aquifers, and several other geologic events. Earth scientists might also want to include dust storms, heat waves, cold waves, tornadoes, and hurricanes, but we would probably be more specific and group them as meteorological hazards.

The second, third, and fourth articles cover three geologic hazards in a superficial manner. Serious students are urged to consider the original longer forms of the second and the fourth articles. Geologic hazards are expected to increase in the future if proper planning to avoid the utilization of hazardous areas is not implemented. As population increases and is forced to settle on *less suitable land, loss of property and life will rise as these areas are affected by hazards of natural or man-induced origin.*

The fifth article, on disposal of radioactive wastes, is important in light of some estimates which indicate we may be obtaining 50 percent of our electrical energy from nuclear power by the year 2000. Although it was decided several years ago to dispose of nuclear wastes in Kansas salt mines, the Atomic Energy Commission's decision to do so was apparently based on incomplete geological information, particularly concerning abandoned wells in the area of waste storage. Salt mines in a tectonically stable area may well be the answer for storage of most nuclear wastes; another possibility includes unsaturated zones in the southwest, such as plateaus and mesas that would isolate the wastes from the hydrosphere for tens of thousands of years. Currently, wastes are being stored in stainless steel and concrete tanks on the surface where they can be monitored. Some of the containers last only about 20 years and then must be replaced. This method of disposal of radioactive wastes imparts a serious responsibility to future generations.

The geologic aspects of health and disease, a field of study also known as medical geology and environmental geochemistry, are covered in the last three articles in this section. The lead article by Warren and Delavault, two of the early researchers in the field, summarizes the origin and importance of the discipline. It is becoming more and more apparent that there is a complex relationship between the natural abundance of some elements in bedrock and the soils of an area and the incidence of certain deficiencies and diseases in that area. One of the best-known examples of this relationship occurs in the Great Lakes region where some types of goiter are related to deficiency of iodine in the soils. For some trace elements, there is a limited range within which animals utilize the elements to maintain health. If the intake and retention of the element is too little, a deficiency in the body is created; too much results in toxicity; and both situations can cause disease. As noted by Pettyjohn (1970), poison is in everything and no thing is without poison.

One aspect of environmental geology only briefly covered in the second article is the relationship between geology and regional planning and urbanization. Although the physical environment has been effective in guiding the establishment and growth of cities for generations, only recently have planners and some engineers become aware of the importance of planning to determine suitable areas for various resource uses, based on an inventory of the physical environment. Such an inventory can provide the necessary data on geologic hazards and resources that could prevent waste and maximize utilization of our resources. In regional planning, the environmental geologist has an opportunity to bring together knowledge of his subject and then apply it for the benefit of man. The delayed emergence of geology as an important aspect of planning can be attributed to lack of communication between planners and geologists—a credit to neither profession. Because of decreasing resources and increasing potential for geologic hazards, the importance of environmental geology in planning will increase as more planning agencies seek to incorporate into their planning process data from reports, files, and the Earth Resources Technology Satellites (ERTS).

15 A Geologist Views the Environment

John C. Frye

John C. Frye is chief of the Illinois State Geological Survey at Urbana, Illinois.

When the geologist considers the environment, and particularly when he is concerned with the diversified relations of man to his total physical environment, he takes an exceptionally broad and long-term view. It is broad because all of the physical features of the earth are the subject matter of the geologist. It is long term because the geologist views the environment of the moment as a mere point on a very long time-continuum that has witnessed a succession of physical and biological changes—and that at present is dynamically undergoing natural change.

Let us first consider the time perspective of the geologist, then consider the many physical factors that are important to man's activity on the face of the earth, and third, turn our attention to specific uses of geologic data for the maintenance or development of an environment that is compatible with human needs.

The Long View

The earth is known to be several billion years old, and the geologic record of physical events and life-forms on the earth is reasonably good for no more than the most recent 500 million years. Throughout this span of known time the environment has been constantly changing—sometimes very slowly, but at other times quite rapidly. Perhaps a few dramatic examples will serve as illustrations. Less than 20,000 years ago the area occupied by such North American cities as Chicago, Cleveland, Detroit, and Toronto was deeply buried under glacial ice. The land on which New York City is now built was many miles inland from the seashore. And part of the area now occupied by Salt Lake City lay beneath a fresh-water lake. Twelve thousand years ago glacial ice covered the northern shores and formed the northern wall of what was then the Great Lakes, and much of the outflow from those lakes was to the Gulf of Mexico rather than to the Atlantic Ocean through the St. Lawrence River, as it is now. Although firm scientific information is not available to permit equally positive statements about atmospheric changes during the past few tens of thousands of years, deductions from the known positions of glaciers and from the fossil record make it clear that the atmospheric circulation patterns were quite different from those of the present, and studies of radiocarbon show that the isotopic content of the atmosphere changed measurably through time.

The purpose of listing these examples is to emphasize that the environment is a dynamic system that must be understood and accommodated by man's activities, rather than a static, unchanging system that can be "preserved." The living, or biological systems of the earth are generally understood to be progressively changing, but much less well understood by the public is that the nonliving, physical aspects of the earth also undergo change at an equal or greater rate.

Clearly, a dynamic system is more difficult to understand fully, and it is more difficult to adapt man's activities to a constantly changing situation than to an unchanging or static system. On the other hand, the very fact of constant change opens many avenues of modification and accommodation that would not be available in a forever constant and unchanging system.

The Problem

Although it is important that we have in mind the long-term facts concerning earth history, modern man has become such an effective agent of physical and chemical change that he has been able to produce major modifications, some of which run counter to the normal evolution of our earthly environment, and to compress millennia of normal evolutionary changes into days. These rapid modifications are, almost without exception, made by man with the intention of producing improvements and advantages for people. Problems result from the fact that by-products and side effects do occur that are neither desirable nor pleasing, and at some times and places may be hazardous or even calamitous. In some cases the undesirable side effects are unknown or are unpredictable; in other cases they are tolerated as a supposed "necessary price" to pay for the desirable end result. It is our intent to examine the role of the earth scientist in defining some of these problems and in devising ways of minimizing or eliminating them.

The ways in which man treats his physical surroundings, produces and uses the available nonliving resources, and plans for his future needs are, of course, social determinations. However, in order that social decisions can be made in such a way that we, and our children, will not find reason to regret them, they should be made in the light of all the factual information that it is possible to obtain. If we, collectively, decide to use the available supply of a nonrenewable resource—for example, petroleum—at a particular rate, we should know how long it will last and what substitute materials are available to replace it when the supply is exhausted; if we decide to dam a river, we should know what the side effects will be in all directions, how long the facility will last, and what the replacement facility might be; if we develop huge piles of discarded trash, we should know whether or not they will cause pollution of water supplies or the atmosphere, and whether or not the terrane is sufficiently stable to retain them; if we substitute one fuel for another with the desire to abate air pollution, we should know if it will make a net over-all improvement in the pollution problem, and if it will be available in the quantity required so that man's needs can be met; and if we plan expanding metropolitan areas we should have full information on the terrane conditions at depth and on the raw material resources that will be rendered unusable by urbanization.

Role of Earth Science in Solving Problems

When we consider the role of earth science in solving problems, we see that the earth sciences can and should develop answers to all of the questions we have asked. Many of the problem areas overlap one another, but it will be easier to discuss them if we class the contributions of the earth scientist to environmental problem solving in five general categories. The first of these is collecting data for planning the proper use of the terrane, or perhaps we should say the most efficient adjustment of man's use of the earth's surface to all of the physical features and characteristics at and below the surface—particularly in expanding urban areas. Second is determination of the factors that influence the safety and permanence of disposal of waste materials and trash of all kinds—both in the rocks near the surface and at great depth in mines and wells. Third is providing information for the planning and development of safe, adequate, and continuing water supplies in locations that will serve populated areas. Fourth is the identification of rock and mineral resources to provide for future availability of needed raw materials, or of appropriate substitute materials. And, fifth is the recognition of man as a major geologic agent by monitoring the changes he has caused in his environment and by providing remedies where these changes are, or may become, harmful.

Proper Use of Land

The first of these general categories—procuring data for physical planning of the proper use of the land surface and of the rocks below the surface—covers data provided by topographic and geologic maps, by engineering geology and soil mechanics investigations, by predictions of potential landslides and other geologic hazards, and by a complete inventory of available mineral resources and future potential water supplies. Much of the geologic data needed in this category can be produced by conventional methods of research, but, to be effective, the research program must be oriented toward environmental applications. Furthermore, the results must be presented in a form that is readily understandable to, and usable by, planners and administrators who often are unfamiliar with science and, particularly, do not have a working knowledge of the geological sciences. Perhaps the best way to explain what I mean by environmental orientation is to cite a few examples.

The first example of the use of geologic research oriented for planning is a laboratory study involving clay mineralogy, petrography, and chemistry. It was prompted by numerous incidences of structural failure of earth materials. In rapidly expanding suburban residential areas there has been a great increase in the use of septic tanks and, simultaneously, a rapid increase in the household use of detergents and water softeners. A laboratory research program was initiated to study the changes induced in the clay minerals by these chemical substances when they were introduced into the near-surface deposits by discharge from septic tanks. Preliminary results showed that the materials in septic tank effluent did, indeed, produce significant and undesirable changes in the properties of some earth materials. The data made it possible to predict changes that could occur in the structural characteristics of common surficial deposits and, thus, to prevent structural failure. Therefore, the conclusions were presented to planners, health officials, architects, and other groups that might have need of the information.

In strong contrast to such a sharply focused and specific research project is a second example provided by a study of a county at the northwest fringe of the Chicago metropolitan area, into which urbanization is spreading from that metropolis. Because of impending problems, the county government organized a regional planning commission, which called upon the State Geological Survey and other agencies to collect data on the physical environment that were essential to wise, long-range planning. Where some of the fields of activity of the agencies overlapped, they cooperated informally so that they could most effectively work with the planning commission. Essential to the project were modern topographic maps of the county, and, even though much of the county had been geologically mapped previously, several man-years of geologist time were devoted to the project.

This project to characterize the physical environment involved many types of geological study. These included (1) analysis of the physical character of the major land forms within the county; (2) interpretation of the relation between geologic units near the surface and the agricultural soil units; (3) establishment of the character of the many layers of rocks and glacial deposits penetrated at depth by drilling below the surface; (4) definition of the occurrence and character of water-bearing strata in the near-surface glacial deposits and the deeper bedrock layers; (5) determination of the geologic feasibility of water-resource management programs; (6) determination of the geologic feasibility of waste management programs; (7) delineation of the geographic occurrence and description of the characteristics of construction material resources; (8) location of commercial mineral resources and assessment of their economic value; (9) determination of the engineering characteristics of the geologic units near the surface; and (10) geologic evaluation of surface reservoir conditions and proposed reservoir sites.

The approach to such development of data includes field work by surficial geologists, engineering geologists, ground-water geologists, stratigraphers, and economic geologists. In the laboratory, chemists, mineralogists, and stratigraphers conduct studies of the subsurface by use of cores and cuttings of the deposits at all depths; make chemical, mineralogical, and textural analyses of all deposits and rocks; determine physical properties; compile statistics; and make economic analyses of the many mineral resource situations. The results of these studies are compiled on interpretative maps, which the planning commission combines with the results of studies by specialists of other agencies and by the commission itself for preparation of maps that show the recommendations for land use. These maps, together with explanatory, nontechnical text, serve as a basis for county zoning and long-range development planning.

Development of Waste Disposal Facilities

Our second category of environmental geologic information includes those geologic data needed for proper and safe development of waste disposal facilities. Man has the propensity to produce toxic and noxious waste materials in progressively increasing quantity and in an ever-increasing variety and degree. Waste products result from manufacturing, processing, and mining—but of even greater concern is the concentrated production of waste by the inhabitants of our large cities. Traditionally, man has used fresh water to dilute liquid waste and the atmosphere to dilute the gaseous waste products of combustion. He has often indiscriminately used the land or large bodies of water for disposal of solid waste. However, even the general public is now aware of the fact that we are exhausting the capacity of fresh waters and the atmosphere to absorb our waste products. Along the sea coasts there is still the ocean—although even the ocean is being restricted as a waste disposal medium—but in the vast region of the continental interior we have no ocean in which to dump our wastes. Instead, our choices are limited to (1) selective recycling, accompanied by essentially complete purification of the residue of waste materials; (2) selective recycling accompanied by land disposal of nonrecyclable residues; or (3) the use of the rocks of the earth's crust for the total future expansion of waste disposal capacity. Geologists, who traditionally have been concerned with the discovery of valuable deposits of minerals and their extraction from the crust, now also must concern themselves with the study of the rocks of the earth's crust as a possibly safe place for the disposal and containment of potentially harmful waste products.

Petroleum geologists were introduced to one aspect of the problem of large quantity underground disposal of waste material more than a quarter century ago when it became necessary to find methods for injecting into deep wells the increasing quantities of brines produced with petroleum in oil fields. However, it was not until population densities approached their present levels that we became genuinely concerned with the most critical problems of the future—that is, the safe disposal of industrial and human waste materials in large quantity, other than by dilution. As some of these undesirable materials are destined to increase at an exponential rate in the future, it is obvious that we must devote our best geologic effort to solution of the problems of disposal. The problem of disposal of high-level and intermediate-level radioactive wastes has already attracted a major research effort, probably because these radioactive materials are obviously so highly dangerous for such a long period of time, and a body of scientific data now exists concerning their safe management.

In my own state of Illinois, solid waste disposal is generally accomplished by sanitary landfill, and frequently the State Geological Survey is asked by state and local departments of health to make geologic evaluations of proposed sites. However, precise and universally accepted criteria for this type of evaluation are only now being developed. Several disposal sites of differing geologic character are being intensively studied by coring and instrumentation of test holes, and analyses are being made of the liquids leached from the wastes and the containing deposits. In addition to laboratory study of the obvious characteristics of the containing deposits—such as texture, permeability, strength, clay mineralogy, and thickness of the units that do not transmit water (and thus protect the aquifers)—studies must be made of the less obvious effects of the seepage of liquids on the structural character of the deposit, the removal of objectionable chemicals from water solutions by the clay minerals of the containing deposits, and the microflow patterns of water in earth materials surrounding the wastes. For some of these determinations it is necessary to know the chemical composition of the liquids that pass through the wastes, as well as the chemistry and mineralogy of the deposits that contain them.

Disposal near the surface by landfill or lagooning methods requires detailed studies of the earth deposits at and immediately below the surface, with only minor data required on the deeper bedrock. On the other hand, disposal of industrial wastes in deep wells requires studies of the character and continuity of all rock layers down to the crystalline basement. Geologic data needed for deep disposal involve a combination of the types of information needed for the exploration for both oil and water, plus a knowledge of the confining beds of shale or clay. It should also enable us to predict possible changes that might be produced by the injected wastes.

A different type of pollution problem is represented by sulfur compounds and other undesirable materials released by the burning of coal, oil, and gas and discharged into the atmosphere. The earth scientist contributes to the solution of this problem by studies of the mineral matter in the coal, studies of methods of processing the coal before it is burned, studies of means of removing the harmful materials from the effluent gases produced by combustion, and research on the conversion of coal into gaseous or liquid fuels from which much of the objectionable material can be extracted.

Planning Water Supplies

The third category of concern for the earth scientist is water, its occurrence, quality, continuing availability, and pollution; its use as a resource, as a diluent for waste materials, as a facility for recreation, and as an aesthetic attribute. Many of the problems and areas of data collection for water resources fall mainly in the province of the engineer, the chemist, the biologist, or the geographer. But, it would be unrealistic to exclude water from the subjects requiring significant and major data input by the geologist, because to a greater or lesser degree geologic data is needed for the proper development and management of each of the above aspects. The occurrence, quality, availability, development, and replenishment of ground water require major attention by the geological scientist because all of these aspects are directly controlled by the character of the rocks at and to considerable distances below the surface.

Of the categories of environmental data we are discussing, water resources and water pollution are the most widely discussed in the news media and most generally recognized by the public and by municipal planners and administrators. Furthermore, an extensive cadre of earth scientists specializing in water-related problems has developed within governmental agencies and industry. There are many examples of the application of geologic data to management of water resources, ranging from dam site evaluations to the mapping of aquifers and determination of areas suitable for artificial recharge. Before we leave this category, however, I should point out that, in contrast to the rock and mineral resources of the earth's crust, water is a dynamic and renewable resource and, therefore, is subject to management. Even the long-range correction of the effects of unwise practices in the past may be possible in some cases.

Future Availability of Rock and Mineral Materials

Our fourth category is, perhaps, the most complicated aspect of environmental geology. It is the problem of assuring adequate supplies of mineral and rock raw materials for the future, and especially of assuring their availability near densely populated areas where they are most needed. We have an increasing shortage of raw material resources to meet the needs of the increasing world population, and also a mounting conflict of interest for land use in and adjacent to our urbanizing regions. Conflict exists in populated regions because buildings and pavements commonly remove the possibility of extracting the rock or mineral raw materials underneath them. While the producer of rock or mineral products is exploring for the best deposit available from both a physical and economic standpoint, the urban developer may be planning surface installations without regard for the presence or absence of rock and mineral deposits, which he may be rendering unavailable. These unavailable resources may be urgently needed for community developments in the future, and it is important that the attitudes of the planner and mineral producer be brought into harmony, and that compatible working relations be evolved so that mineral resource development can move forward as an integral part of the urban plan. The geological sciences can supply an accurate and detailed description and maps of all of the rock and mineral resources, not only in but also surrounding urbanizing areas for a distance reasonable for transporting bulk commodities to the metropolitan centers. Grades of deposits that might have utility in the coming 25 to 50 years as well as grades currently being developed should be included in the study.

An equally important role of the earth scientists now, and more particularly in the future, lies in regions remote from the cities where exploration is needed for the raw materials and fuels required by modern society. Although this topic has been discussed in detail in other talks in this series, it is mentioned here because it is a vital element of environmental application. Furthermore, research by the earth scientist, directed toward the identification of substitute materials and toward meeting more exacting and different specifications for future needs, is called for if we are to keep pace with expanding human needs. Information about natural resources is just as essential a part of needed environmental data as are flood hazard maps, physical data maps for engineering projects, and aquifer maps indicating the occurrence of groundwater supplies.

Man as a Geologic Agent

Our fifth area is the recognition of man as a geologic agent, and here we have the culmination of the problems of Environmental Geology. Man's changes in his physical surroundings are made in order to obtain some real or imagined advantage for people. Some of these changes are designed to prevent natural events from happening and include levees and detention reservoirs to prevent flooding of land, revetments and terracing to prevent erosion, and irrigation to prevent the effects of droughts. But many other changes are intended to produce effects that are not in the natural sequence of things. In both cases, nature is liable to provide unplanned and undesirable side effects. It is the role of the earth scientist to determine the effects of man-made physical changes on all aspects of the physical environment before structural changes are made so that provision can be made to negate undesirable by-product effects—or, if the side effects are too severe, so that a decision against the environmental changes can be made.

A special facet of man-made changes is in the area of mineral and fuel resources. The public need and demand for energy, and for products based on mineral raw materials, is constantly increasing at the same time that increasing populations require that land resources be maintained at maximum utility. Here again, the earth scientist must add to his traditional role of finding and developing sources of energy and mineral resources the equal or more difficult role of devising methods of producing these resources in such a way that land resources have a maximum potential for other human uses. This has led to the concept of planning for multiple sequential use—that is, designing a mineral extraction operation in advance so that the land area will first have a beneficial use before the minerals need to be extracted, then be turned over to mineral extraction, and, finally, be returned to a beneficial use, perhaps quite different from the original use.

Conclusions

The earth scientist is concerned with the physical framework of the environment wherever it may be, with the supply of raw materials essential to modern civilization, and with the management of the earth's surface so that it will all have maximum utility for its living inhabitants. It is this last item that is a relatively new role for the earth scientist, and one in which he must work cooperatively with the engineer, the biologist, and the social scientist. The earth scientist must become the interpreter of the physical environment, and he must do it in the long-term context of a dynamically changing earth so that "architectural" designs will be in harmony with natural forces 50, 100, or 500 years from now, as well as with the conditions of the moment.

16 Hydrology for Urban Land Planning

Luna B. Leopold

Luna B. Leopold is a professor in the Department of Geology and Geophysics at the University of California at Berkeley.

Of particular concern to the planner are those alternatives that affect the hydrologic functioning of the basins. To be interpreted hydrologically, the details of the land-use pattern must be expressed in terms of hydrologic parameters which are affected by land use. These parameters in turn become hydrologic variables by which the effects of alternative planning patterns can be evaluated in hydrologic terms.

There are four interrelated but separable effects of land-use changes on the hydrology of an area: changes in peak flow characteristics, changes in total runoff, changes in quality of water, and changes in the hydrologic amenities. The hydrologic amenities are what might be called the appearance or the impression which the river, its channel and its valleys, leaves with the observer. Of all land-use changes affecting the hydrology of an area, urbanization is by far the most forceful.

Runoff, which spans the entire regimen of flow, can be measured by number and by characteristics of rise in streamflow. The many rises in flow, along with concomitant sediment loads, control the stability of the stream channel. The two principal factors governing flow regimen are the percentage of area made impervious and the rate at which water is transmitted across the land to stream channels. The former is governed by the type of land use; the latter is governed by the density, size, and characteristics of tributary channels and thus by the provision of storm sewerage. Stream channels form in response to the regimen of flow of the stream. Changes in the regimen of flow, whether through land use or other changes, cause adjustments in the stream channels to accommodate the flows.

The volume of runoff is governed primarily by infiltration characteristics and is related to land slope and soil type as well as to the type of vegetative cover. It is thus directly related to the percentage of the area covered by roofs, streets, and other impervious surfaces at times of hydrograph rise during storms.

A summary of some data on the percentage of land rendered impervious by different degrees of urbanization is presented by Lull and Sopper (1966). Antoine (1964) presents the following data on the percentage of impervious surface area in residential properties:

Lot size of residential area (sq ft)	Impervious surface area (percent)
6000	80
6000–15,000	40
15,000	25

The percentage decreases markedly as size of lot increases. Felton and Lull (1963) estimate in the Philadelphia area that 32 percent of the surface area is impervious on lots averaging 0.2 acre in size, whereas only 8 percent of the surface area is impervious on lots averaging 1.8 acres.

As volume of runoff from a storm increases, the size of flood peak also increases. Runoff volume also affects low flows because in any series of storms the larger the percentage of direct runoff, the smaller the amount of water available for soil moisture replenishment and for ground-water storage. An increase in total runoff from a given series of storms as a result of imperviousness results in decreased ground-water recharge and decreased low flows. Thus, increased imperviousness has the effect of increasing flood peaks during storm periods and decreasing low flows between storms.

The principal effect of land use on sediment comes from the exposure of the soil to storm runoff. This occurs mainly when bare ground is exposed during construction. It is well known that sediment production is sensitive to land slope. Sediment yield from urban areas tends to be larger than in unurbanized areas even if there are only small and widely scattered units of unprotected soil in the urban area. In aggregate, these scattered bare areas are sufficient to yield considerable sediment.

A major effect of urbanization is the introduction of effluent from sewage disposal plants, and often the introduction of raw sewage, into channels. Raw sewage obviously degrades water quality, but even treated effluent contains dissolved minerals not extracted by sewage treatment. These minerals act as nutrients and promote algae and plankton growth in a stream. This growth in turn alters the balance in the stream biota.

Land use in all forms affects water quality. Agricultural use results in an increase of nutrients in stream water both from the excretion products of farm animals and from commercial fertilizers. A change from agricultural use to residential use, as in urbanization, tends to reduce these types of nutrients, but this tendency is counteracted by the widely scattered pollutants of the city, such as oil and gasoline products, which are carried through the storm sewers to the streams. The net result is generally an adverse effect on water quality. This effect can be measured by the balance and variety of organic life in the stream, by the quantities of dissolved material, and by the bacterial level. Unfortunately data describing quality factors in streams from urban versus unurbanized areas are particularly lacking.

Finally, the amenity value of the hydrologic environment is especially affected by three factors. The first factor is the stability of the stream channel itself. A channel, which is gradually enlarged owing to increased floods caused by urbanization, tends to have unstable and unvegetated banks, scoured or muddy channel beds, and unusual debris accumulations. These all tend to decrease the amenity value of a stream.

The second factor is the accumulation of artifacts of civilization in the channel and on the flood plain: beer cans, oil drums, bits of lumber, concrete, wire—the whole gamut of rubbish of an urban area. Though this may not importantly affect the hydrologic function of the channel, it becomes a detriment of what is here called the hydrologic amenity.

The third factor is the change brought on by the disruption of balance in the stream biota. The addition of nutrients promotes the growth of plankton and algae. A clear stream, then, may change to one in which rocks are covered with slime; turbidity usually increases, and odors may develop. As a result of increased turbidity and reduced oxygen content desirable game fish give way to less desirable species. Although lack of quantitative objective data on the balance of stream biota is often a handicap to any meaningful and complete evaluation of the effects of urbanization, qualitative observations tend to confirm these conclusions.

* * * * *

References

Antoine, L. H. 1964. Drainage and best use of urban land. *Public Works* (New York) 95:88-90.

Felton, P. N., and Lull, H. W. 1963. Suburban hydrology can improve watershed conditions. *Public Works* 94:93-94.

Lull, H. W., and Sopper, W. E. 1966. *Hydrologic effects from urbanization of forested watersheds in the northeast.* Upper Darby, Pa.: Northeastern Forest Experimental Station.

17 Earthquake "Briefs"

U. S. Geological Survey

The scientific study of earthquakes is comparatively new. Until the 18th century, few factual descriptions of earthquakes were recorded, and the natural cause of earthquakes was little understood. Many people believed that an earthquake was a massive punishment and a warning to the unrepentant. An Italian scholar in the 16th century, for example, suggested that statues of Mercury and Saturn be placed on building walls to protect against earthquakes. Those who sought natural causes often reached fanciful conclusions; one popular theory was that earthquakes were caused by air rushing out of caverns deep in the earth's interior.

The first earthquake for which there is detailed descriptive information occurred on November 1, 1775 in the vicinity of Lisbon, Portugal. An estimated 30,000 people were killed, most of them crushed to death under the ruins of buildings in Lisbon. Shocks from the quake were felt in many parts of the world, including the American Colonies, where "chandeliers rattled." After the quake, Portuguese priests documented their observations, and their records—still preserved—represent the first systematic attempt to investigate an earthquake and its effects. Since then, detailed records have been made of almost every major earthquake that has occurred.

The most widely felt earthquake in the recorded history of North America occurred in 1812 near New Madrid, Missouri. This quake was felt over an area of 2 million square miles—from Canada to the Gulf of Mexico, and from the Rocky Mountains to the Atlantic Ocean. Because the most intense effects were in a sparsely populated region, the destruction to human life and property was slight. If this earthquake occurred in the same area today, it probably would cause severe damage in many cities of the central Mississippi Valley.

The San Francisco earthquake of 1906 was one of the most destructive in the recorded history of North America—the earthquake and the fires that followed killed nearly 700 people, and left the city in ruins. The Alaska earthquake of March 27, 1964, however, was of greater magnitude than the San Francisco earthquake. Releasing perhaps twice as much energy, it was felt over an area of almost 500,000 square miles. Loss of life and property would have been far greater had Alaska been more densely populated.

Most of the world's earthquakes occur in near-continuous seismically active "belts" that coincide with zones in which mountain-building processes are still taking place. About 80 percent of all earthquakes—as well as most of the world's active volcanoes—occur in areas bordering the Pacific Ocean. This circum-Pacific belt, called the "ring of fire," includes the Pacific coasts of North and South America, the Aleutians, Japan, Southeast Asia, and Australasia.

It is estimated that 3 to 5 million lives have been lost in the past 1000 years from earthquakes and volcanoes. Records of past events suggest that a major destructive earthquake will strike the earth at least once a year. With the world's population expected to double in another 40 years, many major cities are becoming increasingly vulnerable to earthquake hazards.

An earthquake is the oscillatory and sometimes violent movement of the earth's surface that follows a release of energy in the earth's crust. This energy can be generated by a sudden dislocation of segments of the crust or by a volcanic eruption. Most of the destructive earthquakes are caused by dislocation of the crust.

When subjected to deep-seated forces (whose origins and natures are largely unknown) the earth's crust may first bend and then, when the stress exceeds the strength of the rocks, the crust breaks and "snaps" to a new position. In the process of breaking, vibrations called "seismic waves" are generated. These waves travel from the source of the earthquake to more distant places along the surface and through the earth at varying speeds depending on the medium through which they move. Some of the vibrations are of high enough frequency to be audible, while others are of low frequency—actually many seconds or minutes between swings. These vibrations cause the entire planet to quiver or ring like a bell or tuning fork.

Geologists have found that earthquakes tend to reoccur along *faults*—fractures in the earth's crust along which two blocks of the crust have slipped with respect to each other. One crustal block may move horizontally in one direction while the block facing it moves in the opposite direction. Or, one block may move upward while the other moves downward. Along many faults, movement is both horizontal and vertical. Faults represent zones of weakness in the earth's crust, but the fact that a fault zone has recently experienced an earthquake offers no assurance that enough stress has been relieved to prevent another quake.

The location of an earthquake is commonly described by the geographic position of its *epicenter*, and by its *focal depth*. The epicenter of an earthquake is the point on the earth's surface directly above the point (focus) where an earthquake's energy originates. The focal depth is the depth from the epicenter to the focus. Focal depths for shallow earthquakes range from the surface to about 60 km (38 miles). Intermediate earthquakes range from 60 to 300 km (38 to 188 miles). The focus of deep earthquakes may reach depths of 700 km (440 miles).

The focuses of most earthquakes are concentrated in the crust and in the upper mantle of the earth. Compared to a depth of about 4000 miles to the center of the earth's core, earthquakes thus originate in relatively shallow parts of the earth's interior. Earthquakes in California along the San Andreas and associated faults have shallow focal depths; for most the depth is less than 10 miles. During the past 100 years, earth movements have occurred along more than half the entire length of the San Andreas fault, and the rupture itself is visible at the land surface in many places.

Very shallow earthquakes are probably caused by fracturing of the brittle rock in the crust, or by internal stresses that overcome the frictional resistance locking opposite sides of a fault.

Earthquakes beneath the ocean floor commonly generate immense sea waves or "tsunamis" (Japan's dread "huge wave"). These waves travel across the ocean at speeds as great as 600 miles an hour, and may be 50 feet high or higher by the time they reach the shore. During the 1964 Alaska earthquake, tsunamis engulfing coastal areas caused much of the destruction at Kodiak, Seward, and other Alaskan communities. They also caused severe damage elsewhere along the west coast of North America, particularly at Crescent City, California. Some waves raced across the ocean to the coasts of Japan.

Water levels in artesian wells fluctuate as seismic waves travel through the rock layers that hold the water. During passage of seismic waves from a large earthquake, water levels in some wells fluctuate wildly, not only in the immediate vicinity of the earthquake, but also at great distances from it. In some wells, the water level change may be long-lasting or even permanent. The Alaskan quake appears to have caused water changes in wells in many areas both local and remote. For example, hydrologists of the U. S. Geological Survey noted that water levels at New Orleans, Louisiana, rose and fell as a result of the Alaska earthquake of 1964.

Landslides triggered by earthquakes often cause more destruction than the earthquake shocks themselves. During the 1964 Alaska quake, shock-induced landslides devastated the Turnagain Heights residential development and many downtown areas in Anchorage. The cause of the landslides and slumps that hit the Anchorage area can be traced to an unstable material called "Bootlegger Cove clay" that underlies much of the city. During the 1964 earthquake, failure in the "Bootlegger Cove clay" led to the landslides that caused most of the severe damage in the Anchorage area.

The vibrations produced by earthquakes are detected, recorded, and measured by instruments called *seismographs*. The zig-zag line made by a seismograph, called a "seismogram," reflects the varying amplitude of the vibrations by responding to the motion of the ground surface beneath the instrument. From the data expressed in seismograms, the time, the epicenter, and focal depth of an earthquake can be determined, and estimates can be made of the amount of energy that was released.

The two general types of vibrations produced by earthquakes are *surface waves*, which travel along the earth's surface, and *body waves*, which travel through the earth. Surface waves usually have the strongest vibrations and probably cause most of the damage done by earthquakes.

The severity of an earthquake can be expressed in several ways. The *magnitude* of an earthquake as expressed by the *Richter Scale*, is a measure of the amplitude of the seismic waves, and is related to the amount of energy released—an amount that can be estimated from seismograph recordings. The *intensity*, as expressed by the *Modified Mercalli Scale*, is a partly subjective measure which depends on the effects of a shock at a particular location. Damage or loss of life and property is another—and ultimately the most important—measure of an earthquake's severity.

The Richter Scale, named after Charles F. Richter of the California Institute of Technology, is the best known scale for measuring the magnitude of earthquakes. The scale is logarithmic, so that a magnitude of 7, for example, signifies a disturbance 10 times as large as a magnitude of 6. A quake of magnitude 2 is the smallest quake normally felt by humans. Earthquakes with a Richter value of 6 or more are commonly considered major in magnitude.

The Modified Mercalli Scale measures the intensity of an earthquake's effects in a given locality in values ranging from 1 to 12. The most commonly used adaptation covers the range of intensity from the condition of "1—Not felt except by very few, favorably situated," to "12—Damage total, lines of sight disturbed, objects thrown into air."

Earthquakes of large magnitude do not necessarily cause the most intense surface effects. The effect in a given region depends to a large degree on local surface and subsurface geologic conditions. An area underlain by unstable ground (sand, clay, or other unconsolidated materials), for example, is likely to experience much more noticeable effects than an area equally distant from the earthquake's epicenter, but underlain by firm ground such as granite.

The United States has experienced less destruction than other countries located in the "ring of fire," but millions of Americans live in potential quake areas, particularly in the western part of the United States. Increasing amounts of construction in many places where the danger of major earthquakes is ever present has created an urgent need for more information on the nature, causes, and effects of earthquakes.

The prediction of earthquakes is one of the objectives of current studies by the U. S. Geological Survey and other scientific groups. At present, it is not possible to predict the time, place, and size of a specific earthquake in advance, but considerable progress has been made in formulating the statistical probability that an earthquake of given magnitude will occur in a region within a specific span of time.

For the present, the best protection against earthquakes is to avoid construction in high-risk areas. For example, the Port of Valdez, Alaska, destroyed by the 1964 earthquake, was found to have been built on incipient landslide blocks. These blocks did not provide stable building sites. Because little property of value remained standing after the earthquake, and because the town was not heavily populated, the people of Valdez chose to rebuild their town at a new site—one recommended on the basis of geologic study. In large metropolitan areas, on the other hand, the emphasis in earthquake protection will inevitably have to be on land use, planning, and improved construction methods.

Earthquakes on the Island of Hawaii, site of the active volcanoes Kilauea and Mauna Loa, appear to be associated with volcanic activity. Abrupt increases in quake activity, at times, herald an eruption, and the location of swarms of tremors can indicate where lava may break out. Also, prior to an eruption, the volcano "swells" measurably in response to the upward movement of molten rock. Continuous seismic and tiltmeter (a device that measures earth movements) records are maintained at the Geological Survey's Hawaiian Volcano Observatory, located on the rim of Kilauea Volcano where study of these records enables specialists to make short-range prediction of volcanic eruptions.

Scientific understanding of earthquakes is of vital importance to the nation. As the population increases, expanding urban development and construction works encroach upon areas susceptible to earthquakes. Earthquake research efforts have been intensified, and special studies of such earthquake-prone zones as the San Andreas fault in California by geologists, geophysicists, and seismologists of the U. S. Geological Survey's National Center for Earthquake Research are yielding new insights into the nature and mechanisms of earthquakes, which, hopefully, may make earthquake predictions possible and which certainly will increase man's ability to live safely in earthquake zones.

18 Land Subsidence

Frank Forrester

Frank H. Forrester is information officer at the National Center of the U. S. Geological Survey in Reston, Virginia.

Although several hundred square miles of land have experienced subsidence (sinking) in the Santa Clara Valley south of San Francisco, California, the rate has decreased markedly in the past few years. Between 1934 and 1967, part of the area sank more than eight feet as a result of excessive pumping of ground water (Figure 1). Since 1966, however, because of a rise of ground water levels, the rate of subsidence has slowed down dramatically.

Results of continuous measurements of subsidence made by the U. S. Geological Survey at San Jose and Sunnyvale since 1960 show that the rate of subsidence has decreased by about 95 percent. At San Jose, the rate decreased from about 1 foot per year in 1961, to about 0.07 foot per year in 1970 (Figure 2). At Sunnyvale, it decreased from about half a foot per year in 1961 to 0.02 foot per year in 1970.

Subsidence is likely to develop when a large amount of water is removed from unconsolidated artesian aquifers (subsurface water-bearing rocks). As ground water is pumped from the aquifer the artesian pressure decreases, and more of the weight of the overburden must be carried by the granular structure of the sediment. The additional weight compacts the unconsolidated sediments and squeezes water out of the pore spaces in the fine-grained clays and silts into the coarse-grained aquifer. The resulting reduction in porosity of the fine-grained sediments causes a reduction in volume which causes subsidence of the land surface.

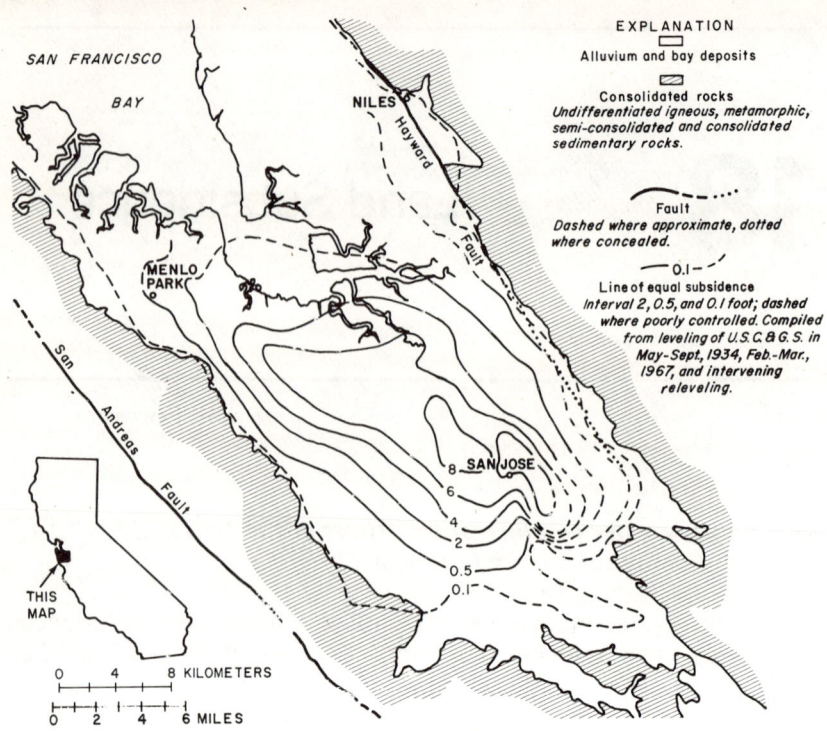

Figure 1. Land subsidence from 1934 to 1967, Santa Clara Valley, California. Modified from an open-file map of the U.S. Geological Survey.

In the Santa Clara Valley the pumping of ground water for a variety of uses increased from about 13 billion gallons (40,000 acre feet) in 1916 to about 58 billion gallons (180,000 acre feet) a year during the early 1960s. As a result, the artesian pressure "head" fell as much as 250 feet. With the loss of this artesian support, sediments in the aquifer compacted and the land settled. Between 1934 and 1967, this compaction amounted to a volume loss of about 20 billion cubic feet (500,000 acre feet). Compaction takes place slowly over a period of years so there is no violent movement of land such as occurs during an earthquake. However, land subsidence remains an important and costly problem.

The Santa Clara County Flood Control and Water District estimates that about $9 million of public funds has been spent on levee construction and other remedial work on stream channels to prevent flooding as a result of land subsidence, especially along the southern shore of San Francisco Bay. Other valley officials have

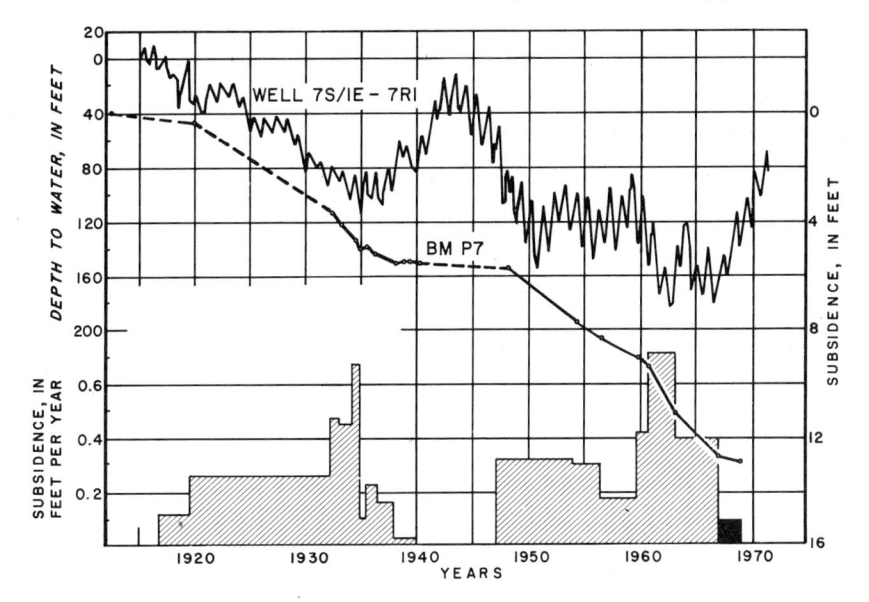

Figure 2. From top to bottom, the graphs show change in artesian head in water well near bench mark P7, land subsidence at bench mark P7 in San Jose, and rate of subsidence in feet per year. Note the marked change in rate of subsidence from 1966 to 1969. Prepared by U.S. Geological Survey.

estimated that the cost of repairing or redrilling several hundred damaged wells amounts to at least $4 million.

Although subsidence can't be fully reversed, it can be stopped. The most dramatic decrease in the rate of subsidence has occurred since 1966 as a result of imports of larger quantities of surface water. These large surface-water imports have permitted a decrease in the amount of ground water that has been withdrawn. As a result, the artesian water level, sometimes called the piezometric surface, has risen about 60 feet. This rise in the artesian water level has reversed the trend of increasing grain-to-grain stress on the sediments and thus has slowed the rate of subsidence. If the artesian head could be raised and maintained at least 20 feet above present levels, then subsidence would be stopped.

The Santa Clara Valley was the first area in the United States where land subsidence caused solely by excessive ground water removal was recognized. The extent and amount of subsidence was first realized in 1933 when bench marks established in 1912 were resurveyed by the Coast and Geodetic Survey, and were found to have subsided as much as 4 feet. Between 1934 and 1967, the year of the most recent survey of bench-mark altitudes, the amount of subsidence increased southeast from Menlo Park and Niles to San Jose, where it exceeded 8 ft.

Local agencies have been working since the 1930s toward the goal of obtaining water supplies adequate to stop the overdraft of ground water and raise the artesian head. From 1964-65 to 1969-70, the amount of water imported from outside the valley increased fourfold—from 30,000 to 125,000 acre-feet a year. To pay for these efforts and to encourage use of imported water, Santa Clara County levies a county-wide ad valorem tax and a charge for water extracted from the ground-water basin. The rise in water level that has nearly eliminated the subsidence is due to this great increase in the use of imported water.

Land-subsidence problems are not limited to the Santa Clara Valley area of California. Similar problems have either developed or exist potentially in such areas as the San Joaquin Valley; the Houston, Texas area; in Mexico City, Mexico; and in Venice, Italy.

The most serious problems are in Japan where more than ten areas have subsided as a result of ground-water overdraft. In Tokyo, two million people live in an area that has sunk below high tide level, and in Osaka, over 600,000 people live under similar conditions.

The U. S. Geological Survey is working in close cooperation with the California Department of Water Resources and various federal, state, and local agencies in studying land-subsidence problems, and in evaluating methods and techniques that might be applied to avoid or alleviate subsidence damage. Subsidence can be reduced or even partially reversed through water management. A reduction or reversal can be achieved only by increasing the artesian pressure either by reduced ground-water withdrawal, or by increased ground-water recharge.

19 Supplemental Areas for Storage of Radioactive Wastes in Kansas

Charles K. Bayne (compiler)

Charles K. Bayne is associate director of the Kansas Geological Survey and an associate state geologist. Ernest E. Angino is chairman of the Department of Geology at the University of Kansas. John C. Halepaska is chief of the Water Resources Section of the Kansas Geological Survey.

In December 1971 the State Geological Survey of Kansas prepared for the U. S. Atomic Energy Commission (A.E.C.) a report on areas potentially suitable for storage of high-level radioactive wastes in the salt beds in Kansas. These areas are supplemental to the federally proposed Lyons Repository site at Lyons, Kansas, on which the Survey also reported in detail.

The report on supplemental areas for the A.E.C. was published in limited numbers and the maps included in the report were larger than desirable for general use. The present report summarizes the data and results of the original report and is produced in a format more suitable for general distribution. In the preparation of the original report no field work was done; however, publications and unpublished data of the Kansas Geological Survey and other state agencies were searched for pertinent information.

Factors considered important in determining the suitability of salt beds for waste disposal are: geology and hydrology, oil and gas developments, pipelines, thickness of salt beds, depth to salt beds, population centers, and potential mineral resource development of the area. Buffer zones having a radius of 6 miles around each population center are shown on the geologic maps for each area studied. The 6-mile radius is an arbitrary figure used in the original report for the A.E.C. It is based on no particular criteria other than the importance of safety to the population centers. No distinction is made as to size of the population centers.

Potential development is concerned principally with the future development of oil and gas. Also to be considered under this factor and related to population is recreational development around large reservoirs in and near potential repository sites. Although these recreational developments would have few permanent residents, large numbers of people could be present at times.

Eight areas (Figure 1) in the state were selected for study. Selection of representative areas was based on a general knowledge of each area, the presence of at least 150 feet of salt at a depth of less than 2000 feet, and a minimum of oil and gas development. In Kansas thick beds of salt occur in Permian rocks at three horizons: the Hutchinson Salt Member of the Wellington Formation, the Cimarron salt which occurs in the Ninnescah Shale just below the Stone Corral Formation, and the Blaine salt which occurs in the Flower-pot Shale just below the Blaine Formation. "Cimarron salt" and "Blaine salt" are informal names used in this report. Salt has been mined in Kansas only from the Hutchinson Salt. The location and status of mining operations is shown in Figure 1. In seven of the areas (Areas 1-7) the suitability of the Hutchinson Salt was studied and in Area 8 the Blaine salt was studied.

The quality and thickness of the Hutchinson Salt is known through the mining operations and the study of cores and many gamma-ray-neutron logs.

The salt beds which occur in the Ninnescah Shale (Cimarron salt) in association with the Stone Corral Formation are known through study of sample logs and geophysical logs. These salt beds were not investigated during this study.

The salt beds which are associated with the Blaine Formation are known from sample logs, geophysical logs, and core description.

Salt in the Permian rocks of Kansas does not occur as a single massive bed of pure salt, but as a succession of layers of relatively pure salt interbedded with shale or anhydrite. Shale is more common near the margins of the salt basins. Key beds within the salt section can be correlated over only relatively short distances, indicating a continual shifting of the locus of deposition within the basin and consequently a change in the thickness of the different lithologies occurring within the salt section.

Quality of a salt section is judged on the basis of the unique

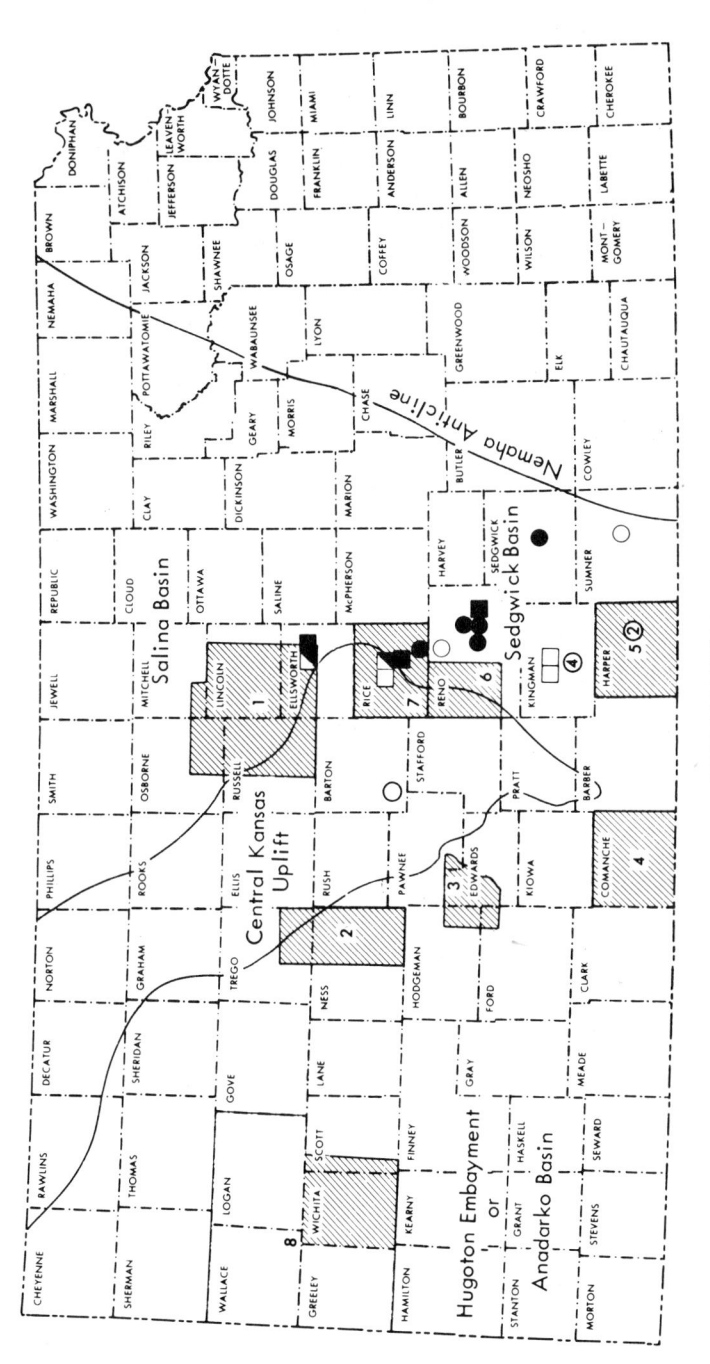

Figure 1. Map showing supplemental study areas, structural features, and salt mines.

signature of each lithology (kind of rock) on radioactive logs. Absolute percentages cannot be determined but the interbedding of shale and anhydrite with salt is clearly displayed and, in addition, dispersed impurities within the salt may be indicated by the displacement of the radioactive curve in its entirety. Each lithology within the salt section may have special properties in reference to the engineering problems which may occur in the salt section of a repository site. The engineering problems associated with quality of the salt section are not considered in this report. Instead, the relative quality of the salt section, as expressed in this report, takes into consideration primarily the overall purity of the section and the definition of a thick mineable salt unit.

* * * * *

Discussion and Conclusions

By John C. Halepaska and Ernest E. Angino

In the matter of suitability for radioactive waste storage, the Kansas Geological Survey has not evaluated the importance of those criteria enumerated in the introduction. However it is the opinion of the Survey staff that these criteria can be ordered generally. The first consideration is the safety of people. Adequate safety buffers are absolutely necessary. Second in importance are engineering and geologic considerations which include depth of salt, salt thickness, salt quality, subsurface geologic structure, old oil and gas holes, and probably most importantly, ground water. Obviously, a most vigorous attempt to minimize breaching the integrity of the salt vault associated with storage is in order. Third would include transportation (assuming inherent danger remains fixed regardless of site), pipelines, and potential recreational and oil and gas development.

Owing to the nature of this study, unfavorable aspects of an area may be more explicitly described and defined than favorable aspects. In areas where information is sparse a broader regional interpretation is required. Obviously, focus must be on regionally good areas where a maximum amount of reliable information exists.

On a regional basis, Area 1 deserves close scrutiny. It is distant from regional oil and gas development and appears to have adequate buffer space. The area fits reasonably well the structural engineering criteria (e.g., thickness, depth, and quality of salt).

The likelihood of unknown hydraulic connection between the salt and an aquifer above or below the salt in Area 1 caused by an old drill hole or fracture pattern is less. The potential for oil and gas development is considerably less than in any of the other seven areas considered. All factors being equal, Area 1 is the best prospect for future study.

Area 2 lies on the western side of an area of regional oil and gas development. It is now undergoing active exploration and, in addition, has good potential for future oil and gas development. The area is marginal in terms of structural engineering criteria in that the salt is near the 2000-foot depth limit. The potential for water problems above and below the salt, in addition to old holes and depth to salt, makes this area a poor candidate for further study.

Area 3 can be categorized much the same as Area 2 and is a poor prospect for future study.

Area 4 lies to the west of regional oil and gas development. It lies east of a large area that appears to have undergone some salt solutioning and faulting. The study area has been, and is now, actively explored for oil and gas. This area, like Areas 2 and 3, is marginal in meeting the structural engineering criteria. The salt is about 1700 feet deep. Water problems above and below the salt, existence of old holes, depth to salt, and salt solutioning immediately to the west make Area 4 a poor prospect for future study.

Area 5, slightly south of the regional oil and gas development, is marginal in terms of structural engineering criteria. Oil and gas exploration has been active; however, sparsely drilled areas do exist. The salt is of poor quality although it is within the limits for thickness and depth. This area does, however, have considerably less ground water above the salt than any other area. A trade-off between thin, poor-quality salt and salt with little or no water above it may make Area 5 worth consideration.

Area 6 lies to the south of the Lyons site on the southeast flank of the regional oil and gas development trend. The area has been, is being, and will continue to be actively explored. The salt fits the structural engineering criteria. Water problems above and below the salt beds will be as great as in any area studied. Because of old oil and gas holes and copious quantities of ground water in the area, Area 6 ranks below Areas 1 and 5, but above Areas 2, 3, and 4.

Area 7 lies on the east flank of the regional oil and gas development. This area has the most producing wells, the greatest history of exploration, and the highest potential for future development of any area considered. The salt fits the structural engineering criteria well. Medium to low quantities of ground water exist above the salt. Large numbers of old oil and gas holes, water above and below the salt, possible deep-seated structural problems, high potential for development in the future, and an inadequate buffer zone make Area 7 the poorest candidate for future study.

Area 8 lies in extreme western Kansas, far to the west of the major Kansas regional oil and gas producing areas, but on the eastern fringe of an active area of oil and gas development in eastern Colorado. The western part of Area 8, though largely undrilled, will probably undergo vigorous exploration. The salt in this area is more than 2000 feet deep. The salt in the eastern part of Area 8 is of high quality, is quite thick, and is 1700 feet deep. Lost-circulation problems and occasional absence of the salt in the western half of this area (Wallace and Greeley counties, Kansas, and eastern Colorado) raise a caution light. A regional hydrodynamic analysis indicating high head differentials across the salt, together with the problem of a regional fault system, may account for the missing salt section. In addition, locally high saturated thickness in the Ogallala Formation indicates possible anomalous structural conditions. Sandstone both above and below the salt may contain large quantities of water. While this area has high-quality salt, its suitability ranks below that of Area 1 and possibly Area 5.

In summary, Areas 2, 3, 4, 6, and 7 rank far below Areas 1, 5, and 8. Area 7 is least suitable. A trade-off between poor quality of salt and little or no ground water in Area 5 may be worth considering. The western part of Area 8 has unanswered questions, but the eastern part of Area 8, with high-quality salt, is worthy of consideration. It is, however, definitely rated below Area 1 and above Area 5. In conclusion Area 1 appears to be the best candidate for future study.

References

Angino, E. E., and Hambleton, W. W. 1971. *Geology and hydrology of the proposed Lyons, Kansas, radioactive waste disposal site (final report)*. Kansas Geological Survey.

Angino, E. E., Hambleton, W. W., Bayne, C. K., and Halepaska, J., eds. 1972. *Preliminary geological investigations of supplemental radioactive waste repository areas in the state of Kansas*. Kansas Geological Survey.

State Geological Survey of Kansas. 1964. Geologic map of Kansas, Map M-1.

20 Medical Geology

Harry V. Warren and Robert Delavault

Harry V. Warren and Robert E. Delavault are professors in the Department of Geology and Geography at the University of British Columbia.

At first thought one might be tempted to assume that there can be little to relate the sciences of geology and medicine. However, a few moments of reflection should be enough to permit the most skeptical to realize that an association between geology and medicine is a natural one. After all, most people are prepared to accept, at least as a hypothesis, the possibility that man's health is determined, to some extent, by the food he eats. The quality of food reflects the makeup of the soil which, in turn, is determined, in part, by the chemistry of the rocks and vegetal life on the earth's crust, which between them, contribute much to the makeup of soil.

It is not difficult to find medical men who can point out how different communities have different ailments. Some of these ailments probably have little or nothing to do with geology, but a few of them may. For generations, the more fortunate members of our society have enjoyed a change of air, food, and water by going each year for a holiday. For some of those who are not too sophisticated, a change of food may come as a delight. Perhaps a change of trace elements is involved in a holiday. At all events, as society gets more and more removed from close contact with the soil, and more and more closely tied to the supermarkets of today, the ties that surely do exist between man and the earth's crust will become more and more tenuous. Even today, it is difficult to find communities suitable for simple epidemiological studies in which trace elements may be involved. However, with care it is still possible to find some communities suitable for trace element epidemiological studies: such communities do exist in some of the undeveloped countries, but they can also be found in conservative and comparatively long-established settlements in Western Europe and North America.

It has long been known that an iodine deficiency can affect whole communities, although we now realize that the problem of goiter is much more involved than was originally realized. Human beings can be affected by poisonous materials in soils when those soils contribute to a large proportion of the diet of a community. In parts of the Dakotas, not only cattle but humans were affected by the selenium that was found in some specific soils. Drinking water may also be the vehicle by which undesirable elements may be introduced into a diet: arsenic and fluorine are two elements that may be present in harmful amounts in local water supplies.

To most geologists, lead is merely lead when it is reported in food or drinking water. Actually, there is one isotope of lead, namely lead 210, derived from uranium by way of radon 222, which is a gas slightly soluble in water. Lead 210 is radioactive with a half-life of some 22 years. However, lead 210 itself breaks down and yields polonium 210, which is a strongly radioactive and poisonous element, particularly dangerous because it finds a resting place in the soft tissues of human bodies.

There may be some good reason, but superficially it seems strange that agriculturists have not done more than they have done to draw doctors and geologists together. It is literally an article of faith with agronomists that soils are related on the one hand to their parent material, and, on the other, to the quality of the food they produce. Nevertheless, doctors and nutritionists alike still publish books that contain tables showing the amounts of various elements, such as copper and iron, that are to be found in different foodstuffs. One may search these medical journals, and all too rarely find any evidence that the significance of geology in relation to food is appreciated. It is equally true that one can search geological literature for a long time before finding any analyses of foodstuffs in relation to their geological background. It is easy to forecast that all this will change before long. Already, in Scandinavia, in Australia, in Great Britain, and in Canada, foresters are realizing more and more how the productivity of their forests is governed, in some areas to a significant extent, by the bedrock involved. It may surprise some geologists to learn that there are areas in the world where foresters would not dream of planting new trees before consulting with their geological confreres.

What makes biogeochemistry such a fascinating subject for a geologist is that appropriate quantities of such elements as copper, zinc, and molybdenum are now known to be just as vital to the health of some plants and animals as they are to the success of mining operations: the concentrations of a specific element needed for the fulfillment of these different needs naturally varies considerably. However, zinc, in appropriate concentrations, is as essential to a diabetic patient as it is for the farmer trying to produce a maximum poundage of pig from each pound of food, and for a forester who is attempting to cultivate a forest. Humans, pigs, and trees would all die if no zinc were available, and equally they can all be adversely affected if they are given too much zinc. The problem yet to be solved in most instances is that of knowing just what is the optimum concentration of zinc needed by a human, a pig, or a tree, as the case may be.

Unfortunately, it is much more difficult to solve equations where there are two variables instead of one, and with each additional unknown element, the complexity of the inquiry increases. A comparatively short time ago only some dozen elements were known to be essential to plant and animal health, but today it may be said that nearer 40 than 30 are thought to play some part in nutrition. Some elements that are always present in plants and animals are not yet known to serve any useful purpose: in the fullness of time it may be found that even these elements, such as lead, mercury, and silver, which are now considered as inhibitors or poisoners of enzyme systems, may actually, if present in appropriate concentrations, act as useful moderators, or regulators, of specific metabolic processes. A word coming more and more into use by biogeochemists is "imbalance." For maximum health in both plants and animals, it is desirable not only that the elements should be present in appropriate concentrations, but also that there should be harmonious relationships between the concentrations of the different elements. It has taken hundreds of millions of years for life to evolve as we now have it on our planet. Surely it is reasonable to assume that life today represents a complex synthesis of the energy and the elements present on the crust of the earth? Thus we might expect to have an infinite variety of living cells representing responses to the different facies extant on the earth's crust. Some forms of life are able to adapt to a sharply different environment, but others perish. Some species of grass have been found to adapt in two generations to a high lead content in soil, but sheep eating this grass could not adapt themselves so promptly: fortunately, it seems that the shepherds in the area involved with this lead soon learned just how long their sheep could tolerate this leaded herbage. Interestingly enough, sheepherders for generations seasonally have moved their flocks. Sometimes this was done merely to ensure a supply of food, but occasionally these moves were dictated by the necessity of avoiding poisonous concentrations in herbage of such elements as molybdenum and selenium.

If it be accepted that variety is the spice of life, then, indeed, biogeochemistry should attract many geologists in the years ahead. However, the number of subjects with which a biogeochemist may become involved is large. Biology, botany, agriculture, bacteriology, virology of the life sciences, and physics and chemistry of the physical sciences all have much to offer to the biogeochemist who must sample them with prudence and wisdom if he is not to be overwhelmed with a mental flatulence likely to be neither congenial nor stimulating but merely frustrating.

At the 1964 meeting of the AAAS in Montreal, there was a symposium on Medical Geology and Geography. The participants included a geologist, a pharmacognosist, a biochemist, a general practitioner, and a geological engineer. It might be a moot question as to whether each one of these persons appreciated fully all the finer points raised by their fellow participants in the panel. Nevertheless, the participants, themselves, seemed to find the interdisciplinary exercise well worthwhile and, judging from the reactions of the press and the audience, cross-fertilization experiments of a similar nature may be expected to attract increasing attention in the future. Biogeochemistry is, to many, an uninspiring name: perhaps we should change the name to "Geological ecology," which would relate this field of study to that of life in relation to its geological environment.

21 A Progress Report on Mercury

John M. Wood

John M. Wood has recently been appointed professor of biochemistry and ecology at the University of Minnesota, where he will be director of the Freshwater Biology Institute.

It is now almost two years since mercury pollution was recognized as a serious problem in the United States and Canada. It is four years since scientists determined why mercury losses to the environment pose a threat to public health. Although the problem is now well defined, legal and political safeguards have proven inadequate to generate solutions in two important respects. First, huge amounts of mercury remain in sediments downstream from industrial polluters, even though the industries have generally halted or drastically reduced the amount of mercury discharged. The mercury in the sediments is in the inorganic, less biologically harmful form, but natural processes are steadily converting the material to methyl mercury, the most lethal organic mercury compound. The second danger is that methyl mercury compounds are still widely used in U. S. agriculture, even though such use leads to serious contamination of the food chain.

As to the first problem, Table 1 shows a published compilation of some of the companies responsible in the past for mercury pollution of inland and coastal waterways in the U. S. Most of these companies have stopped discharging mercury, but abundant deposits of inorganic mercury are present in sediments in the inland and coastal waterways listed in the table. Many local areas of heavy contamination with mercury exist in the U. S. Surveys in many states are still underway, so that the polluted areas listed represent the minimum at the present time.

TABLE 1. MERCURY POLLUTION OF INLAND AND COASTAL WATERWAYS

State	Source	Polluted Waters
Alabama	Diamond Shamrock	Pond Creek to Tennessee River
	Olin Mathieson	Tombigbee River
	Stauffer Chemical Company	Mobile River
Arizona	Pioneer Paint and Varnish Co.	Santa Cruz River
California	Garret-Callahan Company	San Francisco Bay
	Quicksilver Products	San Francisco Bay
Delaware	Diamond Shamrock	Delaware River
Georgia	Olin Mathieson	Savannah River
Kentucky	Goodrich Chemical Company	Tennessee River
	Pennwalt Chemical Company	Tennessee River
Louisiana	Dow Chemical	Mississippi River
	Monochem, Inc.	Mississippi River
	PPG Industries	Bayou d'Inde
	Wyandotte BASF	Mississippi River
Maine	Oxford Paper Company	Androscoggin River
	International Mining and Chem. Company	Penobscot River
Michigan	General Electric Co.	Cedar Lake to Pine River
	Wyandotte BASF	Detroit River
New Jersey	General Aniline and Film Corporation	Arthur Kill
	Woodbridge Chem.	Berrys Creek to Hackensack River
New York	Allied Chemical Company	Buffalo River to Lake Erie
	Allied Chemical Company	Onondaga Lake
	Chesebrough-Ponds, Inc.	Black River to Lake Ontario
	Hooker Electrochem	Niagara River
	Olin Mathieson Chem.	Niagara River
	Williams Gold Refining Co.	Niagara River
North Carolina	Riegel Paper Company	Cape Fear River
Ohio	Dextrex Chem., Inc.	Lake Erie
	General Electric Chem.	Lake Erie
	NASA, Lewis Research Center	Rocky River
	Reactive Metals, Inc.	Ashtabula River
Pennsylvania	Mallinckrodt Chem.	Lake Erie
	NOSCO Plastics	Lake Erie

TABLE 1. (Continued)

State	Source	Polluted Waters
Tennessee	Buckeye Cellulose	Wolfe River to Mississippi River
	Buchman Labs	Mississippi River
	Chapman Chem. Company	Mississippi River
	Olin Mathieson	Hiwassee River
Texas	Aluminum Company of America	Lavaca Bay
	Diamond Shamrock Chem.	Houston Ship Canal
	Monsanto Chem. Company	Galeston Bay
	Tenneco Chem. Company	Houston Ship Canal
Virginia	Olin Mathieson Chem.	North Fork, Holston River
Washington	Georgia-Pacific	Puget Sound
	Weyerhaeuser Company	Columbia River
Wisconsin	Wyandotte BASF	Wisconsin River

Source: Montague, Katherine and Peter, Mercury, a Sierra Club Battlebook, Sierra Club, New York, 1971, pp. 131-148.

The extent of pollution from the second problem, agricultural use of methyl mercury compounds, is even less precisely known, since the compounds can be dispersed in the environment in a number of ways. We do know that methyl mercury compounds are still in widespread use in the United States. About 3 percent of the total mercury used in the United States is applied to seeds and crops. The methyl mercury compounds used as fungicides have entered the food chains of seed-eating birds, carrion, and predators. The hunting season for Hungarian partridge and pheasant was cancelled in Alberta, Canada, in 1969 because of mercury contamination, and mercury-polluted pheasants have been found in California, Montana, and Ohio.

In Sweden, by contrast, steps were taken in 1965 to reduce drastically the agricultural use of methyl mercury compounds. One interesting feature of the decrease of methyl mercury usage in 1965 was that this decrease did not affect crop yields in Sweden (1).

Mercury to Methyl Mercury

At one time, industrial discharge of inorganic mercury compounds did not appear to pose much of a health problem, since inorganic mercury is readily excreted in the urine and is not absorbed by the body in large amounts. In 1966, however, Gunnel Westöö, an employee of the Swedish government, made a disturbing discovery. She reported that the total mercury accumulating in the flesh of fish which she analyzed was in the form of methyl mercury, the organic form readily absorbed by the body and poisonous to human beings. At the time, Dr. A. Jernelöv in Stockholm and my own research group at the University of Illinois realized that some mechanism must be available for the methylation of inorganic mercury, either by microorganisms in sediments or in the fish themselves.

Two years later, both the Stockholm and University of Illinois research groups reported that microorganisms in sediments were capable of producing methyl mercury and dimethyl mercury from inorganic mercury compounds. Both laboratories issued warnings that inorganic mercury pollution would lead to serious public health problems because methyl mercury compounds were very potent neurotoxins (poisons of the central nervous system).

Since 1968 Mr. E. Lien and Dr. R. E. DeSimone (2) of my research group at Illinois have worked out in detail the fundamental chemistry involved in the formation of methyl mercury compounds from inorganic mercury. Furthermore, the elements important in the synthesis of methyl mercury in sediments have been determined (3). The route for the synthesis of methyl mercury compounds is presented schematically in Figure 1. Carbon, nitrogen, phosphates, and trace metals provide microorganisms with the food they require to grow and divide. This supply of food determines the size of the population of bacteria and molds that lives in sediments. The bacteria and molds that live in sediments convert inorganic mercury to methyl mercury compounds. The rate at which methyl mercury is produced fundamentally depends upon the concentration of inorganic mercury present in sediments and also upon the number of bacteria and molds present.

All microorganisms capable of vitamin B_{12} synthesis are capable of methyl mercury synthesis. Methyl mercury does not form stable coordination complexes with organic chemicals in the sediment, and so the compound is released into the water. This reaction can be viewed as detoxification of the microorganisms' environment at the expense of our environment, since the methyl mercury is taken up in the food chain and becomes concentrated in fish (4). Fish also take in methyl mercury directly from the

ENVIRONMENTAL GEOLOGY

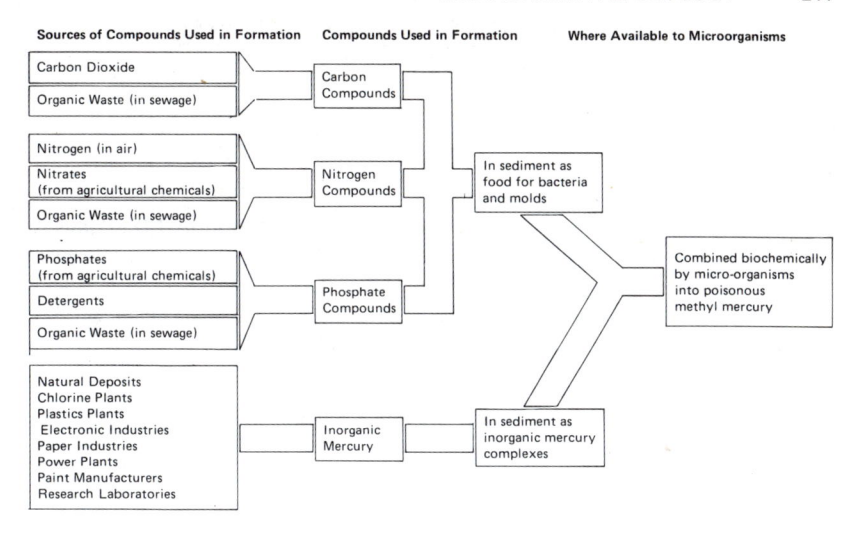

Figure 1. Formation of methylmercury in water.

water: Dimethyl mercury and some monomethyl mercury compounds (for example, methyl mercury chloride) can be directly concentrated in fish by diffusion across the gills.

Many species of fish concentrate methyl mercury because this compound is metabolized (excreted) very slowly. Professor J. K. Miettenen at the University of Helsinki, Finland has studied the retention of radioactive methyl mercury in many species of fish. Table 2 shows the time required for different species of fish to lose half the radioactive methyl mercury administered to them under laboratory conditions.

Clearly, methyl mercury production does not have to be a very rapid process in sediments for fish to accumulate dangerous levels of the substance. When the rate at which methyl mercury is produced, then released into the water and taken up by fish exceeds the rate of metabolism of methyl mercury in fish, then methyl mercury accumulates in fish in quantities that render the fish extremely dangerous for human consumption. The most striking example of the importance of food chains in concentrating methyl mercury is provided by K. Henriksson in *A Study of Seals in Finland* (see Table 3). The average concentration of methyl mercury in various tissues of seals increases significantly in mercury-contaminated waters.

TABLE 2. METHYL MERCURY HALF-LIFE* IN FISH

Variety of Fish	
Flounder	400 to 700 days
Perch	500 days
Pike	500 to 700 days
Eels	900 to 1,000 days

*The time required to eliminate half of the methyl mercury.

Source: Keckes, S., and J. K. Miettenen, "Mercury as a Marine Pollutant," Proceedings of the F.A.O. Conference, Rome, Dec. 9-18, 1970.

TABLE 3. METHYL MERCURY IN THE TISSUES OF FINNISH SEALS

Location	Muscle	Liver	Kidney
Gulf of Finland (nonpolluted)	0.9 ppm*	11.8 ppm	3.2 ppm
Finish Lakes (polluted)	62.4 ppm	137.8 ppm	46.3 ppm

*Parts per million.

Source: Henriksson, Nordiskt Symposium Kring Kuicksilver Problematiken, Stockholm, 1968.

World Problem

To trace briefly the now familiar sequence of events in the mercury story: After 116 people had been poisoned by the effluent of a vinyl chloride plant (vinyl chloride is used in the manufacture of plastics) in Minamata, Japan (1953) (5), and 26 people had been poisoned in an identical situation in Niigata, Japan (1965), scientists in Sweden and Finland (6) realized that mercury losses to the environment pose a serious threat to public health.

Industrial chemists in the U. S. and Canada were not well informed of the problems in Japan and Sweden. Those companies familiar with the literature noted that the Japanese disasters were caused by methyl mercury. Since U. S. industries generally discharged the less harmful inorganic mercury compounds, it was decided that there was no need to change industrial waste-disposal practices. In September 1969, Dr. O. Nielsen (7) of the Department of Veterinary Medicine at the University of Saskatchewan at

Saskatoon analyzed fish taken near the Saskatoon Cooperatives Chlorine plant on the Saskatchewan River. Elevated levels of mercury were discovered in many species of fish, and Dr. Nielsen's research prompted a further investigation of the Saskatchewan River System in the Provinces of Manitoba and Saskatchewan. While this survey was underway, Norvald Fimriete, a graduate student at the University of Western Ontario, collected fish downstream from the Dow Chlorine Plant at Sarnia, and on March 20, 1970 reported elevated levels of mercury in fish taken from the St. Clair River System. The publicity associated with Fimriete's discovery resulted in the establishment of crash programs for fish analysis by both the Canadian and the U. S. governments. Analyses in Saskatchewan and Manitoba indicated that most of the Saskatchewan River System, including Lake Winnipeg, was seriously polluted. Saskatoon Cooperatives at Saskatoon released 66,000 pounds of mercury between 1963 and 1970. Also, the Dryden Chlorine Plant at Lake Wabigoon in Manitoba was shown to be responsible for the pollution of most lakes and rivers in central Manitoba. In the spring of 1970 U. S. Department of the Interior Secretary Walter Hickel rapidly initiated a program for the analysis of fish in the U. S., and the Government of Ontario began extensive studies of the St. Clair River System into which both Dow Chemical Company of Canada and Wyandotte BASF had dumped large quantities of mercury. In the period 1949 to the present, Dow Chemical at Sarnia has released approximately 200,000 pounds of mercury compounds into the St. Clair River System.

Analytic Shortcomings

When Norvald Fimriete announced the results of analyses of fish taken from the St. Clair River System, both the Canadian and U. S. governments, together with the various chemical companies involved, began programs for the rapid analysis of total mercury concentrations in fish. Different methods of analysis were chosen by different laboratories. For some reason, analytical chemists involved in this survey decided to use methods for the analysis of total mercury. No one seemed concerned that Dr. Westöö had clearly demonstrated that methyl mercury is the compound which accumulates in fish. Those who favor total mercury analysis point out that since 95 to 100 percent of mercury in fish is in the methyl mercury form, findings obtained by the total mercury analysis correlate quite closely with the methyl mercury content of the fish. However, all total mercury analytical methods suffer from losses of methyl mercury, dimethyl mercury and methyl mercurythiomethyl. Analytical chemists should be much more concerned with developing methods to analyze for individual compounds rather than for elemental mercury.

At any rate, within a short period of time two methods for total mercury analysis gained popularity for the survey: flameless atomic absorption and neutron activation analysis (8). When samples of the same fish were analyzed by the different methods it became clear that large discrepancies in the findings between the two techniques could be anticipated, with neutron activation analysis giving far more accurate results. However, because the flameless atomic absorption method was cheaper, faster, and easier to apply, it was selected by the U. S. Food and Drug Administration, the Department of Agriculture, and the Federal Water Quality Administration. Thus, the analytical program was crippled from the start by use of the inferior of two testing methods for total mercury; in addition, both methods tested for total mercury rather than for methyl mercury, which is the compound hazardous to life.

The commitment to atomic absorption analysis continued, even though the method consistently gave results lower than those obtained by neutron activation analysis. Recent experiments by D. Thomas and his colleagues have shown conclusively that, for comparisons between atomic absorption analysis for total mercury and electron capture gas chromatography for total methyl mercury, the atomic absorption results are consistently low. Furthermore, there can be as much as a 50 percent error in determinations made by the atomic absorption method. Therefore, since government laboratories generally use the atomic absorption method, one should usually multiply those data published by government agencies by 1.5 in order to compensate for the analytical errors (9).

Recently, a number of laboratories have embarked on projects involving the analysis of museum specimens of fish that were taken from polluted areas before chlorine and other mercury polluting plants began production. In general, these laboratories have analyzed the total mercury content of museum specimens of different species of fish. Once again, since methyl mercury is the compound with which we are concerned, it would be much more realistic to analyze this compound in these old specimens. In addition, since mercuric salts have been used for centuries in the preservation of tissues, it is important that the history of the method of preservation of each fish be well documented. Dr. Frank D'Itri at Michigan State University has analyzed museum specimens of fish caught in the St. Clair River System before the Sarnia plant started production (10). These museum specimens show clearly that the concentrations of total mercury in fish from this area have increased exponentially with the development of the chlorine industry. D'Itri showed that in muskellunge and sea lamprey, mercury concentrations increased threefold from 1965 to 1970. In lake sturgeon, northern pike, sauger, smallmouth bass, walleye, and white crappie, the mean mercury concentration increased about five times since 1965. These fish were all taken from the St. Clair River System.

There has been a great deal of confusion both in the scientific literature and in the popular press concerning the extent of mercury contamination of natural origin in freshwater and saltwater fish. The formation of methyl mercury from inorganic mercury is a reaction catalyzed by microorganisms, and therefore, by definition, the formation and accumulation of methyl mercury in fish is a natural process. Microorganisms undoubtedly had the capacity to synthesize methyl mercury long before the evolution of fish. Examination of samples of tuna fish canned before the widespread use of inorganic mercury in industry clearly demonstrates that certain species of fish naturally contain significant concentrations of methyl mercury in their flesh; for example, samples of tuna and swordfish had 1.0 to 2.0 parts per million (ppm) methyl mercury. Furthermore, some inland lakes that have a naturally high organic content in sediments and a naturally high inorganic mercury content in the lake bed provide conditions for the synthesis of methyl mercury and its accumulation in the fish. Clearly, however, natural methyl mercury contamination of fish is a localized problem; industrial pollution in advanced societies has vastly magnified the problem, making it more widespread. It must be emphasized that the contribution to the mercury problem by industry is to relocate inorganic mercury into areas where rapid synthesis of methyl mercury is assured.

Counteracting Sediment Pollution

In the St. Clair Lake and River System alone, less than 0.2 percent of the total mercury present in sediments has entered food chains.

The sediments in this river system are rich in organic matter, and so the population of bacteria with the capacity to methylate compounds is high. Lake St. Clair's organic content, which provides food for bacteria (see Figure 1), has been increased significantly by the poor sewage treatment practices of the municipality of Sarnia. Therefore, the rate of formation of methyl mercury in these sediments is much more rapid than the rate of excretion of methyl mercury by fish.

If we were to wait for the microbial population in the St. Clair sediments to convert all the inorganic mercury present to methyl mercury, and then wait for the methyl mercury to be flushed out into the sea, it would take approximately 5000 years before the system would clean itself biologically. Of course, during this time most fish taken from inland and coastal waterways would have concentrated so much methyl mercury that they would be totally unfit for human consumption. Furthermore, flushing methyl mercury from inland waterways into the oceans must ultimately lead to similar problems in ocean life. If one considers the hundreds of thousands of pounds of inorganic mercury already present in sediments of the North American continent, something practical must be done to prevent the inevitable loss of fish as a food source for future generations.

In the last two years a number of experiments have been performed with a view to slowing down the rate of formation of methyl mercury from inorganic mercury. The methods that have been tested include: one, dredging heavily polluted areas; two, covering the sediments of heavily polluted areas; and three, improving water quality to a level where microbial populations decrease. Dredging has not been very successful because this procedure has a tendency to mix up and distribute inorganic mercury over a wider area. Minamata Bay, Japan was dredged, and methyl mercury levels in fish from the bay are higher today than they were at the time of the Minamata disaster.

In certain areas, fluorospar tailings have been used to cover polluted sediments and consequently bury the microbial population in the sediments. The bacteria (facultative anaerobes) that are ordinarily active in the methylation process are able to live at much lower than normal levels of dissolved oxygen; however, when tailings cover the sediments, the oxygen concentration is decreased to a level low enough to inhibit even these bacteria, and the rate of methyl mercury formation consequently decreases. A layer of two to three centimeters (three centimeters are equivalent to 1.17 inch) of this covering has been shown to reduce the rate of methyl mercury formation by about 80 percent. The artificial blanket of fluorospar could itself possibly alter normal development of plant and animal communities in the stream and lake bed, but the hazard from methyl mercury released from the sediment seems the far greater danger at this point. Therefore, fluorospar application is recommended as an emergency measure unless there is heavy turbulence with resulting sediment dislocation that would disrupt the cover.

Improvement of water quality really provides the ideal answer. If nitrogen-containing compounds and phosphate can be removed at every water quality treatment plant, and if more effort is made to stop the agricultural abuse of overfertilization with ammonia and phosphate compounds, then the microbial populations in sediments will fall to natural levels and the rate of synthesis of methyl mercury will decrease quite rapidly. Furthermore, improved water treatment facilities will help alleviate similar problems such as the methylation of inorganic arsenic compounds by molds and bacteria (11). In 1933 Challenger showed that the bread mold *Scopulariopsis* will synthesize trimethyl arsine from arsenic salts. In 1970 McBride showed that methane-producing bacteria will synthesize dimethyl arsine from arsenic salts. These two arsines are deadly neurotoxins and undoubtedly are synthesized in the sediments of lakes and rivers polluted with arsenic compounds.

The Legal Aspects

In the summer of 1970 Secretary Hickel of the Department of Interior instructed the Justice Department to use the 1899 Refuse Act for the preparation of law suits against eight companies that had indiscriminately dumped mercury into public waterways. The first company (Oxford Paper Company, Rumford, Maine, a subsidiary of Ethyl Corporation) was to appear in the United States District Court at Portland, Maine a few weeks after Hickel's action. Dr. J. Vostel of Rochester Medical School and myself were called as expert witnesses. Seven lawyers from the Justice Department arrived in Portland. Oxford Paper Company decided to close the chlorine plant at Rumford because the company could not prevent mercury losses if the plant continued production. The Justice Department did not have enough time to work out detailed measures necessary to prevent further problems from the mercury already deposited by the chlorine plant. At this time the mercury lost by this plant still pollutes the Androscoggin River and Casco Bay. In Portland, Dr. Vostel and I prepared affidavits to be used in other law suits. The second company taken to court was the Olin Mathieson Corporation, for pollution of the Niagara River. The company managed to cut mercury losses from this chlorine plant to one-half pound of mercury per day. Nothing has been done yet about the mercury already lost to the Niagara River. Shortly after the Olin Mathieson trial, Secretary Hickel was relieved of his duties in the Department of Interior. Since the fall of 1970 no other lawsuits have been conducted by the Justice Department on this problem.

The 1899 Refuse Act, which applies to solid wastes, is a rather poor law for dealing with pollution problems of this kind. If the law is interpreted strictly, it does not apply to mercury since mercury is a liquid. Hickel's action on the mercury pollution was useful because it caused companies to take steps to prevent further losses of mercury to the environment, but there are no U. S. laws to deal with the mercury already lost to the environment.

In Canada, the laws are much stricter: Companies are held responsible for problems caused by industrial effluents. The Government of Ontario is conducting experiments and making preparations for trials against Dow Chemical of Canada and Wyandotte BASF. The Canadian lawsuits are thorough, and are supported by a law which can be directly applied to pollution problems.

While the Canadians work toward solving their problems, the United States is struggling to pass new laws. Senator Philip Hart has prepared a bill (S1528) which is designed to protect U. S. citizens against the consumption of poisonous fish. Also, Senators McGovern and Hart are trying to get approval for a new environmental protection act. Legislation moves slowly, and it is frightening to consider that it may take 20 years for new laws to be passed and for each company accused of polluting to be tried. Twenty or 30 years from now, methyl mercury concentrations in most species of fish in inland waterways will either be lethal to fish or will render them unfit for human consumption.

An additional problem is that some members of the public are less concerned about mercury pollution than about related problems, such as the loss of jobs resulting from curtailment of industries that pollute. In Saskatoon, a number of people who dislike Indians believe that the mercury pollution problem is beneficial to their community because the Red Indians in that area can no longer make a living from commercial fishing, and so many of them may have to leave the province in the next few years.

Who is responsible for mercury pollution? Since the formation of methyl mercury is a natural phenomenon, it is difficult to place responsibility for a problem which has been present on this planet since before the evolution of mankind. However, the problem has been aggravated by our poor sanitation practices, wasteful agricultural practices, and by a tendency for industries to neglect those aspects of production which do not yield quick profit. Therefore, the mercury pollution problem is a product of advanced industrial societies.

Notes

1. Löfröth, Goran, *Ecological Research Committee Bulletin No. 4,* Swedish Natural Science Research Council, 1969.
2. Dunlap, R., "The Anatomy of a Pollution Problem," *Chemical and Engineering News,* July 5, 1971. pp. 22-34. Wood, J. M., *Enzymes and the Environment*, Bogden and Quigley, New York, 1971. On the subject of the mechanism of methyl-transfer to mercury, see: *International Atomic Energy Commission* [IAEC] *Handbook on Mercury,* IAEC, Geneva, 1971; "Mechanism of Methyl Transfer to Mercury." *Proceedings of 162nd American Chemical Society Meeting*, Washington, D. C., 1971; Wood, J. M., "Environmental Pollution by Mercury," *Advances in Environmental Science and Technology,* vol. 2, Wiley Interscience, 1971; De Simone, R. E., M. W. Penley, and J. M. Wood, "Mechanism of Cobalamin Dependent Methyl Transfer to Mercuric Ion," *Biochemistry*, in press; Wood, J. M., and D. G. Brown, "Vitamin B_{12} Enzymes," *Structure and Bonding*, vol. 11, Springer Verlag, 1971.
3. Langley, D. G. (T. W. Beak Associates, Toronto), "Mercury Methylation in the Aqueous Environment," *Proceedings of the 162nd American Chemical Society Meeting*, Washington, D. C., 1971.
4. Keckes, S., and J. K. Miettenen, "Mercury as a Marine Pollutant," *Proceedings of F. A. O. Conference*, Rome, Dec. 9-18, 1970.
5. Ui, Jun, (Dept. of Engineering, University of Tokyo), "Mercury Pollution of Sea and Fresh Water," *Proceedings of Fourth Colloquium for Medical Oceanography*, Naples, Oct. 2-9, 1969.
6. Keckes and Miettenen, *loc. cit.* Larsson, J. E., "Environmental Mercury Research in Sweden," Swedish Environment Protection Board, June 1970.
7. Nielsen, O., *Mercury in Fish from the Saskatchewan River System*, A report to the Saskatchewan Ministry of Fisheries, Saskatchewan, Canada, 1969.

8. Thomas, D. G. (Olin Mathieson Corporation) et al., "A New Method for the Analysis of Mercury in Tissues," *Analytical Chemistry*, in press.
9. Dunlap, *loc. cit.*
10. D'Itri, Frank, "Mercury Accumulation in the Aquatic Environment," *Proceedings of the 162nd American Chemical Society Meeting*, Washington, D. C., 1971.
11. Challenger, F., "Biosynthesis of Methyl-Arsenic Compounds," *Chem. Reviews*, 1945. McBride, B. C., and R. S. Wolfe, "Heavy Metal Pollution and Biochemistry," *Chemical and Engineering News*, July 19, 1971, p. 33.

22 Good Coffee Water Needs Body

Wayne A. Pettyjohn

Wayne A. Pettyjohn is a professor in the Department of Geology at Ohio State University and an attorney-at-law.

Crosby is a small village, formerly a railhead center, in the northwestern part of North Dakota, a few miles south of the International Boundary. As in many places in North Dakota, water is not in abundant supply. When Crosby was first established, surface water was unavailable, and dug wells were used to supply homesteaders and the community at large. Most of these hand-dug wells, commonly several feet in diameter, were relatively shallow. They were replaced in recent years with a municipal water system supplied by two drilled wells more than 150 feet deep.

In the central part of town, however, there remains an old large-diameter dug well, about 38 feet deep, covered by boards and concrete. Water was formerly withdrawn from the well by a handpump and used by a great number of people who did not like the taste of the water from the city's deep-well supply. Water from the dug well was used specifically in making coffee; it reportedly produced "the best coffee in the state."

During a ground-water investigation of the Crosby area by the U. S. Geological Survey in the early 1960s, a sample of water was collected from the dug well and analyzed (Armstrong, 1965, p. 111). The water contained the following constituents:

Sulfate	846 mg/l
Chloride	164 mg/l
Nitrate	150 mg/l

Dissolved solids   2176 mg/l
Hardness   1300 mg/l

It is evident that this water is highly mineralized. It contains more than four times the limit of 500 mg/l of dissolved solids recommended by the Public Health Service (1962). The local people referred to it as having "body."

Actually, the water had more body than these people realized. The high concentration of nitrate, which far exceeds the Public Health Service limit of 45 mg/l, coupled with the higher-than-normal concentration of chloride, suggested that the well was contaminated by sewage wastes. The apparent contamination was brought to the attention of the local city health official, who declared the well unsafe and removed the pump handle. Immediately many people became angry because their supply of good coffee water had been terminated, apparently for no good reason. After all, they had been drinking the water for years, and no one had ever died from it, or even gotten sick—as far as they knew. Nevertheless, the pump handle was not reinstalled.

Some of the old timers reported that this ancient well had been dug near the site of a former livery stable, which had been built sometime in the late 1800s and had been operated until about 1930. Waste products from the horses evidently contaminated the ground-water supply. These wastes also provided much of the "body" that patrons of Crosby's dug well prized so highly.

E. R. Wood (1962) described an interesting case of ground-water contamination caused by the infiltration of waste materials from an ancient gasworks in Norwich, England. In the early 1950s, a 36-inch diameter well was drilled 150 feet into chalk deposits underlying the city. Although water from the well was of acceptable quality, both biologically and chemically, the yield was insufficient and an adit was driven horizontally from the bottom of the well into the chalk in an attempt to provide more water.

After construction of the adit, the water was sampled and found to contain phenols in the amount of about 0.2 mg/l, causing the well to be abandoned. (The U. S. Public Health Service has recommended a phenol limit of 0.001 mg/l, since at higher concentrations this substance creates significant taste problems.) Mr. Wood, his curiosity aroused, entered the well by means of a bucket and upon examining the roof of the adit found several zones seeping black tarry material. Analyses of this material showed it to consist of tarry carbon with a trace of phenols and volatile matter.

Although the well site was several miles from an existing gasworks plant, the composition of the tarry materials suggested that they had originated from the distillation of coal or similar material. Further investigation brought to light the fact that the first gasworks plant in the city had been constructed around 1815 at the exact site of the contaminated well. In March 1830 the original plant was abandoned. Thus, Wood concluded, the tarry material that contaminated the well had been present in the chalk for at least 120 years.

Perhaps one of the most unique cases of ground-water pollution was provided by G. J. C. Nash (1962) during a discussion of Wood's report. Nash stated that officials of a gasworks plant on the south coast of England once claimed that the presence of hydrogen sulfide in the plant well was due to drainage from a Black Plague burial pit. Apparently, the well was bored through a seventeenth-century graveyard.

In 1925 Boy Scouts in the central Ohio city of Delaware conducted "an outhouse" survey (Bill Rice, oral communications, March 1972). The purpose of the investigation was to determine the distance between each homeowner's outhouse and his well from which a rough estimate could be made of whether the water supply was biologically safe. The data were given to the local health agent, and principal investigator, Dr. B. F. Higley.

Dr. Higley was concerned about the relation of well water to public health in Delaware, because the town was constructed on a nearly flat, till-covered area. Shallow sandy zones, a few inches to a few feet thick, served as a source of water to wells. Although the sandy zones were of limited areal extent, it was possible and in fact highly probable, that some of these permeable layers were acting both as a source of drinking water and as a receptacle for human wastes, thus forming an early and rather primitive example of water recycling and reuse.

One homeowner who had built a privy only a few feet from his shallow well, remained unconvinced by Dr. Higley's most eloquent arguments and would neither dig a new well nor move the outhouse. To prove that there might be a hydrologic connection between the privy and the well, Dr. Higley poured 5 to 10 pounds of salt into the privy. Within two weeks the well water began to taste salty and the homeowner threatened to sue the good doctor for contaminating his well. Ultimately, a new well was drilled on the other side of the house.

In early 1961, members of the Division of Water, Ohio Department of Natural Resources, investigated an unusual, but by no means unique, water pollution problem in the northern Ohio community of Bellevue (Ohio Division of Water, 1961). Contamination of a highly-permeable limestone aquifer had resulted from the dumping of household, municipal and industrial wastes into scores of sink holes and drilled wells. In many instances septic tanks were used, but overflow from the tanks was allowed to discharge into wells. The sewage effluent contaminated the ground water in an area 5 miles wide and more than 15 miles long as it slowly moved down the water-table gradient toward Lake Erie (Figure 1).

Bellevue's first municipal water supply was from a surface reservoir constructed in 1872. By 1919 the practice of disposing of sewage into disposal wells and sink holes was already well established and many wells had been contaminated. To augment the surface water supply, municipal wells were drilled in 1932, outside the known area of contamination. Following a major flood in June, 1937, caused by 10 inches of rain during a three-day period, many sink holes overflowed, pouring raw sewage on the ground throughout several square miles and contaminating many more wells and cisterns. Additional water-supply wells were drilled by the city in the early 1940s. The average yield per well was 500 gpm (gallons per minute) from depths ranging from 137 to 200 feet. By 1944 several of the municipal wells were contaminated, and plans to build a soybean processing plant were abandoned because a well, 230 feet deep, drilled at the prospective plant site was contaminated. All industrial and municipal water wells were abandoned by 1946. On July 4, 1969 the city was again inundated by heavy runoff caused by 10 inches of rain falling over a 16-hour period. The city's main street was covered by 16 feet of water. Infiltration filled the underground reservoir and the waste poured over the ground throughout the city and adjacent areas. About 1100 basements were flooded and surface water supplies were contaminated. Plans to construct a sewage treatment plant were shelved year after year due to "excessive" cost. Only in 1971 was a sewage treatment plant finally completed with Federal assistance.

In the decades during which waste materials were dumped into drilled wells in and around the city, wells occasionally would

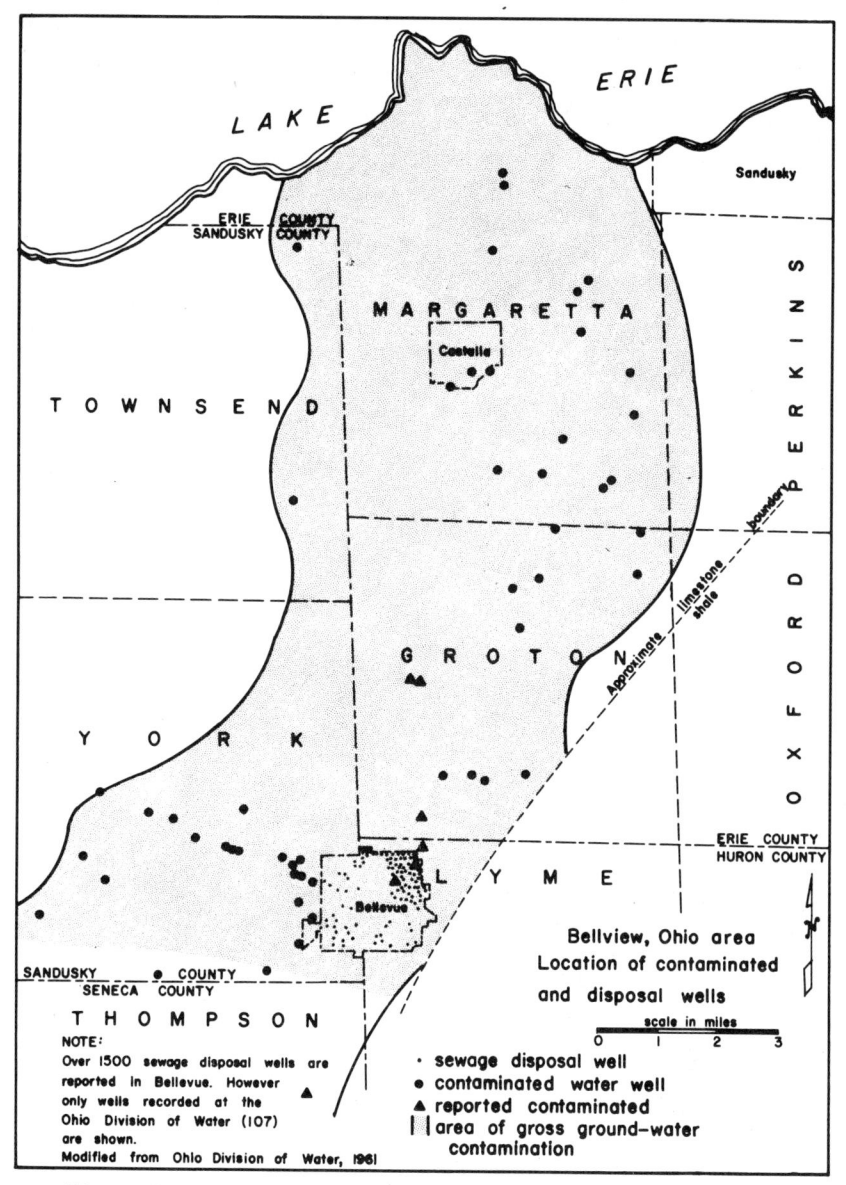

Figure 1. Location of contaminated area and disposal wells in Bellevue, Ohio.

become plugged with sewage, necessitating redrilling. With the increased use of detergents, plugging ceased to be much of a problem. Division of Water investigators found more than 1500 disposal wells within Bellevue's city limits—an area less than 4 square miles (Figure 1). Of 32 samples of ground water collected in the vicinity of Bellevue, 27 contained ammonia. Detergents (alkyl benzene sulfonate or ABS) were found in 22 samples, and all contained nitrate and phosphate. Examples of concentrations found in samples collected in February, March, and April, 1961 are shown (in mg/1) in Table 1.

TABLE 1. CONCENTRATIONS OF ABS, AMMONIA, NO_2 AND NO_3 FOUND IN BELLEVUE, OHIO WELLS IN FEBRUARY-MARCH-APRIL, 1961 (IN MG/L)

Well Owner	Date of Collection	ABS	Ammonia as NH_4	NO_2	NO_3
Weiland	2- 6-61	0.1	0.1	0.5	29
	4-18-61	0.1	0.2	0.05	52
Thomas	3- 7-61	0.1	0.2	0.00	92
	4-18-61	0.3	0.1	.00	158
Neill	3- 7-61	0.1	0.2	1.0	41
	4-18-61	0.1	0.1	.05	38
Andrews	2- 6-61	0.1	0.2	1.5	26
	4-18-61	0.2	0.2	.00	36
Adams	3- 5-61	0.2	0.1	0.05	77
	4-18-61	0.2	0.3	0.20	91

The ground-water resources in the Bellevue area, and in areas downgradient from the town, are obviously grossly contaminated and have been for more than a half century. The Division of Water report states: "Stories have been related to us during this investigation, of wells which yielded easily recognizable raw sewage (including toilet tissue) while being drilled. Others have foamed because of high detergent content, and still others, the contents of which are best left to the reader's imagination."

The attitude of some of the local officials, however, was rather curious at least during 1961. One official stated the cry of ground-water pollution has "been made every three or four years for the last 20 years, and has never been proven" (*Sandusky Register*, June 28, 1961). A commissioner of the Bellevue Health Department said that city officials, since the original inception of the use of sink holes, had made studies of the area and could find no evidence of water contamination (*Fremont News*, June 24,

1961). Bellevue's city engineer reported that the city had wells which it used for drinking water, and the water was not contaminated. However, examination of water from two of these wells proved to be rather enlightening. The location of the wells and the poor condition of the casings had led the Ohio Department of Health to speculate that these wells probably could not be approved as a source of public water supply. Chemical analyses showed one of the wells (212 feet in depth) to be grossly contaminated; the chemist described the water as "foul." Fortunately, neither of these wells had been used since 1946.

Literally hundreds of ground-water contamination cases, such as those described herein, are to be found in the literature. All of them tend to point out the fact that inadequate disposal of waste materials can result in contamination, even though the contaminants have been abandoned and forgotten for decades.

Of all that was done in the past,
you eat the fruit, either rotten or ripe.

T. S. Eliot

References

Armstrong, C. A. 1965. *Geology and ground-water resources of Divide County, North Dakota, part 2, ground-water basic data.* North Dakota State Water Commission County Ground Water Studies, 6.

Nash, G. J. C. 1962. Discussion of paper by E. C. Wood. *Proc. Soc. Water Treatment and Examination* 11:33.

Ohio Division of Water. 1961. *Contamination of underground water in the Bellevue area.* Ohio Department of Natural Resources, Mimeo Report.

U. S. Public Health Service. 1962. *Public health drinking water standards.* Publication 956.

Wood, E. C. 1962. Pollution of ground water by gasworks waste. *Proc. Soc. Water Treatment and Examination* 11:32-33.

Additional Readings

American Petroleum Institute, Committee on Public Affairs. 1969. *Offshore petroleum and the environment.* New York: American Petroleum Institute.

Cannon, H. L., and Hopps, H. C., eds. 1971. *Environmental geochemistry in health and disease.* Geological Society of America Memoir 123.

———, eds. 1972. *Geochemical environment in relation to health and disease.* Geological Society of America Special Paper 140.

Coates, D. R., ed. 1972. *Environmental geomorphology and landscape conservation.* Vol. 1: prior to 1900. Stroudsburg, Pa.: Dowden, Hutchinson, and Ross.

Committee on Geological Sciences, Division of Earth Sciences (NRCNAS). 1972. The earth and human affairs. San Francisco: Canfield Press.

Council on Environmental Quality. 1970. *Environmental quality: the first annual report of the council on environmental quality.* Washington, D. C.: U. S. Government Printing Office.

―――. 1971. *Environmental quality: the second annual report of the council on environmental quality.* Washington, D. C.: U. S. Government Printing Office.

Crandell, D. R., and Waldron, II. H. 1969. Volcanic hazards in the Cascade range. In *Geologic hazards and public problems, conference proceedings,* ed. R. Wilson and M. Wallace, pp. 5-18. Washington, D. C.: U. S. Government Printing Office.

Detwyler, T. R., ed. 1971. *Man's impact on the environment.* New York: McGraw-Hill.

Detwyler, T. R., and Marcus, M. G., eds. 1972. *Urbanization and the environment.* Belmont, Calif.: Duxbury Press.

Fairbridge, R. W., ed. 1968. *The encyclopedia of geomorphology.* Encyclopedia of Earth Sciences Series, vol. 3. New York: Reinhold.

―――, ed. 1972. *The encyclopedia of geochemistry and environmental sciences.* Encyclopedia of Earth Sciences Series, vol. 4A. New York: Reinhold.

Federal Water Quality Administration, Department of the Interior. 1970. *Clean water for the 1970's.* Washington, D.C.: U. S. Government Printing Office.

Flawn, P. T. 1970. *Environmental geology: conservation, land-use planning, and resource management.* New York: Harper and Row.

Grava, S. 1969. *Urban planning aspects of water pollution control.* New York: Columbia University Press.

Gross, D. L. 1970. *Geology for planning in De Kalb County, Illinois.* Illinois Geological Survey, Environmental Geology Notes 33.

Hackett, J. E. 1968. *Geologic factors in community development at Naperville, Illinois.* Illinois Geological Survey, Environmental Geology Notes 22.

Hackett, J. E., and McComas, M. R. 1969. *Geology for planning in McHenry County.* Illinois Geological Survey Circular 438.

Hafen, B. Q., ed. 1972. *Man, health and environment.* Minneapolis: Burgess.

Hambleton, W. W. 1972. The unsolved problem of nuclear wastes. *Technology Review* 74(5):15-19.

Jackson, W., ed. 1971. *Man and the environment.* Dubuque, Iowa: Wm. C. Brown.

Jennings, P. C. 1971. *Engineering features of the San Fernando earthquakes of February 9, 1971.* Pasadena, Calif.: Earthquake Engineering Research Laboratory, California Institute of Technology.

Legget, R. F. 1973. *Cities and geology.* New York: McGraw-Hill.

Mason, A. C., and Foster, H. L. 1953. Diversion of lava flows at O Shima, Japan. *American Journal of Science* 251:249-58.

McKenzie, G. D., and Utgard, R. O., eds. 1972. *Man and his physical environment: readings in environmental geology*. Minneapolis: Burgess.

Murdoch, W. W., ed. 1971. *Environment: resources, pollution and society*. Stamford, Conn.: Sinauer Associates, Inc.

Pettyjohn, W. A., ed. 1972. *Water quality in a stressed environment*. Minneapolis: Burgess.

Shacklette, H. T., Hamilton, J. C., Boerngen, J. G., and Bowles, J. M. 1971. *Elemental composition of surficial materials in the conterminous United States*. U.S. Geological Survey Professional Paper 574-D.

State Earthquake Investigation Commission. 1908-1910 (reprinted 1969). 2 vols. *The California Earthquake of April 18, 1906*. Washington, D. C.: Carnegie Institution.

Steinbrugge, K. V., and Zacher, E. G. 1960. Creep on the San Andreas Fault, art. 1. In Fault creep and property damage. *Bulletin of the Seismological Society of America* 50(3):389-96.

Steinhart, C. E., and Steinhart, J. S. 1972. *Blowout; a case study of the Santa Barbara oil spill*. Belmont, Calif.: Duxbury Press.

Strahler, A. N., and Strahler, A. H. 1973. *Environmental geoscience: interaction between natural systems and man.* Santa Barbara, Calif.: Hamilton.

Tank, R. W. ed. 1973. *Focus on environmental geology*. New York: Oxford University Press.

Thomas, W. L., ed. 1956. Reprinted 1971 in 2 vols. *Man's role in changing the face of the earth*. Chicago: University of Chicago Press.

Turner, D. S. 1969. *Applied earth science*. Dubuque, Iowa: Wm. C. Brown.

U. S. Geological Survey and National Oceanic and Atmospheric Administration. 1971. *The San Fernando, California, earthquake of February 9, 1971*. U. S. Geological Survey Professional Paper 733.

Vedder, J. G., et al. 1969. *Geology, petroleum development, and seismicity of the Santa Barbara Channel region, California*. U. S. Geological Survey Professional Paper 679.

Part Six

Earth Resources

Go, my sons, buy stout shoes, climb mountains, search the valleys, the deserts, the sea shores, and the deep recesses of the earth. Mark well the various kinds of minerals, note their properties and their mode of origin.

Petrus Severinus, 1571

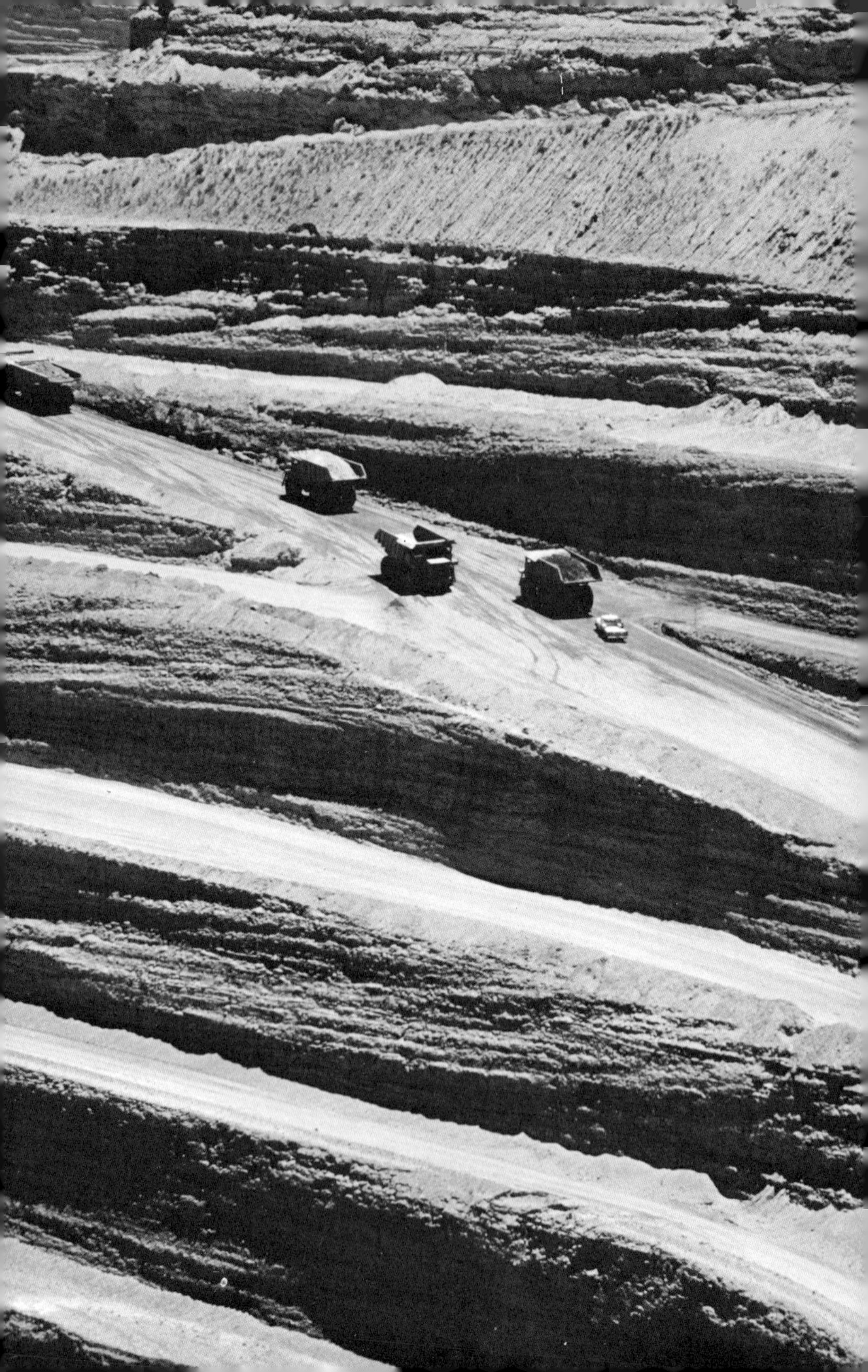

Although this section is entitled Earth Resources, it is restricted mainly to resources of interest to earth scientists. Thus, mineral resources, both fuel and non-fuel, are covered as well as soils. Forest, agricultural, and human resources are omitted, although we realize that they are very important in our total resource picture. Before coal and other fossil fuels came into use, forest products long supplied man with energy. In a way, we can consider fossil fuels to be a supply of buried sunlight which has been stored by biological and geological action for millions of years. We are now using our supply of buried sunlight at a tremendous rate. One estimate indicates that we could, on a world scale, use all of our supply in a total time span of less than 2000 years. At first this appears to be a significantly long time; however, even in the recorded history of man, it is relatively short. The availability of world oil and gas may not extend beyond 2030, and domestic supplies for energy use will probably be near exhaustion by 2000.

Will we have sufficient resources to support continued growth in materials consumption or even enough to maintain consumption at the current level? In seeking the answers to this fundamental question, we must inventory our resources and make an estimation of possible reserves. The article by McKelvey summarizes the ways in which we can determine our mineral supply. If we search the literature, we can find examples of the expected dates for exhaustion of many minerals. These dates have come and gone, and we are still using the materials that we once predicted would be unavailable to man. What has happened? The estimates were made without complete knowledge of the geology of low-grade materials and without any knowledge of the technological improvements that later permitted the extraction of these lower grades of materials.

The continued availability of minerals has led some to think that our resources are inexhaustible and that economics alone, through an increase in price or increase in substitution, will provide the incentive to process more material of lower grade or to search harder for new supplies. Some mineral resources respond to this formula; however, this does not apply to all of them. We face the possibility of global exhaustion of some, such as mercury, gold, and silver, which are rare or geographically restricted.

Abundant elements in the oceans and in crustal rocks may fulfil future mineral requirements, provided that unlimited supplies of inexpensive energy are available in the future for processing raw elements. Such a panacea seems very remote during the present energy crisis. Nuclear fusion, the process in which atomic nuclei combine to form an atomic nucleus of higher atomic number and weight, can yield large quantities of energy and appears unlimited with respect to fuel materials. Fuel for nuclear fusion could be either tritium and deuterium or the combination of two deuterium isotopes. The former combination was the fuel for the hydrogen bomb. However, the problem of maintaining a controlled and sustained nuclear reaction for generation of electricity remains. The use of lasers to produce the necessary temperature of about 100 million degrees Celsius, in combination with experiments to utilize a magnetic field to hold the plasma in a fusion reactor, may provide the technologies needed. Time estimates for developing this source of apparently unlimited energy are anywhere between 30 and 100 years. If we do achieve the ability to process crustal rocks for needed materials, would we have the land resources available for disposal of the waste products?

Energy is a key factor in our standard of living. With about 5 percent of the world's population, the United States consumes about one third of the world's energy production. For many nations, the amount of energy consumed is proportional to the Gross National Product (GNP) per capita; however, there are a few nations with relatively high GNPs that have significant differences in their use of commercial energy. These differences are probably due to the concentration of energy-intensive industry in some countries, although the degree of wastefulness in energy use might be a factor, too. In general, if we accept the premise that the standard of living or the availability of goods and services is indicated by GNP, then it appears that nations with significant quantities of energy will exhibit a high standard of living. If the reverse holds, within the limits described above, then a reduction in the availability of energy could significantly change the quality of life in a nation that has traditionally used large quantities of energy. A similar relationship between steel consumption and GNP is shown in the article by McKelvey, and the consumption of sulfur has been determined to be an indicator of the degree of industrialization of a nation.

Although improved technology for the discovery, extraction, and processing of minerals and advances in recycling, substitution, synthesis, and utilization of materials are important factors in determining the availability of mineral resources for this country, another factor, the world availability of a resource, is becoming increasingly important. The United States is now completely dependent on foreign supplies for 22 of the 74 non-energy mineral commodities considered essential for a modern industrial society. Forecasts indicate that over 50 percent of our primary raw materials will come from foreign sources by the year 2000—even with increased domestic production and recycling. The effects of the increasing dependence of this country on foreign oil resources, with the resultant problem of a poor balance of payments and the availability of billions of excess American dollars in a few countries, are now regularly making the news as part of the energy crisis.

Imaginations run wild in trying to determine the future, and exercises in futuristics are increasing. What effect will the decreased availability of commodity X have on our business? Can the country continue to export lumber or grain? How will we obtain strategic mineral Y? Will armed conflict be needed to get resource Z? (We might ask, have we gone to war to obtain mineral advantages in the past, and, also, have recent warmed relations with the Soviets and the Chinese been precipitated by an expected need for some of their resources?)

One approach to the resource problem, and particularly the energy crisis in this country, is to improve our conservation of resources and energy. An overview of this solution is provided in the article by Abelson. There are many technological improvements and changes in life style that, if implemented, could result in conservation of energy. Inexpensive energy has been one of the factors in this country's current world position. Because the price of energy has been low, wastefulness has been encouraged in manufacturing and commercial operations. In addition, other factors, such as marketing trends, have encouraged the development of a life style in which large, powerful cars, throwaway bottles and cans (which use more energy per unit of beverage than returnable bottles), and planned obsolescence of appliances and cars have contributed to the waste of energy. Some energy companies are now providing suggestions on how to conserve energy, although some electrical utilities persist in advertising to increase sales. Guidelines for voluntary conservation of energy will work with a few conscientious people, but most people will need an economic clout before they respond to the idea that energy should not be wasted. Even then, some people, from habit or because of financial status, will continue to use excessive amounts of energy. Wasting energy may become a new status symbol if it is not declared un-American to do so!

As an exercise, you could speculate on how you might conserve energy as its cost increases. One example of changes in life style with decreased availability of energy could focus on the luxurious activity of showering. With a stepped increase over several years in the cost of heating water, your shower routine might change as follows: 1) shower half as long, 2) shower every other day, 3) shower every other time with cold water, etc. Could it ever come to that extreme (really not an extreme situation by world standards)? It could be that our projected energy demands are too high, particularly for electrical energy, and increases in price and smoothing the daily demand curve could reduce the projected demands upon many electrical utilities. The hills and valleys in the demand curve might be radically altered by development of a metering device for electrical energy that recorded the time of energy use as well as the amount. Very high rates during current peak times would quickly alter our use of this essential commodity, as well as our life styles. It might become fashionable to dine at 10:00 p.m. More food for thought.

A list of possible energy sources would probably surprise the layman. Most people could readily name fossil fuels (oil, coal, natural gas), nuclear energy, and hydroelectric power. With the current increase in energy demand, other less familiar types of energy are now being considered for development or expanded use. These include tidal power, wind, ocean currents, and efficient photosynthesis to produce biomass (possibly by algae, water hyacinths, or sugar cane) which could be burned to produce heat or fermented to produce methane. In this section, two other techniques are considered—geothermal energy and solar energy.

The prospect of solar energy for large-scale electrical energy production is probably several decades away, because technological and production requirements are necessary to make solar cells and solar furnaces economical. The use of solar energy on a small scale for household purposes is probably much closer.

Geothermal energy, long used on a limited basis, is now considered by some to be a possible source of fairly significant quantities of energy. As for any energy source, possible environmental hazards must be considered. At first glance, geothermal energy appears to be a "clean" fuel; however, the brines that are produced with the steam can cause problems if not disposed of properly by processing or reinjection into the ground. But the brines may prove to be a secondary resource if significant quantities of dissolved mineral matter can be obtained from them.

Although we have included no readings on nuclear energy and mentioned it only briefly before, this source of energy will probably be a major one in the next century. Currently, less than 1 percent of our energy needs are met by conventional nuclear reactors using uranium-235 as a fuel. This naturally fissionable material gives off heat in the process of radioactive decay, and the heat is transferred by a coolant to turbines that generate electricity. The process is about 30 percent efficient. The supply of uranium-235 is limited, and with the projected demands for it in the production of nuclear power in this country, the low-cost supply is expected to last only about 30 years.

The use of other isotopes in breeder reactors could extend our supply of nuclear energy from fission by 400 times. The process in which fertile material, such as uranium-238, thorium-232, and plutonium-240, is bombarded with neutrons produced from fissionable uranium-235 in a breeder reactor is known as breeding. The expected time required to double the fuel supply for a reactor by this process is 10 years. Hence, in view of our decreasing supplies of uranium-235, one can see the urgency to develop breeder reactors. There are other factors in addition to need that must be considered in the nuclear program on which we have embarked. A list of these factors includes: disposal of radioactive waste material, accidents in the reactor that allow escape of radiation during daily operation, thermal pollution, sabotage, and plutonium theft. The latter factor and the possibility of accidents are probably the most serious. The problem of thermal pollution from reactor power plants has been reduced by the use of cooling towers. Waste materials will probably be disposed of in a carefully selected salt mine, and the radiation dosage around a properly-operating nuclear power plant is insignificant. In fact, one should be more concerned with the mounting use of medical and dental X-rays and their contribution to the total radiation dose of the population.

Two serious problems remain. The probability of accidents has been calculated by the Atomic Energy Commission to be so low that serious accidents are not a hazard. For some time, most people accepted that this was the case; however, serious questions are now being raised about AEC experiments and reports concerning possible loss-of-coolant accidents in which the cooling mechanism fails and the fuel overheats. As a result, the core of the reactor would become a molten mass of zircaloy and uranium dioxide, and the pressure vessel would probably collapse in about one hour. Although the emergency-core-cooling system has not been proven satisfactory to everyone, this country is pushing ahead with the reactor program. Hopefully, we can solve these technological problems before the 1980s when over 100 reactors will generate 25 percent of our electricity. By then, it would be very difficult to curtail the use of unsafe reactors.

In 1973, as the energy crisis became increasingly obvious, individuals reacted to the situation in proportion to their degree of optimism and/or the amount of information available to them. Thus it was not unusual to hear the energy situation referred to as the "energy crisis," "the so-called energy crisis," "the energy crunch," and even "the energy challenge." To quote A. Weinberg, director of the Oak Ridge National Laboratory, "The present energy crisis is neither theoretical nor long-range: it is real and it is immediate" (1973). Although the expected term of the energy crisis may be debated, most people admit that energy will not get any less expensive in the future.

The last articles in this section describe the techniques for soil investigation and the importance of soil to civilization. Sears points out the value of prairie soils and the importance of fire in the maintenance of the prairie, a factor that has recently been incorporated into forest management. Problems of siltation, ero-

sion, irrigation, and tropical soils are mentioned along with the advantages of floods in replenishing the soils. The importance of improving the fertility of worn-out land as done by Louis Bromfield at Malabar Farm, Ohio, will no doubt gain the approval of members of the growing group of organic gardeners and farmers.

Descriptions of the physical, chemical, and biological properties of soil are useful, not only for the farmer who wants to determine the best crops and fertilizer for his land, but also for the planner who will determine the best use for our important land resources, for the engineer who needs general information for building foundations, and for the geologist for determining possible landslide areas and mineral resources beneath the soils. A detailed account of techniques used in making a soil survey is presented by Himes in the last article in this section.

23 Mineral Resource Estimates and Public Policy

V. E. McKelvey

V. E. McKelvey is director of the U. S. Geological Survey at the National Center in Reston, Virginia.

Not many people, I have found, realize the extent of our dependence on minerals. It was both a surprise and a pleasure, therefore, to come across the observations of George Orwell in his book *The Road to Wigan Pier*. When describing the working conditions of English miners in the 1930s he evidently was led to reflect on the significance of coal:

> Our civilization ... is founded on coal, more completely than one realizes until one stops to think about it. The machines that keep us alive, and the machines that make the machines are all directly or indirectly dependent upon coal ... Practically everything we do, from eating an ice to crossing the Atlantic, and from baking a loaf to writing a novel, involves the use of coal, directly or indirectly. For all the arts of peace coal is needed; if war breaks out it is needed all the more. In time of revolution the miner must go on working or the revolution must stop, for revolution as much as reaction needs coal ... In order that Hitler may march the goosestep, that the Pope may denounce Bolshevism, that the cricket crowds may assemble at Lords, that the Nancy poets may scratch one another's backs, coal has got to be forthcoming.

To make Orwell's statement entirely accurate—and ruin its force with complications—we should speak of mineral *fuels*, instead of coal, and of other minerals also, for it is true that minerals and mineral fuels are the resources that make the industrial society possible. The essential role of minerals and mineral fuels in human life may be illustrated by a simple equation,

$$L = \frac{R \times E \times I}{P}$$

in which the society's average level of living (L), measured in its useful consumption of goods and services, is seen to be a function of its useful consumption of all kinds of raw materials (R), including metals, nonmetals, water, soil minerals, biologic produce, and so on; times its useful consumption of all forms of energy (E); times its useful consumption of all forms of ingenuity (I), including political and socioeconomic as well as technologic ingenuity; divided by the number of people (P) who share in the total product.

This is a statement of the classical economists' equation in which national output is considered to be a function of its use of capital and labor, but it shows what capital and labor really are. Far from being mere money, which is what it is popularly thought to mean, capital represents accumulated usable raw materials and things made from them, usable energy, and especially accumulated knowledge. And the muscle power expended in mere physical toil, which is what labor is often thought to mean, is a trivial contribution to national output compared to that supplied by people in the form of skills and ingenuity.

This is only a conceptual equation, of course, for numerical values cannot be assigned to some of its components, and no doubt some of them—ingenuity in particular—should receive far more weight than others. Moreover, its components are highly interrelated and interdependent. It is the development and use of a high degree of ingenuity that makes possible the high consumption of minerals and fuels, and the use of minerals and fuels are each essential to the availability and use of the other. Nevertheless, the expression serves to emphasize that level of living is a function of our intelligent use of natural resources, and it brings out the importance of the use of energy and minerals in the industrial society. As shown in Figure 1, per capita Gross National Product among the countries of the world is, in fact, closely related to their per capita consumption of energy. Steel consumption also shows a close relation to per capita GNP (Figure 2), as does the consumption of many other minerals.

Because of the key role that minerals and fuels play in economic growth and in economic and military security, the

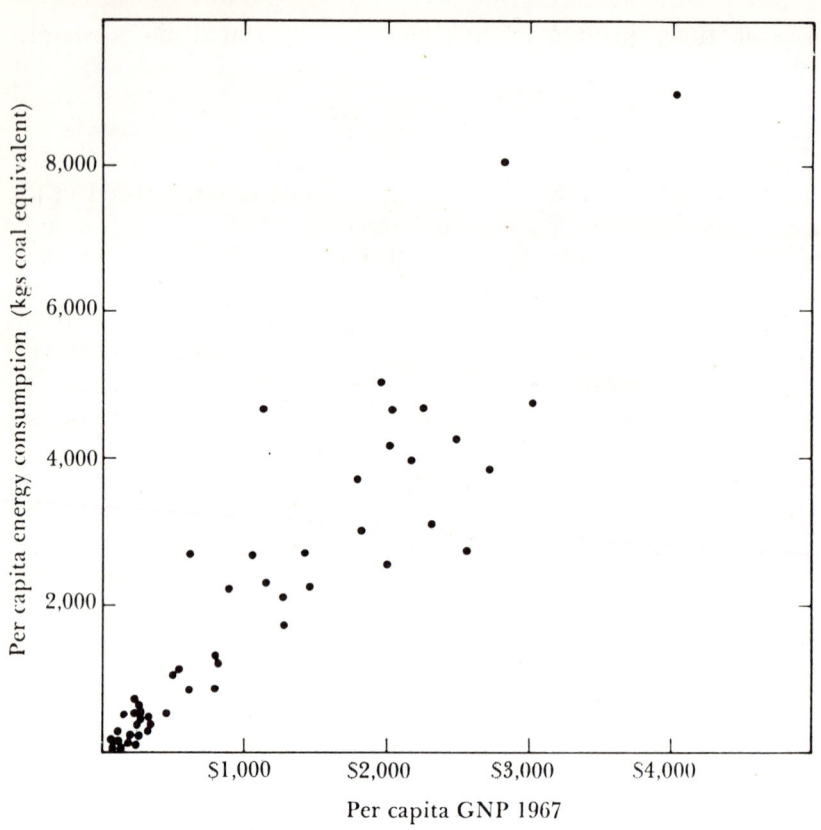

Figure 1. Per capita energy consumption compared to per capita Gross National Product in countries for which statistics are available in the United Nations *Statistical Yearbook* for 1967.

extent of their resources is a matter of great importance to government, and questions concerning the magnitude of resources arise in conjunction with many public problems. To cite some recent examples, the magnitude of low cost coal and uranium reserves has been at the heart of the question as to when to press the development of the breeder reactor—which requires an R & D program involving such an enormous outlay of public capital that it would be unwise to make the investment until absolutely necessary. Similarly, estimates of potential oil and gas resources are needed for policy decisions related to the development of oil shale and coal as commercial sources of hydrocarbons, and estimates are needed also as the basis for decisions concerning prices and import controls.

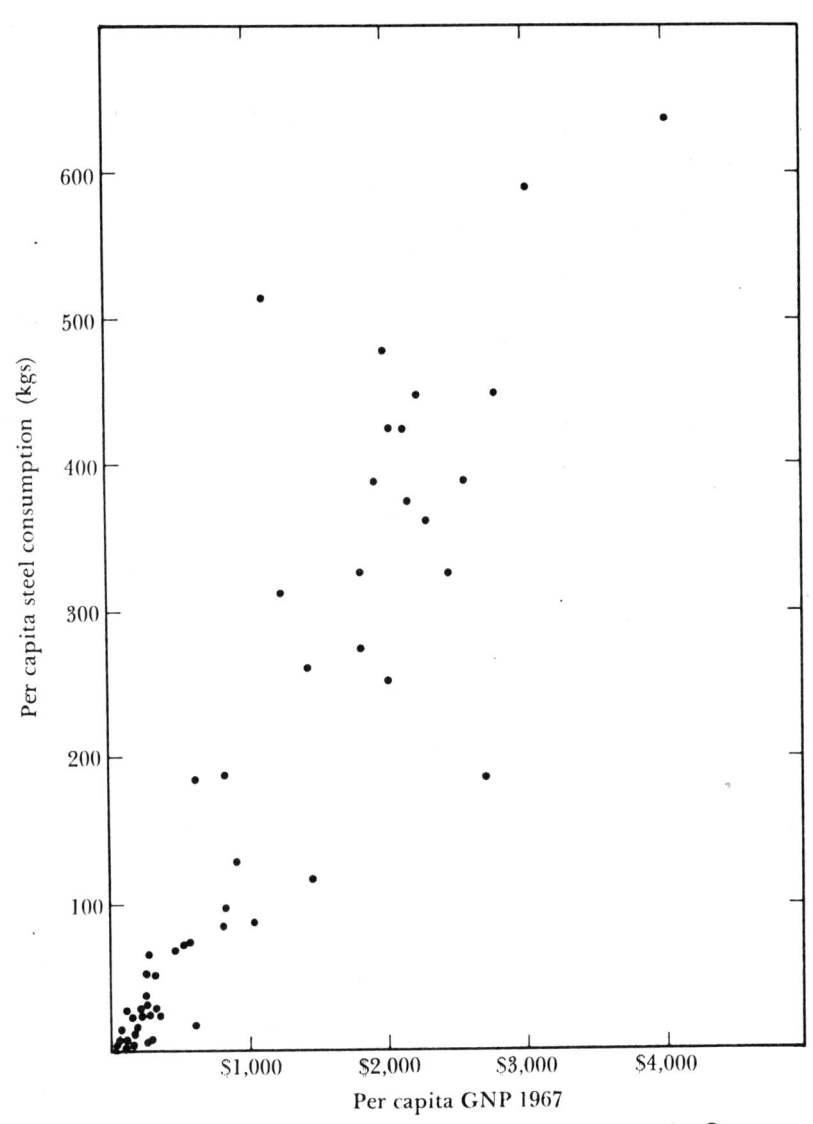

Figure 2. Per capita steel consumption compared to per capita Gross National Product in countries for which statistics are available in the United Nations *Statistical Yearbook* for 1967.

Faced with a developing shortage of natural gas, the Federal Power Commission is presently much interested in knowing whether or not reserves reported by industry are an accurate indication of the amount of natural gas actually on hand; it also wants to know the extent of potential resources and the effect of price on their exploration and development. At the regional or local level, decisions with respect to the designation of wilderness areas and parks, the construction of dams, and other matters related to land use involve appraisal of the distribution and amount of the resources in the area. The questions of the need for an international regime governing the development of seabed resources, the character such arrangement should have, and the definition of the area to which it should apply also involve, among other considerations, analysis of the probable character, distribution, and magnitude of subsea mineral resources.

And coming to the forefront is the most serious question of all—namely, whether or not resources are adequate to support the continued existence of the world's population and indeed our own. The possibility to consider here goes much beyond Malthus' gloomy observations concerning the propensity of a population to grow to the limit of its food supply, for both population and level of living have grown as the result of the consumption of non-renewable resources, and both are already far too high to maintain without industrialized high-energy, and high mineral-consuming agriculture, transportation, and manufacturing. I will say more about this question later, but to indicate something of the magnitude of the problem let me point out that, in attaining our high level of living in the United States, we have used more minerals and mineral fuels during the last 30 years than all the people of the world used previously. This enormous consumption will have to be doubled just to meet the needs of the people now living in the United States through the remainder of their lifetimes, to say nothing about the needs of succeeding generations, or the increased consumption that will have to take place in the lesser developed countries if they are to attain a similar level of living.

Concepts of Reserves and Resources

The focus of most of industry's concern over the extent of mineral resources is on the magnitude of the supplies that exist now or that can be developed in the near term, and this is of public interest also. Many other policy decisions, however, relate to the much more difficult question of potential supplies, a question that to be answered properly must take account both of the extent of undiscovered deposits as well as deposits that cannot be produced profitably now but may become workable in the future. Unfortunately, the need to take account of such deposits is often overlooked, and there is a widespread tendency to think of potential resources as consisting merely of materials in known deposits producible under present economic and technologic conditions.

In connection with my own involvement in resource appraisal, I have been developing over the last several years a system of resource classification and terminology that brings out the classes of resources that need to be taken into account in appraising future supplies, which I believe helps to put the supply problem into a useful perspective. Before describing it, however, I want to emphasize that the problem of estimating potential resources has several built-in uncertainties that make an accurate and complete resource inventory impossible, no matter how comprehensive its scope.

One such uncertainty results from the nature of the occurrence of mineral deposits, for most of them lie hidden beneath the earth's surface and are difficult to locate and to examine in a way that yields accurate knowledge of their extent and quality. Another source of uncertainty is that the specifications of recoverable materials are constantly changing as the advance of technology permits us to mine or process minerals that were once too low in grade, too inaccessible, or too refractory to recover profitably. Still another results from advances that make it possible to utilize materials not previously visualized as usable at all.

For these reasons the quantity of usable resources is not fixed but changes with progress in science, technology, and exploration and with shifts in economic conditions. We must expect to revise our estimates periodically to take account of new developments. Even incomplete and provisional estimates are better than none at all, and if they differentiate known, undiscovered, and presently uneconomic resources they will help to define the supply problem and provide a basis for policy decisions relating to it.

The need to differentiate the known and the recoverable from the undiscovered and the uneconomic requires that a resource classification system convey two prime elements of information: the degree of certainty about the existence of the materials and the economic feasibility of recovering them. These two elements have been recognized in existing terminology, but only incompletely. Thus, as used by both the mining and the petroleum industries, the term *reserves* generally refers to economically recoverable material in identified deposits, and the term *resources* includes in addition deposits not yet discovered as well as identified deposits that cannot be recovered now (e.g. Blondel and Lasky, 1956).

The degree of certainty about the existence of the materials is described by terms such as *proved*, *probable*, and *possible*, the terms traditionally used by industry, and *measured*, *indicated*, and *inferred*, the terms devised during World War II by the Geological Survey and the Bureau of Mines to serve better the broader purpose of national resource appraisal. Usage of these degree-of-certainty terms is by no means standard, but all of their definitions show that they refer only to deposits or structures known to exist.

Thus one of the generally accepted definitions of *possible* ore states that it is to apply to deposits whose existence is known from at least one exposure, and another definition refers to an ore body sampled only on one side. The definition of *inferred* reserves agreed to by the Survey and the Bureau of Mines permits inclusion of completely concealed deposits for which there is specific geologic evidence and for which the specific location can be described, but it makes no allowance for ore in unknown structures of undiscovered districts. The previous definitions of both sets of terms also link them to deposits minable at a profit; the classification system these terms comprise has thus neglected deposits that might become minable as the result of technologic or economic developments.

To remedy these defects, I have suggested that existing terminology be expanded into the broader framework shown in Figure 3, in which degree of certainty decreases from left to right and feasibility of recovery decreases from top to bottom. Either of the series of terms already used to describe degree of certainty may be used with reference to identified deposits and applied not only to presently minable deposits but to others that have been identified with the same degree of certainty. Feasibility-of-recovery categories are designated by the terms *recoverable*, *paramarginal*, and *submarginal*.

Paramarginal resources are defined here as those that are recoverable at prices as much as 1.5 times those prevailing now. (I am indebted to Stanley P. Schweinfurth for suggesting the prefix *para* to indicate that the materials described are not only those just on the margin of economic recoverability, the common economic meaning of the term *marginal*.) At first thought this price factor may seem to be unrealistic. The fact is, however, that prices of many mineral commodities vary within such a range from place to place at any given time, and a price elasticity of this order of magnitude is not uncommon for many commodities over a space of a few years or even months, as shown by recent variations in prices of copper, mercury, silver, sulphur, and coal. Deposits in this category thus become commercially available at price increases that can be borne without serious economic effects, and chances are that improvements in existing technology will make them available at prices little or no higher than those prevailing now.

Over the longer period, we can expect that technologic advances will make it profitable to mine resources that would be much too costly to produce now, and, of course, that is the reason for trying to take account of submarginal resources. Again, it might seem ridiculous to consider resources that cost two or three times more than those produced now as having any future value at all. But keep in mind, as one of many examples, that the cutoff grade for copper has been reduced progressively not just by a factor of two or three but by a factor of ten since the turn of the century and by a factor of about 250 over the history of mining. Many of the fuels and minerals being produced today would once have been classed as submarginal under this definition, and it is reasonable to believe that continued technologic progress will create recoverable reserves from this category.

Examples of Estimates of Potential Resources

For most minerals, the chief value of this classification at present is to call attention to the information needed for a comprehensive appraisal of their potential, for we haven't developed the knowledge and the methods necessary to make meaningful estimates of the magnitude of undiscovered deposits, and we don't know enough about the cost of producing most presently noncommercial deposits to separate paramarginal from submarginal resources. Enough information is available for the mineral fuels, however, to see their potential in such a framework.

The fuel for which the most complete information is available is the newest one—uranium. As a result of extensive research sponsored by the Atomic Energy Commission, uranium reserves and resources are reported in several cost-of-recovery categories, from less than $8 to more than $100 per pound of U_3O_8. For the lower-cost ores, the AEC makes periodic estimates in two degree-

254 MAN'S FINITE EARTH

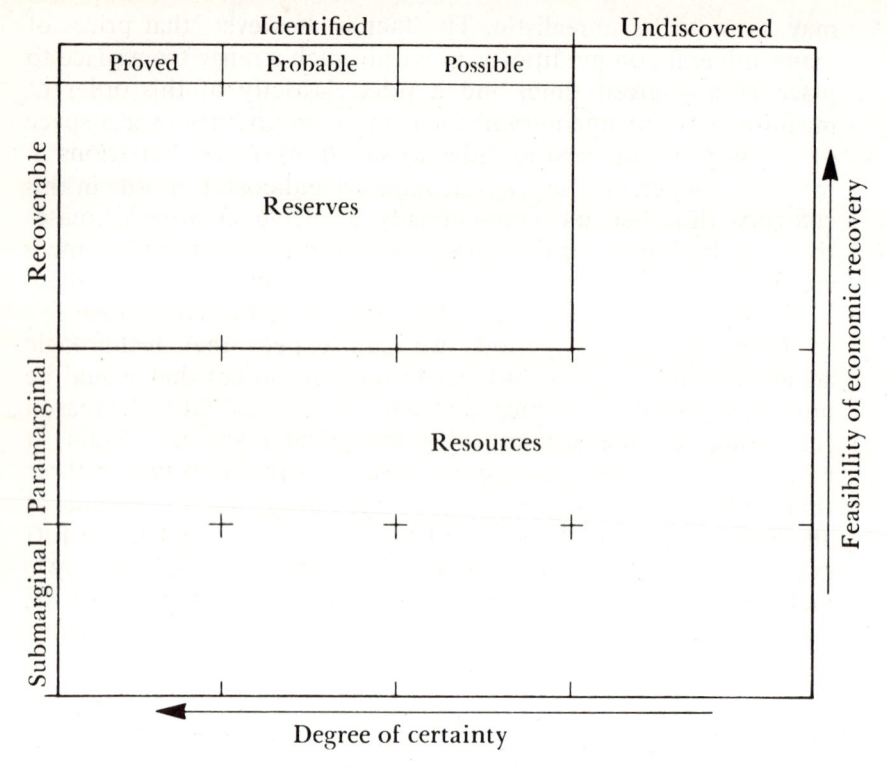

Figure 3. Classification of mineral reserves and resources. Degree of certainty decreases from left to right, and feasibility of recovery decreases from top to bottom.

of-certainty categories, one that it calls *reasonably assured reserves* and the other it calls *additional resources*, defined as uranium surmised to occur in unexplored extensions of known deposits or in undiscovered deposits in known uranium districts. Both the AEC and the Geological Survey have made estimates from time to time of resources in other degree-of-certainty and cost-of-recovery categories.

Ore in the less-than-$8-per-pound class is minable now, and the AEC estimates reasonably assured reserves to be 143,000 tons and additional resources to be 167,000 tons of U_3O_8—just about enough to supply the lifetime needs of reactors in use or ordered in 1968 and only half that required for reactors expected to be in use by 1980. The Geological Survey, however, estimates that undiscovered resources of presently minable quality may amount to 750,000 tons, or about two and a half times that in identified deposits, and districts. Resources in the $8-$30-a-pound category in identified and undiscovered deposits add only about 600,000 tons of U_3O_8 and thus do not significantly increase potential reserves.

But tens of millions of tons come into prospect in the price range of $30-$100 per pound. Uranium at such prices would be usable in the breeder reactor. The breeder, of course, would utilize not only U^{235} but also U^{238}, which is 140 times more abundant than U^{235}. Plainly the significance of uranium as a commercial fuel lies in its use in the breeder reactor, and one may question, as a number of critics have (e.g. Inglis 1971), the advisability of enlarging nuclear generating capacity until the breeder is ready for commercial use.

* * * * *

Quantifying the Undiscovered

Considering potential resources in the degree-of-certainty, cost-of-recovery framework brings out the joint role that geologists, engineers, mineral technologists, and economists must play in estimating their magnitude. Having emphasized the importance of the economic and technologic side of the problem, I want now to turn to the geological side and consider the problem of how to appraise the extent of undiscovered reserves and resources.

* * * * *

Two principal approaches to the problem have been taken thus far. One is to extrapolate observations related to rate of industrial activity, such as annual production of the commodity; the other is to extrapolate observations that relate to the abundance of the mineral in the geologic environment in which it is found.

The first of these methods has been utilized by M. King Hubbert, Charles L. Moore, and Elliott and Linden in estimating ultimate reserves of petroleum. The essential features of this approach are to analyze the growth in production, proved reserves, and discovery per foot of drilling over time and to project these rate phenomena to terminal values in order to predict ultimate production. Hubbert has used the logistic curve for his projections, and Moore has utilized the Gompertz curve, with results more than twice as high as those of Hubbert. As Hubbert has pointed out, these methods utilize the most reliable information collected on the petroleum industry; modern records on production, proved reserves, number of wells drilled, and similar activities are both relatively complete and accurate, at least as compared with quantitative knowledge about geologic features that affect the distribution of petroleum.

The rate methods, however, have an inherent weakness in that the phenomena they analyze reflect human activities that are strongly influenced by economic, political, and other factors that bear no relation to the amount of oil or other material that lies in the ground. Moreover, they make no allowance for major breakthroughs that might transform extensive paramarginal or submarginal resources into recoverable reserves, nor do they provide a means of estimating the potential resources of unexplored regions. Such projections have some value in indicating what will happen over the short term if recent trends continue, but they can have only limited success in appraising potential resources.

Even the goal of such projections, namely the prediction of ultimate production, is not a useful one. Not only is it impossible to predict the quantitative effects of man's future activities but the concept implies that the activities of the past are a part of an inexorable process with only one possible outcome. Far more useful, in my opinion, are estimates of the amounts of various kinds of materials that are in the ground in various environments; such estimates establish targets for both the explorer and the technologist, and they give us a basis for choosing among alternative ways of meeting our needs for mineral supplies.

The second principal approach taken thus far to the estimation of undiscovered resources involves the extrapolation of data on the abundance of mineral deposits from explored to unexplored ground on the basis of either the area or the volume of broadly favorable rocks. In the field of metalliferous deposits, T. B. Nolan pioneered in extrapolation on the basis of area in his study of the spatial and size distribution of mineral deposits in the Boulder Dam region and in his conclusion that a similar distribution should prevail in adjacent concealed and unexplored areas. Lewis Weeks and Wallace Pratt played similar roles with respect to the estimation of petroleum resources—Weeks extrapolating on the basis of oil per unit volume of sediment and Pratt on the basis of oil per unit area. Many of the estimates of crude oil that went into the NPC study were made by the volumetric method, utilizing locally appropriate factors on the amount of oil expected per cubic mile of sediment. Olson and Overstreet have since used the area method to estimate the magnitude of world thorium resources as a function of the size of areas of igneous and metamorphic rocks as compared to India and the United States, and A. P. Butler used the magnitude of sandstone uranium ore reserves exposed in outcrop as a basis for estimating the area in back of the outcrop that is similarly mineralized.

Several years ago, A. D. Zapp and T. A. Hendricks introduced another approach, based on the amount of drilling required to explore adequately the ground favorable for exploration and the reserves discovered by the footage already drilled—a procedure usable in combination with either the volumetric or areal approach. Recently Zimmerman and Long (*Oil and Gas Journal*, 1969) applied this approach to the estimation of gas resources in the Delaware-Val Verde basins of west Texas and southeastern New Mexico, and Haun and others used it to estimate potential natural gas resources in the Rocky Mountain region. In the field of metals, J. David Lowell has estimated the number of undiscovered porphyry copper deposits in the southwestern United States, Chile and Peru, and British Columbia as a function of the proportion of the favorable pre-ore surface adequately explored by drilling, and F. C. Armstrong has similarly estimated undiscovered uranium reserves in the Gas Hills area of Wyoming on the basis of the ratios between explored and unexplored favorable areas.

I have suggested another variant of the areal method for estimating reserves of non-fuel minerals which is based on the fact that the tonnage of minable reserves of the well-explored elements in the United States is roughly equal to their crustal abundance in percent times a billion or 10 billion (Figure 4). Obviously this relation is influenced by the extent of exploration, for it is only reserves of the long-sought and well-explored minerals that display the relation to abundance. But it is this feature that gives the method its greatest usefulness, for it makes it possible to estimate potential resources of elements, such as uranium and thorium, that have been prospected for only a short period. Sekine tested this method for Japan and found it applicable there, which surprised me a little, for I would not have thought Japan to be a large enough sample of the continental crust to bring out this relationship.

The relation between reserves and abundance, of course, can at best be only an approximate one, useful mainly in order-of-magnitude estimates, for obviously crustal abundance of an element is only one of its properties that lead to its concentration. That it is an important factor, however, may be seen not only in

258   MAN'S FINITE EARTH

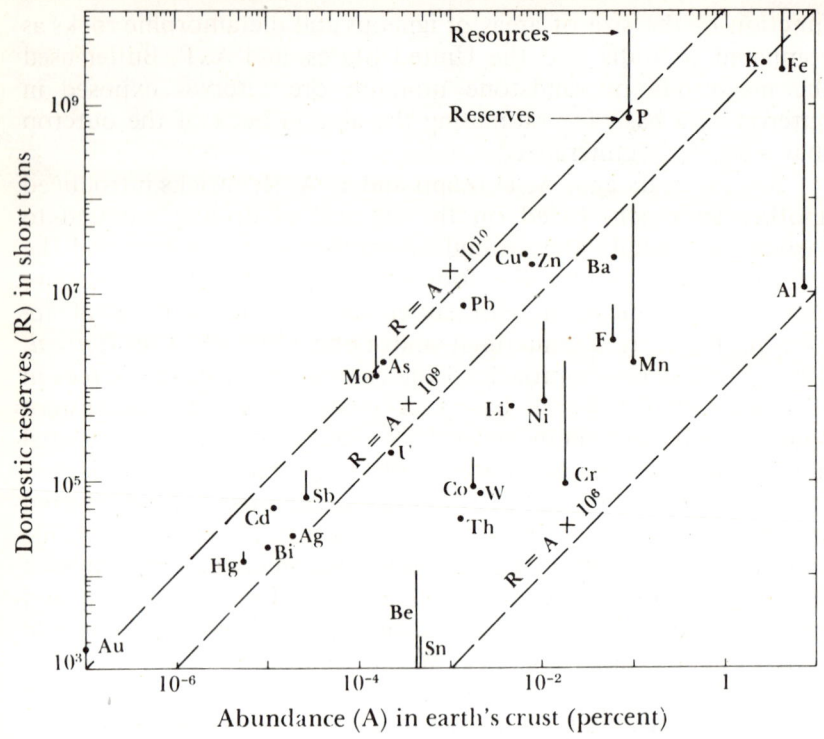

Figure 4. Domestic reserves of elements compared to their abundance in the earth's crust. Tonnage of ore minable now is shown by a dot; tonnage of lower-grade ores whose exploitation depends upon future technological advances or higher prices is shown by a bar.

its influence on the magnitude of reserves but also in other expressions of its influence on the concentrations of the elements. For example, of the 18 or so elements with crustal abundances greater than about 200 parts per million, all but fluorine and strontium are rock-forming in the sense that some extensive rocks are composed chiefly of minerals of which each of these elements is a major constituent. Of the less abundant elements, only chromium, nitrogen, and boron have this distinction. Only a few other elements, such as copper, lead, and zinc, even form ore bodies composed mainly of minerals of which the valuable element is a major constituent, and in a general way the grade of minable ores decreases with decreasing crustal abundance. A similar gross correlation exists between abundance of the elements and the number of minerals in which they are a significant constituent.

Members of a committee of the Geology and the Conservation of Mineral Resources Board of the Soviet Union have described a somewhat similar method for the quantitative evaluation of what they call "predicted reserves" of oil and gas, based on estimates of the total amount of hydrocarbons in the source rock and of the fraction that has migrated into commercial reservoirs—estimates that would be much more difficult to obtain for petroleum than for the elements. Probably for this reason not much use has been made of this method, but it seems likely that quantitative studies of the effects of the natural fractionation of the elements might be of some value in estimating total resources in various size and grade categories.

Some studies of the grade-frequency distribution of the elements have, in fact, been undertaken by geochemists in the last couple of decades, and, taking off from Nolan's work, several investigators have studied the areal and size-frequency distribution of mineral deposits in conjunction with attempts to apply the methods of operations research to exploration (e.g. Allais, 1957; Slichter, 1960; Griffiths, 1964; DeGeoffroy and Wu, 1969; Harris, 1969). None of these studies has been concerned with the estimation of undiscovered reserves, but they have identified two features about the distribution of mineral deposits that may be applicable to the problem.

One is that the size distribution of both metalliferous deposits, expressed in dollar value of production, and of oil and gas, expressed in volumetric units, has been found to be log normal, which means that of a large population of deposits, a few contain most of the ore (e.g. Slichter, 1960; Kaufman, 1963). In the Boulder Dam area, for example, 4 percent of the districts produced 80 percent of the total value of recorded production. The petroleum industry in the United States has a rule of thumb that 5 percent of the fields account for 50 percent of the reserves and 50 percent of fields for 95 percent. And in the USSR, about 5 percent of the oil fields contain about 75 percent of the oil, and 10 percent of the gas fields have 85 percent of the gas reserves.

The other feature of interest is that in many deposits the grade-tonnage distribution is also log normal, and the geochemists have found this to be the case also with the frequency distribution of minor elements.

These patterns of size- and grade-frequency distribution will not in themselves provide information on the magnitude of potential resources, for they describe only how minerals are distributed and not how much is present. But if these patterns are combined with quantitative data on the incidence of congeneric deposits in various kinds of environments, the volume or area of favorable ground, and the extent to which it has been explored, they might yield more useful estimates of potential resources than are obtainable by any of the procedures so far applied. Thus estimates of total resources described in terms of their size- and grade-frequency distributions could be further analyzed in the light of economic criteria defining the size, grade, and accessibility of deposits workable at various costs, and then partitioned into feasibility-of-recovery and degree-of-certainty categories to provide targets for exploration and technologic development as well as guidance for policy decisions.

Essential for such estimates, of course, is better knowledge than is now in hand for many minerals on the volume of ore per unit of favorable ground and on the characteristics of favorable ground itself. For petroleum the development of such knowledge is already well advanced. For example, whereas most estimates of resources have been based on an assumed average petroleum content of about 50,000 barrels per cubic mile of sediment, varied a little perhaps to reflect judgments of favorability, the range in various basins is from 10,000 to more than 2,000,000 barrels per cubic mile. As shown by the recent analysis by Halbouty and his colleagues of the factors affecting the formation of giant fields, the geologic criteria are developing that make it possible to classify sedimentary basins in terms of their petroleum potential. Knowledge of the mode of occurrence and genesis of many metalliferous minerals and of the geology of the terranes in which they occur is not sufficient to support comprehensive estimates prepared in this way. But for many kinds of deposits enough is known to utilize this kind of approach on a district or regional basis, and I hope a start can soon be made in this direction.

Need for Review of Resource Adequacy

Let me return now to the question of whether or not resources are adequate to maintain our present level of living. This is not a new question by any means. In 1908 it was raised as a national policy issue at the famous Governors' Conference on Resources, and it has been the subject of rather extensive inquiry by several national and international bodies since then. In spite of some of the dire predictions about the future made by various people in the course of these inquiries, they did not lead to any major change in our full-speed-ahead policy of economic development. Some of these inquiries, in fact, led to immediate investigations that revealed a greater resource potential for certain minerals than had been thought to exist, and the net effect was to alleviate rather than heighten concern.

Now, however, concern about resource adequacy is mounting again. The overall tone of the recent National Academy of Sciences' report on *Resources and Man* was cautionary if not pessimistic about continued expansion in the production and use of mineral resources, and many scientists, including some eminent geologists, have expressed grave doubts about our ability to continue on our present course. The question is also being raised internationally, particularly in developing countries where concern is being expressed that our disproportionate use of minerals to support our high level of living may be depriving them of their own future.

Personally, I am confident that for millennia to come we can continue to develop the mineral supplies needed to maintain a high level of living for those who now enjoy it and to raise it for the impoverished people of our own country and world. My reasons for thinking so are that there is a visible undeveloped potential of substantial proportions in each of the processes by which we create resources and that our experience justifies the belief that these processes have dimensions beyond our knowledge and even beyond our imagination at any given time.

Setting aside the unimaginable, I will mention some examples of the believable. I am sure all geologists would agree that minable undiscovered deposits remain in explored as well as unexplored areas and that progress in our knowledge of regional geology and exploration will lead to the discovery of many of them. With respect to unexplored areas, the mineral potential of the continental margins and ocean basins deserves particular emphasis, for the technology that will give us access to it is clearly now in sight. For many critical minerals, we already know of substantial paramarginal and submarginal resources that experience tells us should be brought within economic reach by technologic advance. The process of substituting an abundant for a scarce material has also been pursued successfully, thus far not out of need but out of economic opportunity, and plainly has much potential as a means of enlarging usable resources.

Extending our supplies by increasing the efficiency of recovery and use of raw materials has also been significant. For example, a unit weight of today's steel provides 43 percent more structural support than it did only ten years ago, reducing proportionately the amount required for a given purpose. Similarly, we make as much electric power from one ton of coal now as we were able to make from seven tons around the turn of the century. Our rising awareness of pollution and its effects surely will force us to pay even more attention to increasing the efficiency of mineral recovery and use as a means of reducing the release of contaminants to the environment. For similar reasons, we are likely to pursue more diligently processes of recovery, re-use, and recycling of mineral materials than we have in the past.

Most important to secure our future is an abundant and cheap supply of energy, for if that is available we can obtain materials from low-quality sources, perhaps even country rocks, as Harrison Brown has suggested. Again, I am personally optimistic on this matter, with respect both to the fossil fuels and particularly to the nuclear fuels. Not only does the breeder reactor appear to be near enough to practical reality to justify the belief that it will permit the use of extremely low-grade sources of uranium and thorium that will carry us far into the future, but during the last couple of years there have been exciting new developments in the prospects for commercial energy from fusion. Geothermal energy has a large unexploited potential, and new concepts are also being developed to permit the commercial use of solar energy.

But many others do not share these views, and it seems likely that soon there will be a demand for a confrontation with the full-speed-ahead philosophy that will have to be answered by a deep review of resource adequacy. I myself think that such a review is necessary, simply because the stakes have become so high. Our own population, to say nothing of the world's, is already too large to exist without industrialized, high energy- and mineral-consuming agriculture, transportation, and manufacturing. If our supply of critical materials is enough to meet our needs for only a few decades, a mere tapering off in the rate of increase of their use, or even a modest cutback, would stretch out these supplies for only a trivial period. If resource adequacy cannot be assured into the far-distance future, a major reorientation of our philosophy, goals, and way of life will be necessary. And if we do need to revert to a low resource-consuming economy, we will have to begin the process as quickly as possible in order to avoid chaos and catastrophe.

Comprehensive resource estimates will be essential for this critical examination of resource adequacy, and they will have to be made by techniques of accepted reliability. The techniques I have described for making such estimates have thus far been applied to only a few minerals, and none of them have been developed to the point of general acceptance. Better methods need to be devised and applied more widely, and I hope that others can be enlisted in the effort necessary to do both.

References

Allais, M. 1957. Methods of appraising economic prospects of mining exploration over large territories. *Management Science* 2:285-347.

Armstrong, F. C. 1970. Geologic factors controlling uranium resources in the Gas Hills District, Wyoming. 22nd Annual Field Conference. *Wyoming Geological Association Guidebook*, pp. 31-44.

Blondel, F., and Lasky, S. F. 1965. Mineral reserves and mineral resources. *Economic Geology* 60:686-97.

Bush, A. L., and Stager, H. K. 1956. Accuracy of ore-reserve estimates for uranium-vanadium deposits on the Colorado Plateau. *U. S. Geological Survey Bulletin* 1030-D:137.

Buyalov, N. I., Erofeev, N. S., Kalinin, N. A., Kleschev, A. I., Kudryashova, N. M., L'vov, M. S., Simakov, S. N., and Vasil'ev, V. G. 1964. *Quantitative evaluation of predicted reserves of oil and gas*. Authorized translation Consultants Bureau.

DeGeoffroy, J., and Wu, S. M. 1970. A statistical study of ore occurrences in the greenstone belts of the Canadian Shield. *Economic Geology* 65:496-504.

Elliott, M. A., and Linden, H. R. 1968. A new analysis of U. S. natural gas supplies. *Journal of Petroleum Technology* 20:135-41.

Griffiths, J. C. 1962. Frequency distributions of some natural resource materials. 23rd Tech. Conf. on Pet. Prod., September 26-28. *Min. Ind. Expt. Sta. Circ.* 63:174-98.

–––. 1966. Exploration for natural resources. *Operations Research* 14:189-209.

Halbouty, M. T., Meyerhoff, A. A., King, R. E., Dott, R. H., Sr., Klemme, H. D., and Shabad, T. 1970. World's giant oil and gas fields: world's giant oil and gas fields, geologic factors affecting their formation, and basin classification, part 1. *Geology of giant petroleum fields*. American Association of Petroleum Geologist Memoir 14:502-28.

Halbouty, M. T., King, R. E., Klemme, H. D., Dott, R. H., Sr., and Meyerhoff, A. A. 1970. Factors affecting formation of giant oil and gas fields, and basin classification: worlds's giant oil and gas fields, geologic factors affecting their formation, and basin classification, part 2. *Geology of* giant petroleum fields. American Association of Petroleum Geologists Memoir 14:528-55.

Harris, D. P., and Euresty, D. 1969. A preliminary model for the economic appraisal of regional resources and exploration based upon geostatistical analyses and computer simulation. *Colorado School of Mines Quarterly* 64:71-98.

Hendricks, T. A. 1965. *Resources of oil, gas, and natural gas liquids in the U.S. and the world.* U. S. Geological Survey Circular 522.

Hendricks, T. A., and Schweinfurth, S. P. 1966. Unpublished memorandum quoted in *United States petroleum through 1980* (1968). Washington, D.C.: U. S. Department of the Interior.

Hubbert, M. K. 1969. Energy resources. In *Resources and man*, pp. 157-239. National Academy of Sciences–National Research Council. San Francisco: W. H. Freeman & Co.

Inglis, D. R 1971. Nuclear energy and the Malthusian dilemma. *Bulletin of the Atomic Scientists* 27(2):14-18.

Kaufman, G. M. 1963. *Statistical decision and related techniques in oil and gas exploration.* Englewood Cliffs, N. J.: Prentice-Hall.

Lasky, S. G. 1950. Mineral-resource appraisal by U. S. Geological Survey. *Colorado School of Mines Quarterly 45:1-27.* See also his (1950) How tonnage and grade relations help predict ore reserves. *Engineering Mining Journal* 151(4):81-85.

Lowell, J. D. 1970. Copper resources in 1970. *Mining Engineering* 22(April): 67-73.

McKelvey, V. E. 1960. Relation of reserves of the elements to their crustal abundance. *American Journal of Science* (Bradley Vol.) 258-A:234-41.

Moore, C. L. 1966. *Projections of U. S. petroleum supply to 1980.* Washington, D.C.: U. S. Department of the Interior, Office of Oil and Gas.

National Academy of Sciences–National Research Council. 1969. *Resources and man.* San Francisco: W. H. Freeman & Co.

Nolan, T. B. 1950. The search for new mineral districts. *Economic Geology* 45:601-8.

Oil and Gas Journal. 1969. Vast Delaware-Val Verde reserve seen. 67(16):44. (Re: Zimmerman and Long.)

Olson, J. C., and Overstreet, W. C. 1964. Geologic distribution and resources of thorium. *U. S. Geological Survey Bulletin* 1204.

Orwell, G. 1937. *The road to Wigan Pier.* American edition, 1958. New York: Harcourt, Brace & World.

Pratt, W. E. 1950. The earth's petroleum resources. In *Our oil resources*, ed. L. M. Fanning, pp. 137-53. 2nd ed. New York: McGraw-Hill.

Rodionov, D. A. 1965. *Distribution functions of the element and mineral contents of igneous rocks: a special research report*, pp. 28-29. Authorized translation Consultants Bureau.

Sekine, Y. 1963. On the concept of concentration of ore-forming elements and the relationship of their frequency in the earth's crust. *International Geological Review* 5:505-15.

Slichter, L. B. 1960. The need of a new philosophy of prospecting. *Mining Engineering* 12:570-6.

Slichter, L. B., et al. 1962. Statistics as a guide to prospecting. In *Math and computer applications in mining and exploration symposium proceedings*, pp. F-1—27. Tucson: Arizona University College of Mines.

Weeks, L. G. 1950. Concerning estimates of potential oil reserves. *American Association of Petroleum Geologists Bulletin* 34:1947-53.

———. 1958. Fuel reserves of the future. *American Association of Petroleum Geologists Bulletin* 42:431-8.

———. 1956. World offshore petroleum resources. *American Association of Petroleum Geologists Bulletin 49:1680-93.*

Zapp, A. D. 1962. Future petroleum producing capacity of the United States. *U. S. Geological Survey Bulletin* 1142-H.

24 Energy Conservation

Philip H. Abelson

Philip H. Abelson is the editor of Science, *the official publication of the American Association for the Advancement of Science and president of the Carnegie Institution of Washington.*

The series of articles on energy appearing currently in *Science* points up the long-range importance of this topic and many problems connected with it. If we are to solve our energy problems, the public and the government must give these matters an enduring high priority. This is chancy. When there is a dramatic crisis, the public usually behaves well. For example, during some of the recent power shortages, the public responded to pleas for conservation. Once the immediate crisis passed, though, the public returned to its old habits. Consumption of energy continued to grow exponentially. And use of gasoline is growing at a fast rate because emission controls are making automobiles less efficient.

Importation of petroleum and its products has been increasing rapidly. The Bureau of Mines now estimates that by 1985 imports alone will amount to 15 million barrels a day, which is our current total use. Such a volume could only be met by drawing heavily on the Middle East. Demand from Europe and Japan has already created a seller's market. Recently the Oil Producing and Exporting Countries have obtained substantial increases in their take. The most aggressive has been Libya, which in 2 years has doubled its return per barrel.

What will the Middle Eastern countries do with the enormous wealth that they will extract? The example of one is disquieting. Libya has chosen to devote part of its revenues to financing terrorist activities. The largest petroleum reserve is found in Saudi Arabia. That country has a small population and limited demand for goods. It has already begun to move toward controlling interest in some of the great International Petroleum companies. At the moment it seems to be a force for stability in the Middle East. However, who knows for how long? Prudence dictates that we examine alternatives to massive dependence on foreign oil.

One alternative that has not had much attention is conservation of energy. A recent useful 250-page government study* points to many possible measures that could be taken to reduce energy demand without great interference with life styles. It provides data on the various categories of energy consumption—transportation (25 percent), industry (29 percent), electric utilities (25 percent), and residential/commercial (21 percent) as well as the many components of these categories. The report discusses in detail possible short-term and mid-term savings in energy. For example, better insulation of houses provided at nominal cost would save very substantial amounts of both energy and money. The study suggests that energy conservation measures could reduce U.S. energy demand in 1980 by as much as the equivalent of 7.3 million barrels of oil per day. To achieve these economies in energy would require voluntary public cooperation on a scale that has heretofore not been sustained for long.

The surest way of obtaining public cooperation in the expenditure of energy is to make energy costly, and this is likely to occur whether we wish it or not. If present trends continue, a doubling or trebling in cost of oil and gasoline could occur in this decade.

Ultimately we will find that we must rethink our attitude about automobiles. Most of us would be reluctant to part with our mobile castles. But must these castles weigh 2 tons or more? If the government can dictate exhaust standards, safety features and more, why can't it exert pressure for lighter weight and greater mileage. Indeed it is likely that history will record that instead of its push on manufacturers to cut emission, the government should be pressing now for sharply better fuel economy.

---

*The Potential for Energy Conservation: Office of Emergency Preparedness, 1972, Executive Office of the President, U. S. Government Printing Office.

25 Geothermal Energy

L. J. P. Muffler and D. E. White

L. P. J. Muffler and Donald E. White are geologists for the U. S. Geological Survey. Dr. Muffler is coordinator of the Geothermal Research Program at Menlo Park, California, and Dr. White managed U. S. G. S. geothermal resource investigations prior to 1972.

Geothermal energy, in the broadest sense, is the natural heat of the earth. Temperatures in the earth rise with increasing depth. At the base of the continental crust (25 to 50 km), temperatures range from 200°C to 1000°C; at the center of the earth (6371 km), perhaps from 3500°C to 4500°C. But most of the earth's heat is far too deeply buried to be tapped by man, even under the most optimistic assumptions of technological development. Although drilling has reached 7½ km and may some day reach 15 to 20 km, the depths from which heat might be extracted economically are unlikely to be greater than 10 km.

White (7) has calculated that the amount of geothermal heat available in this outer 10 km is approximately 3×10^{26} calories, which is more than 2000 times the heat represented by the total coal resources of the world (1). But most of this geothermal energy is far too diffuse ever to be recovered economically. The average heat content of each gram of rock in the outer 10 km of the earth is only 0.3 percent of the heat obtainable by combusting one gram of coal and is less than 0.01 percent of the heat equivalent of fissionable uranium and thorium contained in one gram of average granite. Consequently, most of the heat within the earth, even at depths of less than 10 km, cannot be considered a potential energy resource.

Geothermal energy, however, does have potential economic significance where the heat is concentrated into restricted volumes in a manner analogous to the concentration of valuable metals into ore deposits or of oil into commercial petroleum reservoirs. At present, economically significant concentrations of geothermal energy occur where elevated temperatures (40°C to >380°C) are found in permeable rocks at shallow depths (less than 3 km). The thermal energy is stored both in the solid rock and in water and steam filling pores and fractures. This water and steam serve to transfer the heat from the rock to a well and thence to the ground surface. Under present technology, rocks with two few pores, or with pores that are not interconnected, do not comprise an economic geothermal reservoir, however hot the rocks may be.

Water in a geothermal system also serves as the medium by which heat is transferred from a deep igneous source to a geothermal reservoir at depths shallow enough to be tapped by drill holes. Geothermal reservoirs are located in the upflowing parts of major water convection systems (Figure 1). Cool rainwater percolates underground from areas that may comprise tens to thousands of square kilometers and then circulates downward. At depths of 2 to 6 km, the water is heated by contact with hot rock (in turn probably heated by molten rock). The water expands upon heating and then moves buoyantly upward in a column of relatively restricted cross-sectional area (1 to 50 km^2). If the rocks have many interconnected pores or fractures, the heated water rises rapidly to the surface and is dissipated rather than stored. If, however, the upward movement of heated water is impeded by rocks without interconnected pores or fractures, the geothermal energy may be stored in reservoir rocks below the impeding layers. The driving force of this large circulation system is gravity, effective because of the density difference between cold, downward-moving recharge water and hot, upward-moving geothermal water.

Many investigators in the past considered the water in geothermal systems to be derived from molten rock at depth. Modern studies of hydrogen and oxygen isotopes in geothermal waters, however, indicate that at least 95 percent of most geothermal fluids must be derived from surface precipitation and that no more than 5 percent is volcanic steam.

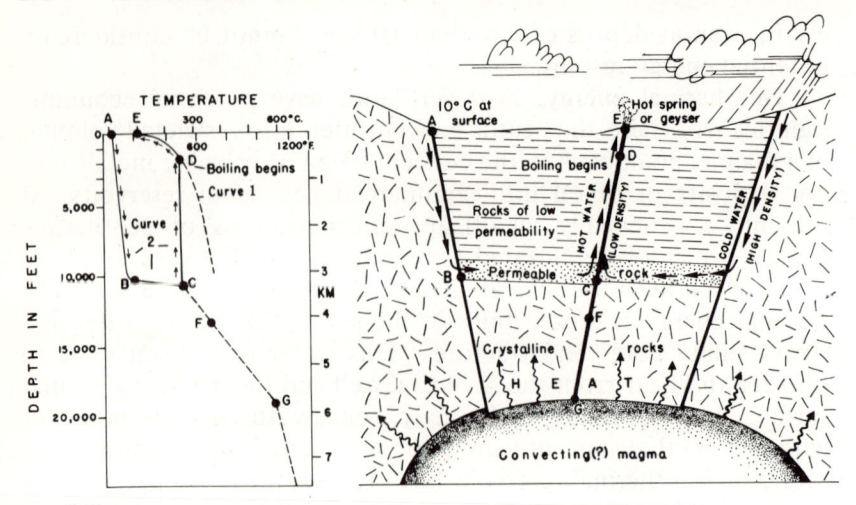

Figure 1. Schematic model of hot-water geothermal system, modified from White (8, 9). Curve 1 shows the boiling point of pure water under pressure exerted by a column of liquid water everywhere at boiling, assuming water level at ground surface. Dissolved salts shift the curve to the right; dissolved gases shift the curve to the left. Curve 2 shows the ground temperature profile of a typical hot-water system.

Location of Geothermal Systems

Geothermal reservoirs are the "hot spots" of larger regions where the flow of heat from depth in the earth is one and one-half to perhaps five times the worldwide average of 1.5×10^{-6} calories per square centimeter per second. Such regions of high heat flow commonly are zones of young volcanism and mountain building and are localized along the margins of major crustal plates (Figure 2). These margins are zones where either new material from the mantle is being added to the crust (i.e., spreading ridges; see Figure 3) or where crustal material is being dragged downward and "consumed" in the mantle (subduction zones). In both situations, molten rock is generated and then moves buoyantly upward into the crust. These pods of igneous rock provide the heat that is then transferred by conduction to the convecting systems of meteoric water.

Figure 2 shows that the geothermal fields presently being exploited or explored occur in three major geologic environments: (1) along spreading ridges, (2) above subduction zones, and (3)

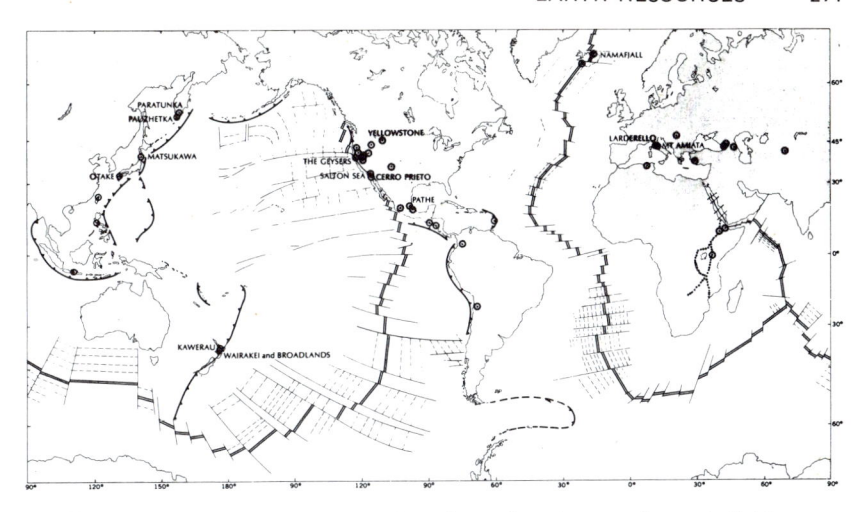

Figure 2. World map showing location of major geothermal fields along plate margins. Heavy double lines represent spreading ridges; heavy lines with barbs represent active subduction zones; heavy dotted lines represent rift valleys. Light lines represent transform faults; dashed light lines represent approximate position of magnetic anomalies. Base map and tectonic features from Coleman (2, Figure 4).

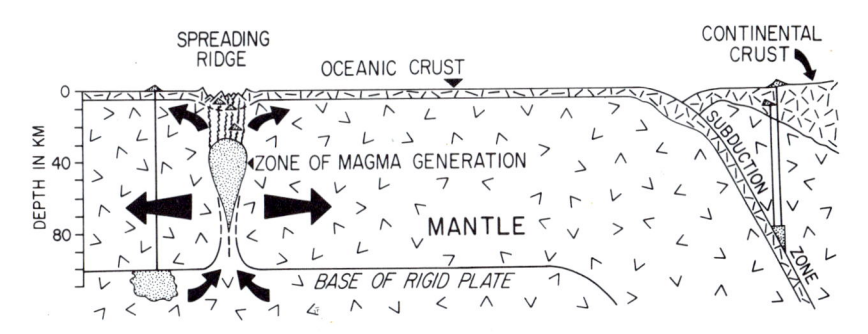

Figure 3. Model of development of oceanic crust at spreading ridges and subduction of oceanic crust at consuming plate margins. (Generalized from Coleman, 2, Figure 6.)

along the belt of mountains extending from Italy through Turkey to the Caucasus. Although this last zone is not a modern subduction zone, it is the zone where the African and European plates are in contact, and it appears to have been a subduction zone in the past. Geothermal fields are absent from the stable, continental shields, which are characterized by lower-than-average heat flow. Although there are no known shallow geothermal reservoirs in the non-volcanic continental areas bordering the shields, hot water has been found at depths of 3 to 6 km in the U.S.S.R., in Hungary, and on the Gulf Coast of the United States (6).

Uses of Geothermal Resources

The primary use of geothermal energy to date is for the generation of electricity. For this purpose, under existing technology, the geothermal reservoir must have a temperature of at least 180°C, and preferably 200°C. Geothermal steam, after separation of any associated water (as much as 90 weight percent of the total effluent), is expanded into a turbine that drives a conventional generator. World electrical capacity from geothermal energy in 1971 was approximately 800 megawatts (Table 1), or about 0.08 percent of the total world electrical capacity from all generating modes. Power from favorable geothermal systems is competitive in cost with either fossil fuel or nuclear power. The production of

TABLE 1. WORLD GEOTHERMAL POWER PRODUCTION, 1971.

Country	Field	Electrical Capacity, MW Operating	Under Construction
ITALY	Larderello	358.6	
	Mt. Amiata	25.5	
U.S.A.	The Geysers	192	110
NEW ZEALAND	Wairakei	160	
	Kawerau	10	
JAPAN	Matsukawa	20	
	Otake	13	
MEXICO	Pathe	3.5	
	Cerro Prieto		75
U.S.S.R.	Pauzhetka	5	
	Paratunka[a]	.7	
ICELAND	Namafjall	2.5	
		790.8	185 $\Sigma = 975.8$

[a]Freon Plant

geothermal power is obviously restricted to areas where geothermal energy is found in sufficient quantity. Unlike coal, oil, gas, or uranium, geothermal steam cannot be transported long distances to a generating plant located near the existing load centers.

Geothermal resources have other uses, but to date they have been minor. Geothermal waters as low as 40°C are used locally for space heating and horticulture. Much of Reykjavik, the capital of Iceland, is heated by geothermal water, as are parts of Rotorua (New Zealand), Boise (Idaho), Klamath Falls (Oregon), and various towns in Hungary and the U.S.S.R. Geothermal steam is also used in paper manufacturing at Kawerau, New Zealand and has potential use for refrigeration. Some geothermal waters contain potentially valuable by-products such as potassium, lithium, calcium, and other metals. Use of geothermal energy to desalt geothermal water itself has been proposed, and the U. S. Bureau of Reclamation and the Office of Saline Water are presently developing a pilot operation for producing fresh water from the geothermal waters of the Imperial Valley, Southern California.

Types of Geothermal Systems

There are two major types of geothermal systems: hot-water systems and vapor-dominated ("dry-steam") systems (11). Among geothermal systems discovered to date, hot-water systems are perhaps twenty times as common as vapor-dominated systems (10).

Hot-Water Geothermal Systems

Hot-water geothermal systems contain water at temperatures that may be far above surface boiling, owing to the effect of pressure on the boiling point of water (curve 1 of Figure 1). A typical hot-water system has temperature-depth relations similar to those of curve 2. Little change in temperature occurs as meteoric water descends from A to B, heat is absorbed from B to C, and from C to D the system contains water at nearly constant temperature (the "base temperature"). From D to E pressure has decreased enough for water to boil, and steam and water coexist. In major zones of upflow, coexisting steam and water extend to the surface and are expressed as boiling hot springs and locally as geysers. Geothermal wells, however, are usually sited in nearby cool, stable ground where near-surface temperatures are controlled by conduction of heat through solid rocks; the temperature-depth curve is therefore initially to the left of curve 1.

Water in most hot-water geothermal systems is a dilute solution (1000 to 30,000 milligrams per liter), containing mostly sodium, potassium, lithium, chloride, bicarbonate, sulfate, borate, and silica. The silica content and the ratio of potassium to sodium are dependent on temperature in the geothermal reservoir, thus allowing prediction of subsurface temperature from chemical analysis of hot springs (3,4).

In hot-water geothermal systems, only part of the produced fluid is steam and can be used to generate electricity with present technology. For example, water at 250°C will produce only about 20 weight percent of steam when the confining pressure is reduced to 6 kilograms per square centimeter, the approximate well-head pressure commonly used in geothermal installations. The steam and water at this pressure are mechanically separated before the steam is fed to the turbine.

Some attention is currently being directed toward a heat exchange generating system. Heat in the geothermal water is transferred by a heat exchanger to a low-boiling-point fluid, such as freon or isobutane, which is then expanded into a turbine. The geothermal water is not allowed to boil and is reinjected as water into the ground. If this binary fluid generating technology proves economically feasible, it will allow more complete extraction of heat from geothermal fluids and will allow use of hot-water geothermal systems of lower temperature than are presently required for direct geothermal steam generation.

The major known hot-water geothermal fields are Wairakei (160 megawatts) and Broadlands (100 megawatts proposed) in New Zealand, Cerro Prieto (75 megawatts under construction; 200 megawatts proposed) in Mexico, the Salton Sea field in California, and the Yellowstone geyser basins in Wyoming. Although the Yellowstone region is the world's most intensive display of hot-spring and geyser phenomena, the area is permanently withdrawn as a National Park and will never be exploited for power.

Whereas the salinities of most hot-water fields are 0.1 to 3 percent, the Salton Sea geothermal reservoir contains a brine with more than 25 percent by weight of dissolved solids, mainly chloride, sodium, calcium, and potassium. In addition, the brine is rich in a variety of metals (8). Although temperatures reach 360°C, development of the field has been hindered by problems of corrosion, deposition of silica, and disposal of unwanted effluent.

Hot, saline brines also occur in pools along the median trench of the Red Sea where geothermal fluids discharge directly onto the sea floor 2 km below sea level (8).

Vapor-Dominated Geothermal Systems

Vapor-dominated geothermal systems, in contrast to hot-water systems, produce superheated steam with minor amounts of other gases (CO_2, H_2S), but no water. The total fluid can therefore be piped directly to the turbine. Within the vapor-dominated geothermal reservoir, saturated steam and water coexist, with steam being the phase that controls the pressure. With decrease in pressure upon production, heat contained in the rocks dries the fluids first to saturated and then to superheated steam, with as much as 55°C superheat at a well-head pressure range of 5 to 7 kilograms per square centimeter. Owing to the thermodynamic properties and flow dynamics of steam and water in porous media, vapor-dominated reservoirs are unlikely to exist at pressures much greater than about 34 kilograms per square centimeter and temperatures much above 240°C (5). Hot brine probably exists below the vapor-dominated reservoirs at depth, but drill holes are not yet deep enough to confirm the presence of such a brine.

Drilling has demonstrated the existence of only three commercial vapor-dominated systems: Larderello, Italy; The Geysers, California; and probably Matsukawa, Japan. Two small fields in the Monte Amiata region of Italy are marginally commercial. Larderello was the first geothermal field to be exploited, starting in 1904, and is still the largest producer of geothermal power (350 megawatts). The Geysers at present produces 192 mw, but plants under construction will boost capacity to 302 mw in 1972, and ultimate potential is in excess of 500 mw.

The Geothermal Energy Resource

White (7) estimated that the total stored heat of all geothermal reservoirs to a depth of 10 km was 10^{22} calories. This estimate specifically excluded reservoirs of molten rock, abnormally hot rocks of low permeability, and deep sedimentary basins of near "normal" conductive heat flow, such as the Gulf Coast of the United States or Kazahkstan in the Soviet Union. The geothermal resources in these environments are at least 10 times greater than the resources of the hydrothermal systems, but they are recoverable only at much more than present costs. Should production of these geothermal resources someday become feasible, the potential geothermal resource in all reservoir types would be at least 10^{23} calories, which is approximately equivalent to the heat represented by the world's potential resources of coal.

For a hot-water geothermal system, approximately 1 percent of the heat stored in the reservoir can be extracted and converted into electricity under present technology. For the far less abundant vapor-dominated geothermal systems, perhaps 2 to 5 percent of the heat in the reservoir can be extracted and converted to electricity. Therefore, if the use of geothermal energy continues to be restricted primarily to electrical generation by proven techniques, then the potential geothermal resource to 10 km is only about 10^{20} calories. To a depth of 3 km (the deepest well drilled to date for geothermal power), the resource for electrical generation by proven techniques is even less, approximately 2×10^{19} calories (7). Use of geothermal resources for other than electrical generation (e.g., heating, desalination, horticulture, etc.) would greatly increase these geothermal resource estimates, perhaps by 10 times, but all these uses involve special geographic and economic conditions that to date have been implemented only on a local scale.

Production of electricity from geothermal energy is presently attractive environmentally, because no solid atmospheric pollutants are emitted and no radiation hazard is involved. But geothermal generation is not without environmental effects. Effluent from either a hot-water or a vapor-dominated system can pollute streams or ground water. Consequently, federal and state regulations require reinjecting objectionable fluids back into a deep reservoir. Thermal pollution is also a problem, particularly in hot-water systems, but it can be solved in part by reinjection of unwanted water and of residual steam condensate. Noise, objectionable gases, visual impact, and subsidence of the land surface due to fluid withdrawal are other problems that are faced in any geothermal energy development.

Geothermal energy is unlikely to supply more than perhaps 10 percent of domestic or world electrical power demand. But in favorable areas, geothermal power may be of major importance, particularly in underdeveloped countries that have few other energy resources.

Although geothermal power was produced in Italy as early as 1904 and in New Zealand by 1955, extensive interest in geothermal resources of the United States has developed only in the past 10 years. Large areas in the western United States appear to be favorable for geothermal exploration, but knowledge of the nature and extent of our geothermal resources is inadequate. Further investigations are necessary, not only of the distribution and characteristics of geothermal reservoirs but also of the various ways in which geothermal energy can be used in the most beneficial and least wasteful manner.

References

1. Averitt, P. 1969. Coal resources of the United States, January 1, 1967. U. S. Geological Survey Bulletin 1275.
2. Coleman, R. G. 1971. Plate tectonic emplacement of upper mantle peridotites along continental edges. *Journal of Geophysical Research* 76:1212-22.
3. Ellis, A. J. 1970. *Quantitative interpretation of chemical characteristics of hydrothermal systems.* United Nations Symposium on the development and Utilization of Geothermal Resources, Pisa, Italy, September, 1970. Paper I/11.
4. Fournier, R. O., and Rowe, J. J. 1966. Estimation of underground temperatures from the silica content of water from hot springs and wet-steam wells. *American Journal of Science* 264:685-97.
5. James, R. 1968. Wairakei and Larderello: geothermal power systems compared. *New Zealand Journal of Science* 11:706-19.
6. Jones, P. H. 1970. *Geothermal resources of the northern Gulf of Mexico basin.* United Nations Symposium on the Development and Utilization of Geothermal Resources, Pisa, Italy, September, 1970. Paper I/24.
7. White, D. E. 1965. *Geothermal energy.* U. S. Geological Survey Circular 519.
8. ———. 1968a. Environment of generation of some base-metal ore deposits. *Economic Geology* 63:301-35.
9. ———. 1968b. *Hydrology, activity, and heat flow of the Steamboat Springs thermal system, Washoe County, Nevada.* U. S. Geological Survey Professional Paper 458-C.
10. ———. 1970. *Geochemistry applied to the discovery, evaluation, and exploitation of geothermal energy resources.* United Nations Symposium on the Development and Utilization of Geothermal Resources, Pisa, Italy, September, 1970. Rapporteur's Report, Section 5.
11. White, D. E., Muffler, L. J. P., and Truesdell, A. H. 1971. Vapor-dominated hydrothermal systems compared with hot-water systems. *Economic Geology* 66:75-97.

26 Solar Energy— Prospects for Its Large-Scale Use

Peter E. Glaser

Peter E. Glaser is vice president of engineering sciences at Arthur D. Little, Inc., Cambridge, Massachusetts. He has served as president of the International Solar Energy Society and is currently the editor-in-chief of the Journal of Solar Energy.

Our present efforts to meet the burgeoning demand for energy are proceeding along three paths: increasing the efficiency of fuel plants, developing advanced nuclear power plants, and attempting to control the nuclear fusion reaction. Each line of investigation is both encouraged and constrained by the need to reduce environmental degradation. So are other less advanced efforts, such as the use of geothermal energy or solar energy.

The two desired qualities, abundant energy and minimal side effects, indicate that we cannot ignore our primary energy source—the sun, which has sustained life and led to the complex interrelationships of the earth's ecology. We have always depended on this most abundant energy source, but only for its natural function of providing the earth with a hospitable biosphere through the processes of atmospheric, hydrologic, and oceanic circulation and through photosynthesis.

The sun's energy is low in density (we receive about one kilowatt per square meter at the earth's surface), and it is subject to diurnal, weather, and geographical availability. Nevertheless, the sun provides 178 trillion kilowatts to the earth—about one-half million times greater than the present U. S. electrical power generating capacity, 5000 times greater than the world's geothermal capacity, and 60,000 times greater than the total tidal energy. We have not tapped more than a minute fraction of this colossal reservoir of energy, even though solar energy is available everywhere, is free of pollution, and costs nothing to supply or distribute.

Solar energy devices if deployed on earth would occupy large areas, which might be a significant diversion of land from alternative uses. The area required depends upon the efficiency of conversion and the quantity of solar energy available in a specific location. If a device with a conversion efficiency of 10 percent were available, one acre of a typical site would produce one million kilowatt hours per year, enough power to supply 200 typical American households. If the 3.2 million acres of land disrupted by strip mining were covered with similarly efficient solar-energy conversion devices, about one million megawatts could be produced when the sun is shining. Thus, the disadvantage of diverting land to solar energy collection would have to be weighed against the benefits to be gained by drawing on this source of energy.

Even if the world's population were 18 billion (a pessimistic projection for the year 2030) and per capita energy consumption were equal to that presently experienced in the United States, the consumption of solar energy would still be very small, only about 0.2 percent of the total solar energy available. The diversion of this small percentage of solar energy to meet human needs would appear to represent little environmental risk. The heat loads imposed on the earth do not increase through the use of solar energy. The concentration of energy consumption, and consequent heat production in large urban centers with high consumption per unit area, may have deleterious local effects, but this problem exists no matter what energy source is employed.

The factors that adversely affect the practicality and cost of the use of solar energy are its high variability because of regular hourly, daily, and seasonal changes and irregular cloud-cover variations. Although these problems can be surmounted through the use of standby power, energy storage schemes, and the like, the task will be difficult. The challenge is to find the means to utilize solar energy on a large scale so that a significant portion of energy demands can be met. This can be accomplished either by constructing large solar power stations with a very high electrical power output or by deploying a large number of individual units to significantly decrease our dependence on other energy sources.

Since ancient times, man has tried to take advantage of the

Above, Artist's rendering of a satellite solar-power station. The diagram below shows a similar station designed to produce 10,000 megawatts of electricity, enough to meet the demands of a city such as New York in the year 2000.

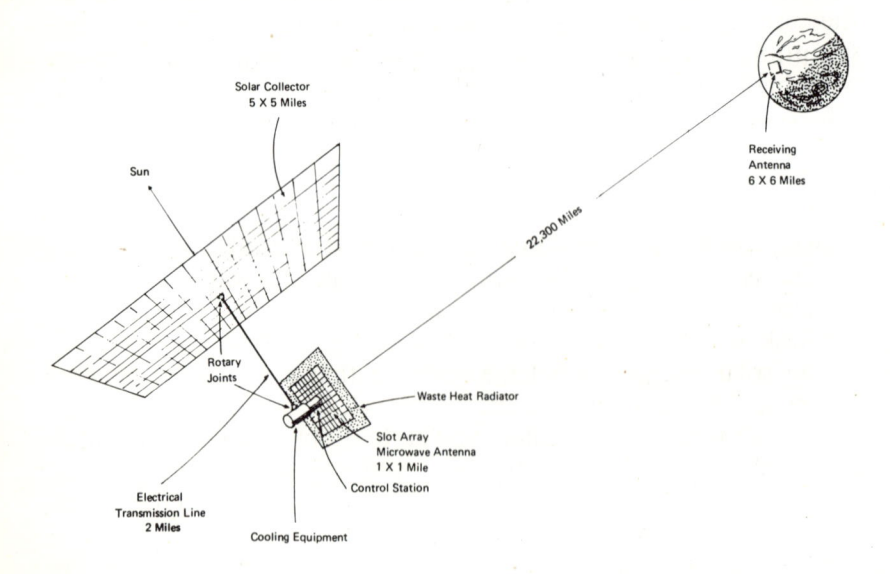

sun's energy. More recently he has tried to convert solar energy directly to useful power, one example being the photovoltaic cell. He has also tried the indirect approach—using the sun's heat to dry foods and chemicals and to heat houses and water. He has always been fascinated with concentrating the sun's rays by lenses or mirrors to achieve high temperatures for use in the laboratory or to generate steam to drive engines.

By and large, however, solar energy has been dismissed as being a less feasible means of generating power than other methods. This judgment was understandable in light of the technology available a decade or so ago, but circumstances now dictate that we reappraise this view.

The sun's radiation is readily convertible to heat; one need only provide a surface on which the solar energy can be absorbed. If a fluid such as air or water is then brought in contact with the heated surface, the energy can be transferred into the fluid and subsequently utilized for practical purposes.

This principle is successfully applied in several million domestic hot-water heaters in use in a dozen countries, including Australia, Israel, and Japan. A solar hot-water heater usually consists of a blackened sheet of metal or plastic in a shallow glass or plastic-covered box occupying 10 to 50 square feet of roof area. Water circulates through tubing attached to the surface of the blackened sheet and picks up the heat. The warmed water is then stored in an insulated tank. The tank can be supplied, if desired, with auxiliary heat from an electric heating element or other heat source.

The principle employed in the solar hot-water heaters can be applied to heating homes—all that is required are more or larger solar-heating panels. Enough heat can be absorbed in the circulating water to provide most of the heating requirements of houses in reasonably sunny climates. Of course, the storage tank must be large enough to provide carry-over capacity during short periods of cloudy weather. Conventional energy sources can be used to supplement the solar heater. Instead of water, air can be used as the heat-transfer medium; it is delivered by an air distribution system either directly throughout the house or to a heat storage tank where the heat from the air is transferred to the water or to stones. Reversal of the air flow during cloudy weather allows heat to be taken from this heat reservoir.

In the past 25 years a number of houses and laboratory buildings in the United States, Australia, and Japan have been heated at least partially with solar energy. Most of these installations have been technically successful, and extensive performance data have been obtained on various modifications of air- or water-heating systems.

In most U. S. locations residential heating with solar energy would be somewhat more costly than present conventional means because of the relatively high cost of equipment still under development. If produced on a large scale, however, solar house-heating systems could be competitive with fossil fuel systems, particularly when all of the hidden environmental costs are accounted for. An economic study of residential solar house-heating indicates that, in a least-cost system, about one-half of the total heat supply should be provided by solar energy.

Energy savings achieved through heating homes with solar energy can have a significant impact on our energy consumption, because the energy requirements of the residential housing are nearly 30 percent of total energy consumption. If the solar heating system were designed to supply a larger percentage of the total heat requirement, average costs would be quite high because the cost of solar heating is almost all in fixed capital investment. Because 100 percent of solar heating is not a rational objective, a combination of solar and conventional heat has to be employed. The optimum combination of solar and electric heat, for example, will be that which is equivalent in cost to the cost of electrical power over a 20-year period.

The least-cost, house-heating system will be applicable to the Southwest and Far West. Higher cost systems will be useful at more northerly latitudes (e.g., New York) and in more humid climates. The Southeast (e.g., Florida) is a poor site for solar house heating because the heating requirements are too low to justify such equipment.

Residential cooling systems that rely on absorption-refrigeration cycles will require more technical development before they can use solar-heated water or air from a roof-mounted collector. However, the solar collector would be the same unit used to heat residences, and the cooling unit would be a somewhat more expensive version of the conventional heat-operated air conditioner. The inherent advantage of solar cooling is that the maximum requirement coincides roughly with the time when the maximum amount of energy is available to operate the system. In addition, the solar collector, which is the most expensive portion of the system, can be employed nearly year-round if cooling is combined with solar heating.

The development of the devices required to cool and heat houses with solar energy has reached the stage where these devices could be available in less than 10 years if a mass market is realized. In mass use, solar heating and cooling devices could easily reduce consumptiom of conventionally generated electricity by 10 percent. Such a reduction in electrical usage normally translates into a 30 percent reduction in the energy requirements at a power plant. In the year 2000, such a reduction would mean about 120,000 fewer megawatts that our power plants would have to generate.

The primary advantage of converting solar energy to power is the inherent absence of virtually all of the undesirable environmental conditions ascribed to present and anticipated means of power generation with fossil or nuclear fuels. The interest in producing power with solar energy goes back at least a century. A solar-powered steam engine was a central attraction at the Paris Exposition of 1878. More recently, several imaginative concepts have been proposed for the large-scale utilization of solar energy on earth. Which of these approaches will be the most feasible alternative to present power-generation methods remains to be established. The important fact is that most of them are based on existing technology and well-known physical principles.

Solar energy conversion to electricity on the surface of the earth is intriguing. For example, a 10 percent solar-energy-conversion efficiency would produce 180,000 kilowatts per square mile while the sun is shining.

One possible approach is based on photovoltaic energy conversion employing solar cells. This method relies on the direct conversion of photon energy to electricity in a semiconductor crystal such as silicon. The state of the art of solar cells, based on single-crystal silicon, is well understood, and conversion efficiencies of about 10 percent are reached routinely. The major drawback is that present solar cells are prohibitively expensive because they have been developed primarily to meet the stringent requirements imposed by their application in various spacecraft missions.

New designs for solar cells include microcrystalline cells based on single-crystal solar cell principles or based on organic compounds which exhibit characteristic semiconductor properties. Such designs will likely be developed for mass production possibilities and the consequent cost savings. Or costs may be reduced by partially concentrating solar radiation by means of mirrors to minimize the number of solar cells.

Another approach would make use of the greenhouse effect. This approach involves the use of thin layers of films that selectively absorb radiation. Such films absorb most of the incident solar radiation but emit very little thermal radiation. In principle it should be possible to reach temperatures in excess of 1000°F if the solar collector is kept under a vacuum.

At present, those thin films that can withstand high temperatures for extended periods have too low an absorptance-to-emittance ratio, less than 20. For efficient conversion the absorptance-to-emittance ratio should be about 40, indicating that an optical concentration of 2 to 4 will be required. A solar power-generating system based on this greenhouse effect has been proposed for the Southwest. For example, a land area of about six square miles covered by such solar collectors would provide the equivalent power output of a conventional 1000-megawatt station.

The location of a solar power station would be crucial. Even in Arizona, January normally has 9 days in which cloud cover is 0.8 or greater, 7 days in which it is between 0.3 and 0.8, and only 15 days in which it is 0.3 or less. This amount of cloud cover would greatly reduce the solar-energy conversion capability. Thus, energy storage for night periods and overcast days would be required. Whether this would be pumped-water storage, storage in a heat transfer medium or in the form of hydrogen and oxygen produced by electrolysis would have to be carefully evaluated. Land costs, power substations, and transmission lines will substantially affect the economics of operation. An alternative would be a location where solar energy would be available 24 hours a day. This implies that the solar power station should be located in orbit, outside the atmosphere as an earth satellite.

The successful missions of unmanned spacecraft in earth orbit and to other planets have demonstrated the feasibility of direct conversion of solar energy in space to produce power in the spacecraft and telemeter signals back to earth. On the basis of this experience and the further advances in space technology which can be projected, the concept of a system of satellite solar-power stations has been advanced. Suggestions to use satellite collectors to concentrate solar energy for use on earth are not new, but the technology now available makes such suggestions practical.

A satellite, when placed in synchronous orbit around the earth's equator, will be exposed to solar energy for 24 hours a day except for short periods near the equinoxes. Lightweight solar cells forming a solar collector of substantial area to produce electricity can be used to power microwave generators. Microwaves in a selected wavelength band can be beamed by an antenna to a receiving station on earth with no significant ionospheric or atmospheric absorption. The receiving station on earth, consisting of a microwave rectifying antenna, converts the microwave to DC electricity which is supplied to the electricity users. A network of such satellite solar-power stations could generate enough power to meet all foreseeable U. S. energy production requirements.

The primary advantage of this approach is that the inefficiencies of the conversion process can be tolerated in outer space because the energy source is inexhaustible. On earth, microwaves could be converted to electrical power with efficiencies of about 90 percent, thereby reducing thermal pollution by a factor of five compared to any other power plants based on any known thermodynamic process.

The power density of the microwave beam would range from one-tenth of the density of solar radiation received on earth at the fringes of the receiving antenna to an amount nearly equal to the solar radiation density in the middle of the array. Beyond the confines of the receiving arrays the beam density would drop to very low values. Since the position of the microwave beam with respect to the receiving array can be carefully controlled, the beam should not represent a health hazard based on presently known biological effects.

A satellite solar-power station of the size envisaged would be capable of producing 10,000 megawatts, enough to meet the demands of a city such as New York in the year 2000. To place such a large satellite in orbit would require an earth-to-orbit transportation system with reusable boosters and space tugs to convey the satellite components to synchronous orbit for assembly.

To develop a system of satellites which could meet future U. S. and eventually world needs will require the application of systems engineering and management techniques which have already been proven in the undertakings associated with the space program. Although the task appears to be immense, so are the opportunities if we can succeed.

We appear to be at the threshold of a new era of energy production. Before we commit ourselves to pursue any one direction, solar energy, along with its potential large-scale application relative to other technically feasible energy production methods capable of meeting future needs, should be evaluated. This evaluation should involve the definition of standards, criteria, and procedures for analysis on technical, economic, social, and political grounds. With such an evaluation, the steps we take over the coming decades can be made consistent with our national goals and broader national purposes.

We can also envisage the possibility of forming a global partnership to develop solar-energy applications to benefit all peoples. Such a partnership might very well mean the realization of the goal so aptly stated by U Thant: "to curb the arms race, to improve the human environment, to defuse the population explosion, and to supply the required momentum to development efforts before the problems facing the world today will have reached such staggering proportions that they will be beyond our capacity to control."[1]

References

Daniels, F. 1964. *Direct use of the sun's energy*. New Haven, Conn.: Yale University Press.
Glaser, P. E. 1968. Power from the sun: its future. *Science* 162:857-61.
———. 1970. Power without pollution. *Journal of Microwave Power* 5 (4). Special Issue on Satellite Solar Power Station and Microwave Transmission to Earth.

---

[1] Quoted by J. Reston, *The New York Times*, October 23, 1970, p. 39M.

27 Soil

Paul B. Sears

Paul B. Sears has served on the faculties of Ohio State University, University of Nebraska, Oklahoma State University, Oberlin College, and as Director of the Yale Conservation Program, 1950-60, his last academic appointment. He is past president of several scientific societies and an honorary president of the National Audubon Society. His present address is Las Milpas, Taos, New Mexico 87571.

> So cream that was richest
> And meat that was rare
> Were free to them always
> As water and air.
> *Palmer Cox*

So went the final lines of a rhyme for children, written less than a lifetime ago. The plot, which involved animals, need not concern us here, but the comparison does. Water is no longer free, while anyone who has gasped and choked in the atmosphere of some of our great urban and industrial centers should know that it will cost money, and a great deal of it, to give us all pure fresh air to breathe.

There is another old expression "cheap as dirt" which is beginning to sound quaint and out of fashion. The word "dirt" does not here refer to filth, of which we have more than ever, some of the worst of it invisible. Instead it means the soil that covers the ground and which the farmer tills. It has become a costly commodity. True, the highest prices are not for the soil itself, but for the space it occupies. But when the space alone is purchased, it is for such structures as highways and buildings that end its usefulness as soil. In the United States not less than a million acres a year are lost in this way to the production of crops. In Holland the amount of land so painfully reclaimed from the shallow sea margin is barely enough to equal that which is required for purposes other than farming.

One cannot have a sound perspective on human history if he ignores the role that soil has played. Nor can he properly interpret today's problems, domestic and international, without taking soil into consideration. Or better, we should speak of soils, for soil exists in great variety, and it is this variety which has been so significant for plants, animals, and man. Yet the relationship is a curiously intricate one, for without living organisms there would be no soils. The stuff that covers the earth becomes soil only by being lived in and upon. It comes about by the interaction of air, water, and life upon rock-stuff, the energy for this complex operation being furnished by the sun.

Rock materials which form the matrix of soils differ greatly from place to place, both in texture and chemical make-up. Climates likewise vary, and the resulting patterns of moisture, temperature, and light influence changes in the exposed rock materials as well as the activities of living organisms. The several sciences, physics, chemistry, geology, and biology, have all been needed to help us find out what is in the various soils and what is going on in them. The result has been an elaborate catalog of types. Rock particles range in size from coarse gravels through sands, silts and to impalpably fine clays, mixed in various proportions. The kinds and amounts of organisms and dead organic material likewise differ, as does the content of air and water. Furthermore—and this is most important—soils differ in the amount and kind of materials that can be dissolved in soil water and so be available for plant nutrition.

A huge catalog of this kind of information about soils was built up during the nineteenth century, yet something was needed to bring order out of this knowledge. For one thing, the soils of Western Europe, then the center of scientific activity, had been greatly changed from their original condition by centuries of heavy agricultural pressure. This fortunately was less true in the vast expanses of Russia and the United States, where natural or virgin soils were still to be found.

In Russia, since the days of Peter the Great, it had been

customary to sponsor men of high talent in the arts and sciences. Though few in number, they were of exceptional ability, and among them was one—later others—who turned his attention to a study of undisturbed natural soils. And as original scientific activity increased in the United States, such work got under way here.

Many great scientific advances seem surprisingly simple after they have been made. Soil science is no exception. Progress came by studying not just what soils are like but how they have been formed. Trenches or cuts extending down into the parent material give what is called a profile. This enables one to trace the changes that have transformed raw rock into fertile soil. If then, as has happened, profiles are examined in different regions whose climate is expressed by vegetation—forest, grassland, or arid steppe and desert—a significant pattern begins to emerge.

Let us take a few examples. Under forest cover, for example, one finds a dark rich layer seldom more than eight or ten inches thick. Below this is normally a light-colored mineral layer poor in such plant nutrients as lime and potash. Then comes a zone of transition to the parent rock. Under clearing and agriculture such soils are usually very productive until the thin black top layer disappears. Unless unusual care is taken, as it has been in Denmark, Britain and France, one is likely to find abandoned farms such as those of New England where once there was forest.

In a forest most of the organic material falls to the surface as leaves, twigs and dead trunks. There are few animals that burrow beneath the surface. Droppings and remains of deer, squirrels and birds accumulate on the forest floor, while insects, worms and organisms of decay are confined to the surface litter. This litter is moist and compact, favoring the production of organic acids by fermentation, a process similar to the making of sauerkraut and ensilage. These acids in turn are carried downward by percolating rain, dissolving out the alkaline nutrients in the layer just below. That is why this leached mineral layer is relatively infertile when at length it is laid bare by the destruction, through practices which fail to renew it, of the rich surface layer of leaf mold.

Roughly speaking, leafy forests are found where rainfall exceeds evaporation. Where the reverse is true, trees will not grow unless there is an extra supply of soil moisture, as from irrigation. As one moves from forest to successively drier conditions, he will first encounter the tall-grass prairie of subhumid climates; passing further towards the continental interior he will find the grasses growing shorter and the vegetation sparser; assuming that he does not encounter mountains on his way, he will next meet with thorny, wide-spaced woody scrub such as one finds in driving through northern Mexico. Beyond this he will pass by degrees into true desert, where evaporation far exceeds rainfall. (Some find this statement confusing. It simply means that the capacity of the air to remove moisture is much greater than the amount of water available.)

So much for the vegetation and climate. What of the soils passed over on such a journey? Those of the tall-grass prairie or subhumid steppe are in dramatic contrast with that formed under the adjoining forest. Instead of less than a foot of dark humus, they exhibit three to five feet or more of rich black topmost layer. Below this instead of a leached zone, there will be a transition to the parent material, often containing rather than lacking nutrient minerals.

The deep organic layer of the prairie is densely occupied by the fine fibrous roots of grasses and owes its character to their activity while alive, the products of their decay after they die. In this zone are also the storage organs of other plants, the burrows and tunnels of countless animals, large and small, and the abode of a rich variety of microorganisms. Both in prairie and forest food is manufactured in the sunlit leaves. But in the forest much of the surplus is stored in great trunks and branches or eaten by browsers and tree dwellers. In the prairie much of this surplus is moved below ground and there stored or eaten and ultimately converted back for re-use by green plants. That which remains available above ground as tender leaves or nutrient seeds serves as food to the herds of grazing animals and flocks of birds which in turn support a population of meat eaters. But ultimately the wastes and remains of animal life are returned to the soil and used again.

In addition there are hosts of microorganisms, some of which, sustained either on organic wastes or on the roots of the many legumes (clovers, beans and the like), draw nitrogen from the air and make it available for the building of plant proteins. Small wonder then that the prairie or moist steppe regions of the world have become its most fertile source of human food when placed under cultivation. The orderly cycles of nature with their beautiful economy of accumulation and re-use through the ages that native vegetation and wildlife have been at work have made this possible. If the United States were suddenly deprived of its prairie soils, our civilization would be profoundly altered, if indeed it did not collapse.

This is equally true of other continents, the sub-humid steppes of Russia, for example, and those of the Argentine, as the ruinous meddling of Peron demonstrated recently. The true wealth of this country lies in its grasslands with their rich black soil. But mesmerized by the thought that the secret of prosperity lay in urbanization and industry, Peron diverted capital and energy away from the land. The result was, as we know, catastrophic.

Because prairie soils are so productive, remnants of their original cover of plant and animal life are much scarcer today in our Midwest than are forest preserves farther east. Such as remain should be protected jealously, both as living museums and for their aesthetic beauty. They are not, as we might suppose, monotonous stretches of grass, but carpets of infinite variety whose colors change throughout the seasons with the growth, flowering, and fruiting of many kinds of herbaceous plants.

In this rich variety lies the secret of that resilience which has enabled the prairie to flourish through the millennia in an environment too harsh to permit the growth of trees. This environment is marked by sudden changes of temperature, daily and seasonal, by strong winds and hot sun. Drought years tend to come in groups, interspersed between groups of moist years. Genial spring weather may be followed by biting frost. To all of this the great grasslands have become adjusted, as they have to the moving herds of grazing animals, and even to fire. For fires were set by lightning long before the invasion of man.

Curiously, grassland completely protected from fire and grazing deteriorates. Dead stems and leaves become matted on the surface, locking up needed nutrients and choking the new growth needed to maintain vigor and production. These remnants above ground do not decay rapidly as do fallen leaves and twigs in the more humid air of the forest. In short, the natural prairie is a beautiful example of a system organized for permanence in the face of recurrent hazards, even fire, and a model to man for intelligent use of the landscape.

In the drier, short-grass steppes which we call the High Plains in North America growth is far less luxuriant. The mat of roots penetrates less deeply than in the prairie. Food production is less, yet remarkably dependable, sustaining animal life both above and below ground. The organic layer is shallow, grey or brown rather than black since there is no great surplus of humus. Whenever the infrequent rains moisten the soil, the subsequent evaporation serves to bring up soluble plant nutrients, making the soil potentially fertile. Only the scarcity of water limits production. If these soils happen to be ploughed up and planted during a time of better than average rainfall, they produce good crops of wheat at low cost. But our cereal crops do not withstand the inevitable groups of dry years. These bring crop failure and destruction of the thin soil by wind, with resultant distress to man. The continued attempts to exploit the short-grass, semi-arid, lands have brought several periods of such distress, while their heavy and economical production of wheat during normal years is a major source of our present costly surplus. Yet this land is admirably adapted to extensive grazing and even permits the growing of forage in selected places. Some wheat growers who have returned to livestock production are doing very well. Our agriculture and general economy might benefit if more did so, but so long as every state has two senators and profits are high during good years, a miracle would be required to bring about such a change through political means.

In desert soils there is practically no production of humus by decay. Plant and animal remains dry out on the surface. Yet here, to an even greater degree than in the semi-arid grassland, nutrient salts are concentrated at the surface by evaporation. Except in low alkaline spots where these salts are too concentrated, desert soil can be highly productive if supplied with water. But irrigation has to be conducted with great skill, lest it result in too great an increase of mineral salts.

This is not the only hazard. When, as often happens, erosion from upstream loads the irrigation channels with silt, fields may be converted into marsh and swamp. This has taken place in the region around Albuquerque, where engineers have become occupied with flushing and draining the soil to keep it reasonably productive. And as one travels north from that city he sees a panorama of arroyos that have been cut into the foothills draining into the upper Rio Grande and contributing their silt. Even if the silt does not create marshy conditions, it may vastly increase the labor and cost of production by clogging the ditches that carry water to the fields. In Mesopotamia this task had become oppressive long before the irrigation works were destroyed by invaders after some thousands of years of use.

The reckless expansion of irrigation at great public expense in a time of food surplus makes heavy demands upon water in regions where it is scarce, increases competition for farmers with heavy costs elsewhere, and has at times resulted in failure from soil deterioration. In discussing a large appropriation to extend irrigation, Senator Frank J. Lausche from Ohio noted that we are now paying farmers as much as $50 an acre to limit production, while at the same time we propose to expend $500 to $2000 an acre on land priced at $7 to bring it into production. His objection was overruled.

Each of the great climatic types of soil owes its character to the orderly economy of nature in which energy is channelled and materials recycled—a process which affords the model for any permanent human activity, a model too often ignored. Within each climatic type there are many variations due to such factors as slope, exposure, drainage, and nature of parent material. This last is important, as we have seen, for if necessary chemical materials are scarce or absent, they will limit production.

Doubtless in nature, before the day of fences and other man-made barriers, these local differences were offset to some degree by the horizontal movement of animals. For animals have a notable ability to make up mineral deficiencies by visiting saltlicks and mineral springs, or migrating to favorable feeding grounds. In this way their droppings and remains have tended to equalize soil differences to some extent.

There are places where this process has gone on through thousands of years. Often around and about them may be found the bones of animals long extinct, as in the bonelicks of Kentucky. Here the rich accumulation of organic matter was exploited for fertilizer before its scientific record could be properly cataloged and interpreted.

Recurring floods and their residues dropped in valley floors have likewise helped to distribute nutrient minerals and organic matter. The Nile is the best known but by no means the only example of this function and it has sustained an intensive human economy for thousands of years. Our present policy of using floodplains for urban and industrial expansion, safeguarded by the building of dykes and the preemption of productive valley land for reservoirs might well be reexamined in the light of natural and human history.

Man is sometimes referred to as a Pleistocene mammal, which means that his existence has been coextensive with the glacial, or Pleistocene Age—approximately one million years. This fact has influenced his activity, even his modern economy, in more ways than we realize. As to soil specifically, the moving glaciers have presented him with mineral material from remote areas. In New England, for example, the ice masses have brought in the weathered products or granites and sandstones containing little of the precious calcium so necessary to a prosperous agriculture. Once the natural soils had been exhausted or washed away as a result of cultivation, New Englanders have had little choice but to give way or invest heavily in fertilizer. Two-thirds of this area was in farms in the early 1800s. Today an equal amount is back in inferior second- or third-growth forest, since the best seed-trees have long since been harvested.

Further west much of the mineral material brought in by the glaciers was better provided with plant nutrients. When the late Louis Bromfield established his farm in Ohio on glacial soil, he took over "worn-out" land whose fertility had apparently been exhausted. He was obliged to restore some of the missing materials by fertilizing. But he made use of deep-rooting crops. Presently these began to bring up the needed mineral elements from the rich storehouse of glacial till below. Then instead of selling off everything he grew, he returned as much of it to the land as possible, patterning his operations after the recycling economy of nature which we have described. The sound practices which he demonstrated have had a wide influence in Ohio and elsewhere. Invidious tongues have complained that his farm was not a money-maker, ignoring the fact that he was an artist, not an accountant, and that throughout his life he lavished baronial hospitality. He was a master farmer who understood the model that nature has given us and which we so generally choose to ignore.

Much has been said about using the tropics to feed the world's expanding population. A short answer is that had it been possible, it would have been done by this time. Here again, the answer lies in the soil. In the moist tropics there is ample water, while temperatures are suitable to luxuriant plant growth. But the tropical forest is about as different as can be from a grain field. It is maintained by an extremely rapid and efficient turnover of a minimum amount of mineral nutrients, under a protecting canopy of many species which shields the ground from driving rain and searing sunshine.

Once a tropical soil is exposed to open cultivation, it is doomed. The combination of heat and heavy rain leaches and oxidizes the nutrient materials. Only dense growths of jungle or useless grasses can take over, until by a long process the forest is restored. In Indonesia, for example, former forestland has been taken over by a dense growth of coarse grass, useful only to furnish fuel. Where the tropics are being used effectively it is by growing tree-crops, or other crops with sheltering trees, as in the coffee plantations of El Salvador, shaded by the well-named Madre de Cacao.

To repeat, the great climatic zones have stamped their character on the soils formed within them. Occasionally, by luck, we find bits of fossil soils preserved and so are able to deduce climatic conditions during times long past. And since climates have shifted measurably, even within the past ten thousand years, we occasionally find soils that are changing their character under the impact of climatic change. This is notably true in the wavering borderline between forest and prairie in the middle states.

Yet most farmers know that even on their limited acres the soils may vary from place to place and that these differences are important to proper use and management. Slight differences in elevation or greater differences in parent material may serve to modify the effects of climate, or even to mask it. So does the history of the soil itself. The chalk downs of England, though formed by the weathering of lime rock, may become deficient in lime through generations of rainfall and pasturing. In time, even the rich phosphate soils of the Blue Grass region or the dairy lands of Wisconsin may have to be restored with phosphates from Florida, to replace that which has gone to market in the form of animals and animal products. There is no escape from the relentless principle that, to keep land productive, materials must either be recycled or restored.

Coal, gas and oil are the legacy to us of past sunshine, stored below the surface of the earth. Soil, the dark carpet upon the earth, is a similar legacy and a more precious one because it is more important that we eat than delegate our work to machines. The generations of plants and animals that form the soil and are sustained by it during their brief existence pass on. What remains, as evidence of their activity in catching and transforming the sun's energy, is the organized soil. It is vulnerable to our blundering, immensely rewarding to good husbandry.

28 Inventorying Soil Resources

F. L. Himes

Frank L. Himes is a professor in the Department of Agronomy at Ohio State University.

There may be almost as many different kinds of soils as there are plants that live upon them. Each soil has a distinctive set of properties which are derived from the starting (parent) material and environment. Soil is more than disintegrated rock; during its formation, materials have been added and others have been removed. Plants and animals have added organic compounds and altered the physical properties of the parent material. Climate influences not only the rates of chemical weathering, leaching, and erosion, but also the type and rate of biological activities. Land relief and time of exposure are the other factors that account for variations among soils.

Why has man divided soils into groups? He has found that one soil is better suited to a particular use than are others. Light-colored soils are usually lower in fertility than are the associated dark-colored soils, but they do not require artificial drainage. Sandy soils have a lower water-holding capacity than clay soils. Light (sandy) soils are easier to cultivate than heavy (clay) soils. Very acid soils are suitable for certain plant species, but near neutral soils are best for most. Well-drained soils are more desirable than poorly drained ones for home construction, road construction, and playgrounds.

Making an inventory of the soils of an area includes locating the boundaries between soil types and measuring properties of samples from each soil. The boundaries between soils frequently

Figure 1. Lines may be drawn on a soil survey map to illustrate horizontal differences among soils. Brookston (Br) is a silty clay loam with high water table part of each year and poor natural drainage. Crosby (CrA) is a silt loam with drainage intermediate between Brookston and Miamian; land slope is 0 to 2 percent. Miamian (MlC3) is a clay loam; severely eroded; good natural drainage; land slope is 6 to 12 percent. (Source: USDA, Soil Conservation Service.)

coincide with changes in colors. These color changes within a field occur when changes in slope (relief) affect soil moisture and erosion. They are also influenced by the parent material (original mineral material) and vegetation or a combination of these. Climate and time, too, contribute greatly to horizontal differences among soils. In the humid climatic regions, soils with poor natural drainage have a darker surface color than that of the well-drained soils. This darker color results from the higher concentration of humus. Soils in depressions or with an impermeable layer are poorly drained. Drainage systems must be installed in these soils before they are satisfactory for growing many species of plants, or for building houses, constructing roads, and developing athletic fields. In arid regions salts, transported from the sloping areas by leaching and erosion, accumulate in the depressions, and the soils often have a white color.

The Soil Profile

A vertical section of soil, or profile, usually shows three layers known as horizons. The horizons can be identified by color changes, by texture, and/or by structure. The topsoil or surface mineral layer is the *A*-horizon, the subsoil is the *B*-horizon, and the parent material is the *C*-horizon.

The *A*-horizon is the layer of soil that supports plant and animal life. The *A*-horizon tends to be sandy or silty if the clay has been transported into the *B*-horizon. Soluble mineral matter likewise leaches downward. Organic material or *humus* from

Figure 2. Vertical soil differences are the result of unequal rates of addition and removal of materials to horizons. Above, Bearden, very fine, sandy loam, showing a fairly deep dark surface soil, brownish upper Bearden horizon, and light gray zone of lime enrichment in the lower Bearden horizon. (Source: USDA, Soil Conservation Service.)

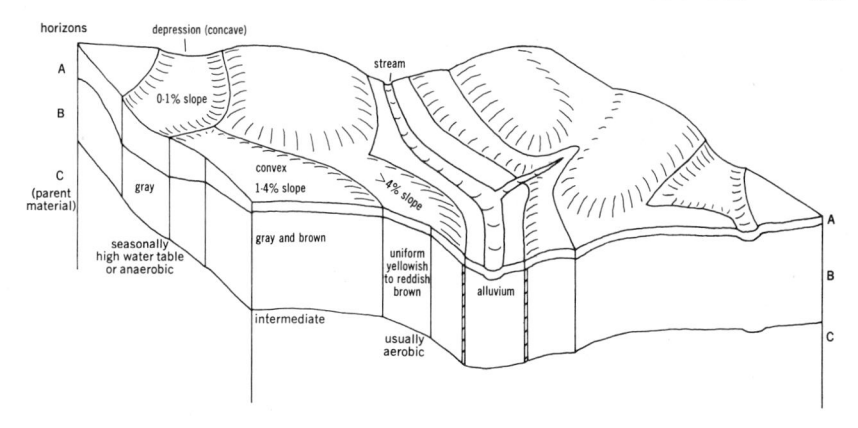

Figure 3. Landscape in a humid, temperate region where forest was the natural cover.

decayed plants and animals makes this layer darker than the *B*-horizon below it.

The internal movement of water and the oxidation conditions of soils in humid regions can be determined by observing the color of the iron oxides in the subsoil or *B*-horizon. Soils with rapid internal water movement have a uniform yellowish-brown to reddish-brown color in the *B*-horizons; those with slow internal water movement have uniform gray colors. Soils between these extremes have mottled (yellow and gray) *B*-horizons. The poorly drained soils also have higher concentrations of organic matter to greater depths than have the better drained soils.

The *B*-horizons of most soils are zones of accumulation. Soils formed under acid conditions have an accumulation of clay and/or iron and aluminum oxides. The accumulation of clay can be detected by feel, as will be described. The increased concentration of iron oxides can be observed by the increased brightness of color. In arid and subhumid regions, calcium carbonate accumulates in the *B*-horizon. This occurs as a white layer and varies from a soft and discontinuous layer to the very hard and continuous crust known as caliche. The accumulation of materials in the *B*-horizon can decrease the rate of water movement, restrict root growth, and increase the energy required to dig holes and foundations. The decrease in water transmission must be considered when designing drainage fields for septic tanks.

The *C*-horizon is the parent material. The parent material

Figure 4. Sites for housing and other uses should be inspected during a wet season. (Source: USDA, Soil Conservation Service.)

may be bedrock (residual) or unconsolidated material. The latter has usually been transported to the site by water, ice, wind, or gravity. At soil age zero, the surface layer was similar to the *C*-horizon. During soil formation, the compositions of the upper layers changed because of various removals and additions. The *C*-horizon contains the material that has undergone little or no change during the development of the *A*- and *B*-horizons.

A soil sample can also be inventoried for physical, chemical, and biological properties. The procedures for determining these properties vary from qualitative to quantitative. Some procedures will be briefly described.

Physical Properties

Soil particles are graded by size (from coarse to fine) as sand, silt, or clay. Since most soils are mixtures, the *texture* of a soil is determined by the relative proportions of these particles. On the basis of texture, soils are divided into large classes: sands, loams, silts, and clays. Soil scientists want to know the textural class of a soil before making a management recommendation.

The textural class can be determined qualitatively by feel. The sample is moistened to the consistency of putty and forced into a thin ribbon between the thumb and forefinger. If the soil has more than 40 percent clay, it will form a long, thin ribbon and is classified as clay. Soils with 27 to 40 percent clay will form shorter ribbons and are called clay loams. Soils containing less than 27 percent clay will not form a thin ribbon and are called loams. Each of these classes can be subdivided; for example, into sandy loam and silt loam. Quantitative methods are described in references 1 and 3.

The arrangement of the soil particles influences the rate of water infiltration and percolation, the movement of gases, and root development. Soil *structure* refers to the arrangement of soil particles into aggregates. The size and arrangement of soil particles and aggregates are important in determining the size and shape of pore spaces. Pores may contain air or water or both. Granular (spherical) structure is preferred for root development, water movement, and water retention. The large pores between aggregates permit rapid movement of water, and the small pores within the granules retain water for later use by the plants.

A soil crust is a thin, close-packed structure on the soil surface which greatly restricts the infiltration of water and the diffusion of gases to lower layers of the soil. Clay pans, fragipans, and caliche occur below the surface (usually in the B-horizon) and restrict water movement and root penetration. Another structure that retards water and root penetration is platy; platy structural units overlap like shingles.

Common structural forms in the B-horizons are blocky, prismatic, and columnar. As the cross-sectional area of the aggregates decreases, the number of vertical pores per unit area increases. The rate of percolation and ease of root penetration increase with the number of vertical pores.

The *quantity of water* in a sample can be determined by weighing it before and after drying. Soil samples are usually dried at 105°C.

$$\text{Percent water} = \frac{\text{wt. of water}}{\text{wt. of dry soil}} \times 100$$

This laboratory procedure can be used to compare the moisture content of many soils (poorly drained vs. well drained, A-horizon vs. B-horizon) and the change in moisture content with time after the addition of water.

The *rates of water movement* can be determined as illustrated in Figure 1.

The rates of movement differ among soils, among horizons, and with past treatment. On sloping sites, water runoff occurs when the rate of rainfall exceeds the rate of infiltration.

In urban areas, construction sites are a major source of sediment in streams. There is increased erosion from such sites because the surface layer has been compacted, decreasing the rate of infiltration and increasing runoff.

The volume of the particles in a sample can be determined by displacement of water. *Particle densities* (D_p) do not vary much from mineral soil to mineral soil. *Bulk densities* (D_B) do vary greatly and are calculated by the following formula:

$$D_B = \frac{\text{wt. of dry sample}}{\text{vol. of sample (or vol. of pores + vol. of particles)}}$$

One common method for determining the bulk density is to force a sharp-edged, metal cylinder or can with both ends removed into the soil. After the soil is trimmed to the volume of the cylinder, it is dried and weighed, and the volume of the cylinder is calculated. Care must be taken to avoid compaction during collection of the sample.

Bulk densities not only vary among soils but also among horizons and with use. For example, compare a sample from a footpath with a sample taken 10 feet away. Most plant roots will not penetrate a layer if the bulk density is greater than 1.6 g/cm^3.

The percent *pore space* can be calculated with the following formula:

$$\text{Percent pore space} = 100\% - (D_B / D_p \times 100)$$

The surface layers of productive agricultural soils have approximately 50 percent pore space, or a bulk density of approximately 1.3 g/cm^3.

Chemical Properties

A useful and easy chemical determination is *soil acidity* or *alkalinity* as measured by pH. The pH of soil samples can be determined by the use of color indicators and pH meters. Variations in pH occur among soils, among horizons, and with past treatments. Soil pH is often changed by farmers and gardeners.

The optimum pH range for many species of plants is near neutral, that is, 6.2 to 7.0. Some of the favorite flowering shrubs (e.g., azaleas) grow best in the acidic range of pH 4.5 to 5.0.

Clay particles and organic colloids are negatively charged and attract the cations. The larger the *cation exchange capacity* (CEC), the greater buffering capacity, or ability to protect from major change in pH. Larger quantities of fertilizers and amendments can be more safely applied to soils with large CEC than to those with low CEC.

The cation exchange capacity of a soil must be known for efficient use of soil and fertilizers. One procedure for determining the CEC of soil is described below:

1. Weigh 10.0 g of soil into a 125 ml Erlenmeyer flask.
2. Add 50 ml of approximately 1 N HCL. Shake for 10 minutes.
3. Let the soil settle, then decant most of the supernatant liquid through a filter paper; discard the leachate; and repeat acid leaching.
4. Shake the soil with 50 ml of distilled water for one minute. Let the soil settle, and decant the supernatant liquid through the filter. Repeat this step two more times. Discard the leachate. Place filter paper and soil in flask.
5. Add 50 ml of 1 N barium acetate. Shake occasionally for five minutes. Let the soil settle and decant the superantant liquid through the filter, collecting the filtrate in a clean 250 ml suction flask. Repeat this step, collecting the filtrate in the same flask.
6. Add 20 ml of distilled water to the soil, shake, let the soil settle, and decant through a filter, collecting the filtrate in the same container as used in step 5.
7. Add a few drops of phenolphthalein indicator solution and titrate with NaOH. Record the volume and concentration of NaOH used on the data sheet and calculate the milliequivalents of exchangeable cations per 100 g of soil.

The concentration of *organic matter* can be determined by loss of weight. The organic matter can be oxidized with hydrogen peroxide or by heating to 550°C. Increasing the concentration of organic matter increases the CEC, the water-holding capacity, the biological activity, and the release of available nitrogen in soil. Organic matter will adsorb many pesticides. It is recommended that some pesticides not be added to soil with more than 5 percent organic matter because the pesticide will not efficiently control the pest.

Available phosphorus and potassium can be determined by the procedures described in soil analyses books and in soil test kits.

Many *minerals* of the sand fraction can be identified by examining the sand grains with a microscope. Mineral soils that shrink upon drying, developing large cracks, usually contain the clay mineral montmorillonite.

Biological Properties

The distribution of plant roots in the soil can be determined by carefully washing the soil particles from the roots. Roots of representative volumes can be compared by taking cores at various distances from the base of the plant. The roots are usually weighed after drying at 70°C. Small animal populations can be determined by using the Berlese funnel technique. The bacteria and fungi populations can be determined by suspending a sample of soil in sterile water, diluting, and adding to the proper agar mixture for culturing. The number and types of organisms vary with such factors as pH, aeration, temperature, and quantity of food.

Summary

Soil varies horizontally and vertically. The horizontal changes are associated with changes of slope, vegetation, parent material, climate, and time. The vertical differences on a soil profile are the results of unequal rates of addition and removal of materials to horizons. For efficient use of a unit area of soil, a manager needs to know the surface and internal water movement characteristics, the textural class, the structure, the bulk density, the pH, the cation exchange capacity, the percent organic matter, and the types of pathogenic organisms. The boundaries and properties of soil types in many counties can be obtained from the detailed Soil Survey Reports published by the Soil Conservation Service of the U. S. Department of Agriculture.

References

1. Black, C. A., ed. 1965. *Methods of soil analysis, parts 1 and 2.* Madison, Wisc.: American Society of Agronomy, Inc.

2. Buckman, H. O., and Brady, N. C. 1969. *The nature and properties of soil.* 7th ed. New York: Macmillan Company.
3. Himes, F. L. 1969. *Audio-tutorial notes.* Minneapolis: Burgess Publishing.
4. Millar, C. E., Turk, L. M., and Foth, H. D. 1966. *Fundamentals of soil science.* 4th ed. New York: John Wiley and Sons, Inc.

Additional Readings

American Petroleum Institute. 1967. *Petroleum facts and figures.* New York: American Petroleum Institute.

Brady, N. C., ed. 1967. *Agriculture and the quality of our environment.* American Association for the Advancement of Science Publication 85.

Brobst, D. A., and Pratt, W. P. 1973. *United States mineral resources.* U. S. Geological Survey Professional Paper 820.

Ciriacy-Wantrup. S. V., and Parsons, J. J., eds. 1967. *Natural resources-quality and quantity.* Berkeley, Calif.: University of California Press.

Cloud, P. E., Jr. 1968. Realities of mineral distribution. *Texas Quarterly* 11(2):103-26.

Committee on Resources and Man, Division of Earth Sciences (NAC/NRC). 1969. *Resources and man.* San Francisco: Freeman.

Cook, E. 1973. Energy for millenium three. *Technological Review* 75(2):16-23.

Dansereau, P., ed. 1970. *Challenge for survival: land, air, and water for man in Megalopolis.* New York: Columbia University Press.

Geological Survey of Alabama. 1971. *Environmental geology and hydrology, Madison County, Alabama, Meridianville Quadrangle.* Geological Survey of Alabama, Atlas Series 1.

Hammond, A., Metz, W., and Maugh, T., II. 1973. *Energy and the future.* Washington, D. C.: American Association for the Advancement of Science.

Heller, A., ed. 1972. *The California tomorrow plan.* Los Altos, Calif.: William Kaufman.

Hibbard, W. R., Jr. 1968. Mineral resources: challenge or threat? *Science* 160:143.

Holman, E. 1952. Our inexhaustible resources. *American Association of Petroleum Geologists Bulletin* 36(7):1323-29.

Hunt, C. B. 1972. *Geology of soils: their evolution, classification, and uses.* San Francisco: W. H. Freeman.

Kline, A. B., Jr. 1972. *The environmental and ecological forum, 1970-1971.* Oak Ridge, Tenn.: U. S. Atomic Energy Commission.

Landsberg, H. H. 1964. *Natural resources for U. S. growth: a look ahead to the year 2000.* Baltimore: Johns Hopkins Press.

Leith, C. K. 1935. Conservation of minerals. *Science* 82(2114):109-17.

McHarg, I. L. 1969. *Design with nature.* Garden City, N. Y.: Doubleday/Natural History Press.

McKelvey, V. E., and Singer, S. F. 1971. Conservation and the minerals industry—a public dilemma. *Geotimes* 16(12):21.

Millar, C. E., Turk, L. M., and Foth, H. D. 1958. *Fundamentals of soil science*. 3rd ed. New York: Wiley.

Murphy, J. J., ed. 1972. *Energy and public policy–1972*. Conference Board Report 575. New York: The Conference Board.

Nace, R. G. 1967. Water resources: a global problem with local roots. *Environmental Science and Technology* 1(7):550-60.

Office of Emergency Preparedness, Executive Office of the President. 1972. *The potential for energy conservation: a staff study*. Washington, D.C.: U. S. Government Printing Office.

Owen, O. S. 1971. *Natural resource conservation: an ecological approach*. New York: Macmillan.

Pederson, J. A., ed. 1973. Future energy outlook. *Quarterly of the Colorado School of Mines* 68(2).

Risser, H. E. 1973. *Energy supply problems for the 1970s and beyond*. Illinois Geological Survey Environmental Geology Note 62.

Schneider, W. J., and Spieker, A. M. 1969. *Water for the cities–the outlook*. U. S. Geological Survey Circular 601-A.

Scientific American. 1971. *Energy and power*. San Francisco: W. H. Freeman.

Wang, F. H. 1970. *Mineral resources of the sea*. New York: United Nations.

Weinberg, A. M. 1968. Raw materials unlimited. *Texas Quarterly* 11(2):90-102.

———. 1973. Some views of the energy crisis. *American Scientist* 61(1):59-60.

Winsche, W. E., Hoffman, K. C., and Salzano, F. J. 1973. Hydrogen: its future role in the nation's energy economy. *Science* 180:1325-32.

Part Seven

Man and the Finite Earth

Once a photograph of the earth, taken from the outside, is available—once the sheer isolation of the earth becomes plain—a new idea as powerful as any in history will be let loose.

Fred Hoyle, 1948
British astrophysicist

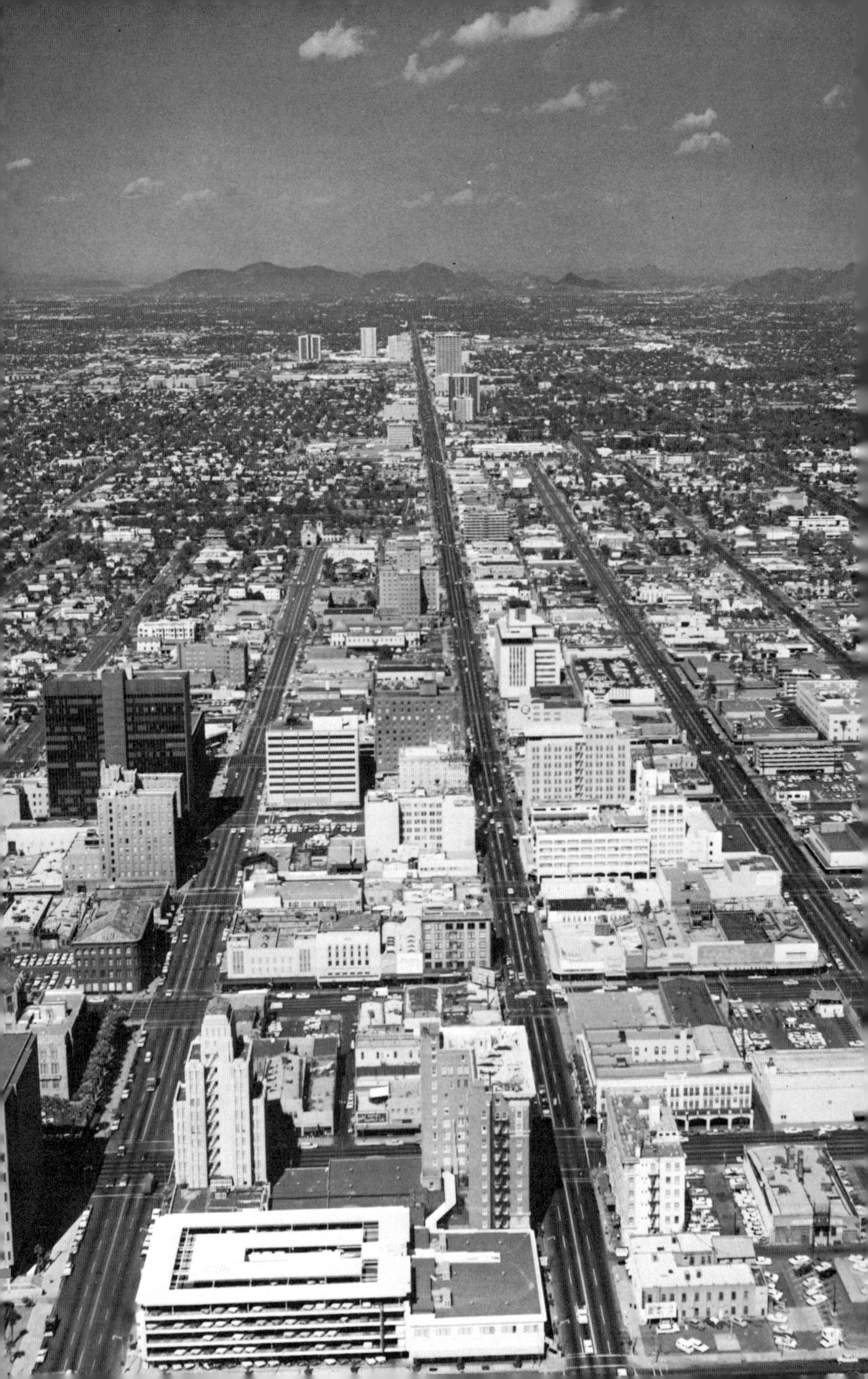

The finiteness of the earth became apparent for the first time to many earthbound observers as they followed the news of the Apollo astronauts landing on the moon. Photographs showing the isolated earth and the astronauts totally dependent upon their portable life-support systems and small spacecraft awakened many to the fact that the earth is indeed finite and that it contains man's only life-support systems. For the first time, many people realized that unlimited increase in population and consumption of resources (the latter partly the result of increased population and partly the result of increased affluence) could have disastrous effects on their spaceship earth.

The quality of life and the quality of the environment on earth are controlled by many factors. The interrelationship of these factors has been summarized by McKelvey in article number 23 with the following equation:

$$L = \frac{R \times E \times I}{P} \qquad \text{where,}$$

L is the average level of living;
R is the consumption of raw materials;
E is the consumption of energy;
I is the consumption of ingenuity (other authors use technology here);
P is the population that shares the total product.

Although it might be argued that population is an important resource and a continually expanding population makes a country richer, there is a point where increase in population reduces more options and individual freedoms than it creates. For example, the choices among recreation styles could become limited. As affluence in the United States continues to increase and there is more leisure time, the use of camping and outdoor recreation areas will increase. The effects of tin-tent camping have already had a major impact on camping regions, especially where the campers bring snowmobiles and motorcycles; and even backpacking areas have felt the impact of more people. In some wilderness areas, the number of hikers has been limited to 2000; more backpackers would ruin the experience and the wilderness. As the public observes the decreasing quality of wilderness recreation because of misuse and overuse, the finite aspect of the earth will continue to take on meaning.

To the portion of the population that enjoys outdoor recreation, the changes in camping and outdoor recreational facilities have already signaled that the earth may be approaching its population capacity. Many more people point to more important indicators, such as shortages of food and energy and the increased loss of life and property resulting from geologic hazards affecting people who were forced to utilize unsuitable land because of population pressure.

Average reproduction in the United States (2.08 children per family) is now slightly below the rate (2.10) needed for zero population growth. This lower birth rate has decreased concern over increasing population in this country, and it has been suggested jokingly that, if this rate were to continue for about 4000 years, there would be no one left in the country. However, the problem of accommodating increased population will continue for several decades, because there is still a large percentage of females in the population who have not yet reached childbearing age. This fact illustrates one long-term aspect of the population problem. If we decided that the population capacity of the world were 7 billion people—the population is now over 3.5 billion—how could we limit it to this number? What plan of action could nations use and what would be the consequences of a rapid change in the number of people in different age groups in a nation's population? Probably a more fundamental problem would be getting all nations to discuss the population question, in addition to other world problems!

The table on page 311, from Ehrlich and Ehrlich (1972), provides doubling times for populations based on different rates of increase. The current world rate of increase in population is 2 percent per year and by using the table, we can see that at this rate the population could reach 14 billion people before the middle of the next century. This is within the lifetime of children born today. The question of determining the earth's population capacity is one that cannot wait long for an answer.

In this section, the first article describes the possible future populations of the United States and, as is true with most projections, becomes out-of-date in a very short time. A tongue-

Annual percent increase	Doubling time (years)
0.5	140
0.8	87
1.0	70
2.0	35
3.0	24
4.0	17

From *Population, Resources, Environment: Issues in Human Ecology,* Second Edition, by Paul Ehrlich and Anne H. Ehrlich. W. H. Freeman and Company. Copyright © 1972.

in-cheek approach to future population and associated factors is provided in the next article by Brown; the final article is a draft declaration on the human environment from the United Nations. It shows the many factors that must be considered in solving environmental and other problems in a finite world. The perception of problems and possible solutions in a developed country are often quite different from those in a developing nation. In fact, within any one country, many different preferences of environments and life styles can be found.

In attempting to solve national and world problems, we will employ large-scale planning techniques including computer simulation of dynamic models of cities, countries, and the world. Technology and new knowledge in the physical sciences will be seen by many (the technological optimists) as the panacea for all our problems; however, others will realize that human attitudes and actions in building a world society will be at least as important. The question remains as to what forces will act to make the necessary commitment for improved technology and the necessary attitudinal changes toward our fellow space travelers.

29 Population

Council on Environmental Quality

Population is a critical environmental factor. Its growth contributes to most other environmental problems. Moreover, the density and distribution of population are intrinsically important in determining the quality of the environment.

The 1970 population of the United States was 204 million. This is double the population in 1920. At the current rate of growth, we will reach 300 million around the year 2008.

The large number of persons now in the child-bearing-age range makes continued population growth all but inevitable. The Commission on Population Growth and American Future reports:

> Even if immigration from abroad ceased and families had only two children on the average—just enough to replace themselves—our population would continue to grow until the year 2037, when it would be a third larger than it is now.

The effects of a three-child family versus a two-child family are shown in Figure 1.

The Commission has also commented on the difficulties of achieving zero population growth now:

> Our past rapid growth has given us so many young couples that they would have to limit their childbearing to an average of only about one child to produce the number of births consistent with immediate zero growth. Ten years from now, the population under 10 years old would be only 43% of what it now is, with disruptive effects on the school system and ultimately on the number of persons entering the labor force. Thereafter, a constant total pop-

MAN AND THE FINITE EARTH 313

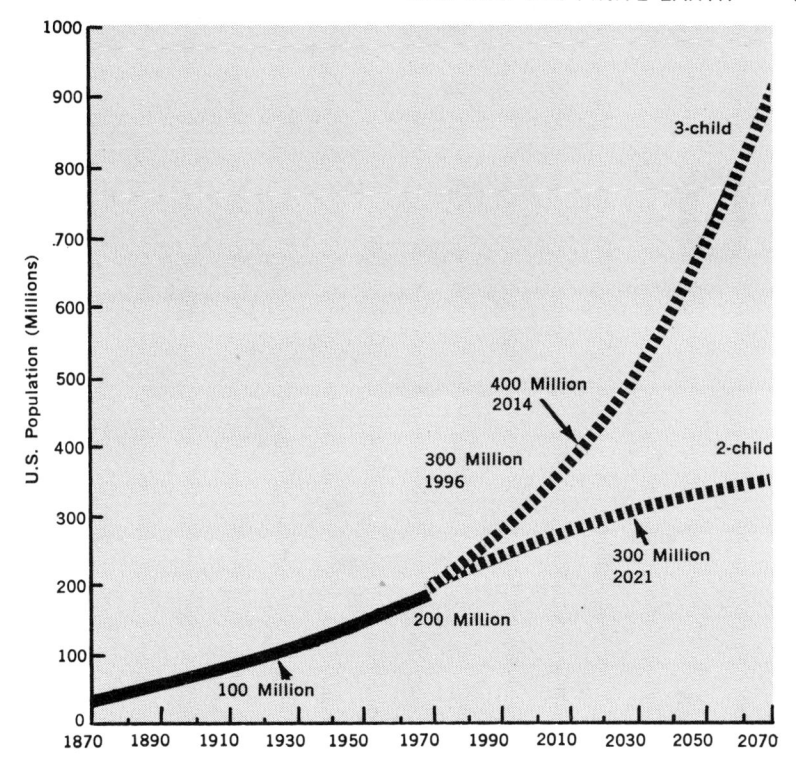

Source: Commission on Population Growth and the American Future.

Figure 1. Projected U. S. Population: Effects of 3-Child and 2-Child Families.

ulation could be maintained only if this small generation in turn had two children and their grandchildren had nearly three children on the average. And then the process would again have to reverse, so that the overall effect for many years would be that of an accordion-like mechanism requiring continuous expansion and contraction.

Although stabilization of population at the current levels does not seem possible, there is widespread recognition that population growth is a problem that must be explicitly considered in government planning and policy. It is particularly essential that population growth be considered with respect to its impact on the availability of natural resources.

30 After the Population Explosion

Harrison Brown

Harrison Brown is a professor of geochemistry, science, and government at the California Institute of Technology. He is also foreign secretary of the National Academy of Science in Washington, D. C.

At one time or another almost all of us have asked: How many human beings can the earth support? When this question is put to me, I find it necessary to respond with another question: In what kind of world are you willing to live? In the eyes of those who care about their environment, we have perhaps already passed the limits of growth. In the eyes of those who don't care how they live or what dangers they create for posterity, the limits of growth lie far ahead.

The populations of all biological species are limited by environmental factors, and man's is no exception. Food supplies and the presence of predators are of prime importance. When two rabbits of opposite sex are placed in a fenced-in field of grass, they will go forth and multiply, but the population will eventually be limited by the grass supply. If predators are placed in the field, the rabbit population will either stabilize at a new level or possibly become extinct. Given no predators and no restrictions on food, but circumscribed space, the number of rabbits will still be limited, either by the psychological and biological effects of overcrowding or by being buried in their own refuse.

When man, endowed with the power of conceptual thought, appeared upon the earth scene, something new was introduced into the evolutionary process. Biological evolution, which had dominated all living species for billions of years, gave way to cultural evolution. As man gradually learned how to control various elements of his environment, he succeeded in modifying a number of the factors that limited his population. Clothing, fire, and crude shelters extended the range of habitable climate. Tools of increasing sophistication helped man gather edible vegetation, hunt animals more effectively, and protect himself from predators.

But no matter how effective the tools, there is a limit to the number of food gatherers who can inhabit a given area of land. One cannot kill more animals than are born or pick more fruit than trees bear. The maximum population of a worldwide food-gathering society was about ten million persons. Once that level was reached, numerous cultural patterns emerged that caused worldwide birth rates and death rates to become equal. In some societies, the natural death rate was elevated by malnutrition and disease; in others, the death rate was increased artificially by such practices as infanticide or the waging of war. In some cases, certain sex taboos and rituals appear to have lowered the birth rate. But, however birth and death rates came into balance, we can be confident that for a long time prior to the agricultural revolution the human population remained virtually constant.

With the introduction of agriculture about 10,000 years ago, the levels of population that had been imposed by limited supplies of food were raised significantly. Even in the earliest agricultural societies, several hundred times as much food could be produced from a given area of fertile land than could be collected by food gatherers. As the technology of agriculture spread, population grew rapidly. This new technology dramatically affected the entire fabric of human culture. Man gave up the nomad life and settled in villages, some of which became cities. Sufficient food could be grown to make it possible for about 10 percent of the population to engage in activities other than farming.

The development of iron technology and improved transportation accelerated spread of this peasant-village culture. Indeed, had new technological developments ceased to appear after 1700, it is nevertheless likely that the peasant-village culture would have spread to all inhabitable parts of the earth, eventually to reach a level of roughly five billion persons, some 500 million of whom would live in cities. But long before the population had reached anything close to that level, the emergence of new technologies leading up to the Industrial Revolution markedly changed the course of history. The steam engine for the first time gave man a means of concentrating enormous quantities of inanimate mechanical energy, and the newly found power was quickly applied.

During the nineteenth century in western Europe, improved transportation, increased food supplies, and a generally improved environment decreased the morbidity of a number of infectious diseases and virtually eliminated the large fluctuations in mortality rates that had been so characteristic of the seventeenth and eighteenth centuries. As mortality rates declined and the birth rate remained unchanged, populations in these areas increased rapidly. But as industrialization spread, a multiplicity of factors combined to lessen the desirability of large families. After about 1870, the size of families decreased, at first slowly and then more rapidly; eventually, the rate of population growth declined.

During the nineteenth and early twentieth centuries, some of the new technologies were gradually transplanted to the nonindustrialized parts of the world, but in a very one-sided manner. Death rates were reduced appreciably, and, with birth rates unchanged, populations in these poorer countries increased rapidly and are still growing.

In spite of the fact that the annual rate of population growth in the industrialized countries has dropped to less than 1 percent, the worldwide rate is now close to 2 percent, the highest it has ever been. This rate represents a doubling of population about every 35 years. The human population is now 3.5 billion and at the present rate of increase is destined to reach 6.5 billion by the turn of the century and 10 billion 50 years from now. Beyond that point, how much further can population grow?

An analysis of modern technology's potential makes it clear that from a long-range, theoretical point of view, food supplies need no longer be the primary factor limiting population growth. Today nearly 10 percent of the land area of the earth, or about 3.5 billion acres, is under cultivation. It is estimated that with sufficient effort about 15 billion acres of land could be placed under cultivation—some four times the present area. Such a move would require prodigious effort and investment and would necessitate the use of substantial quantities of desalinated water reclaimed from the sea. Given abundant energy resources, however, it now appears that in principle this can be done economically.

Large as the potential is for increasing the area of agricultural land, the increases in yield that can be obtained through fertilizers, application of supplementary water, and the use of new high-yielding varieties of cereals are even more impressive. Whereas in the past the growth of plants was circumscribed by the availability of nutrients and water, this need no longer be true. Using our new agricultural technology, solar energy can be converted into food with a high degree of efficiency, and even on the world's presently cultivated lands several times as much food can be produced each year than is now being grown.

To accomplish these objectives, however, an enormous amount of industrialization will be required. Fertilizers must be produced; thus, phosphate rock must be mined and processed, and nitrogen fixation plants must be built. Pesticides and herbicides are needed; thus, chemical plants must be built. All this requires steel and concrete, highways, railroads, and trucks. To be sure, the people of India, for example, might not need to attain Japan's level of industrialization in order to obtain Japanese levels of crop yield (which are about the highest in the world), but they will nevertheless need a level of industrialization that turns out to be surprisingly high.

Colin Clark, the director of the Agricultural Economics Research Institute of Oxford and a noted enthusiast for large populations, estimates that, given this new agricultural land and a level of industrialization sufficiently high to apply Japanese standards of farming, close to 30 billion persons could be supported on a Western European diet. Were people to content themselves with a Japanese diet, which contains little animal protein, he estimates that 100 billion persons could be supported.

To those who feel that life under such circumstances might be rather crowded, I should like to point out that even at the higher population level, the mean density of human beings over the land areas of the earth would be no more than that which exists today in the belt along the Eastern Seaboard between Boston and Washington, D. C., where the average density is now 2000 persons per square mile and where many people live quite comfortably. After all, Hong Kong has a population density of about 13,000 persons per square mile (nearly six times greater), and I understand that there are numerous happy people there.

Of course, such a society would need to expend a great deal

of energy in order to manufacture, transport, and distribute the fertilizers, pesticides, herbicides, water, foodstuffs, and countless associated raw materials and products that would be necessary.

In the United States we currently consume energy equivalent to the burning of twelve-and-a-half short tons of coal per person per year. This quantity is bound to increase in the future as we find it necessary to process lower-grade ores, as we expend greater effort on controlling pollution (which would otherwise increase enormously), and as we recover additional quantities of potable water from the sea. Dr. Alvin Weinberg, director of the Oak Ridge National Laboratory, and his associates estimate that such activities will cost several additional tons of coal per person per year, and they suggest that for safety we budget 25 tons of coal per person per year in order to maintain our present material standard of living. Since we are a magnanimous people, we would not tolerate a double standard of living (a rich one for us and a poor one for others); so I will assume that this per capita level of energy expenditure will be characteristic of the world as a whole.

It has been estimated that the world's total usable coal reserve is on the order of 7600 billion tons. This amount would last a population of 30 billion persons only 10 years and a population of 100 billion only 3 years. Clearly, long before such population levels are reached, man must look elsewhere for his energy supplies.

Fortunately, technology once again gets us out of our difficulty, for nuclear fuels are available to us in virtually limitless quantities in the form of uranium and thorium for fission, and possibly in the form of deuterium for fusion. The Conway granite in New Hampshire could alone provide fuel for a population of 20 billion persons for 200 years. When we run out of high-grade granites, we can move on to process low-grade granites. Waste rock can be dumped into the holes from which it came and can be used to create new land areas on bays and on the continental shelf. Waste fission products can be stored in old salt mines.

Actually, a major shift to nuclear fuel might well be necessary long before our supplies of fossil fuels are exhausted. The carbon dioxide concentration in our atmosphere is rapidly increasing as a result of our burning of coal, petroleum, and natural gas, and it is destined to increase still more rapidly in the future. More than likely, any such increase will have a deleterious effect upon our climate, and if this turns out to be the case, use of those fuels will probably be restricted.

Thus, we see that in theory there should be little difficulty in feeding a world population of 30 billion or even 100 billion persons and in providing it with the necessities of life. But can we go even further?

With respect to food, once again technology can come to our rescue, for we have vast areas of the seas to fertilize and farm. Even more important, we will be able to produce synthetic foods in quantity. The constituents of our common oils and fats can already be manufactured on a substantial scale for human consumption and animal feeds. In the not too distant future, we should be able to synthetically produce complete, wholesome foods, thus bypassing the rather cumbersome process of photosynthesis.

Far more difficult than the task of feeding people will be that of cooling the earth, of dissipating the heat generated by nuclear power plants. It has been suggested that if we were to limit our total energy generation to no more than 5 percent of the incident solar radiation, little harm would be done. The mean surface temperature of the earth would rise by about 6 degrees F. A temperature rise much greater than this could be extremely dangerous and should not be permitted until we have learned more about the behavior of our ocean/atmosphere system.

Of course, there will be local heating problems in the vicinity of the power stations. Dr. Weinberg suggests a system of "nuclear parks," each producing about 40 million kilowatts of electricity and located on the coast or offshore. A population of 333 billion persons would require 65,000 such parks. The continental United States, with a projected population of close to 25 billion persons, would require nearly 5000 parks spaced at 20-mile intervals along its coastline.

Again, I want to allay the fears of those who worry about crowding. A population of 333 billion spread uniformly over the land areas of the earth would give us a population density of only 6000 persons per square mile, which, after all, is only somewhat greater than the population density in the city of Los Angeles. Just imagine the thrill of flying from Los Angeles to New York and having the landscape look like Los Angeles all the way. Imagine the excitement of driving from Los Angeles to New York on a Santa Monica Freeway 2800 miles long.

A few years ago Dr. J. H. Fremlin of the University of Birmingham analyzed the problem of population density and concluded that several stages of development might be possible beyond the several-hundred-billion-person level of population. He conceives of hermetically sealing the outer surface of the planet and of using pumps to transfer heat to the solid outer skin from which it would be radiated directly into space. Combining this with a roof over the oceans to prevent excessive evaporation of water and to provide additional living space, he feels it would be possible to accommodate about 100 persons per square yard, thus giving a total population of about 60 million billion persons. But, frankly, I consider this proposal visionary. Being basically conservative, I doubt that the human population will ever get much above the 333-billion-person level.

Now some readers might be thinking that I am writing nonsense, and they are right. My facts are correct; the conclusions I have drawn from those facts are correct. Yet, I have truthfully been writing nonsense. Specifically, I have given only *some* of the facts. Those facts that I have omitted alter the conclusions considerably.

I have presented only what is deemed possible by scientists from an energetic or thermodynamic point of view. An analogy would be for me to announce that I have calculated that in principle all men should be able to leap 10 feet into the air. Obviously, such an announcement would not be followed by a sudden, frenzied, worldwide demonstration of people showing their leaping capabilities. Some people have sore feet; others have inadequate muscles; most haven't the slightest desire to leap into the air. The calculation might be correct, but the enthusiasm for jumping and the ability to jump might be very low. The problem is the behavior of people rather than that of inanimate matter.

We are confronted by the brutal fact that humanity today doesn't really know how to cope with the problems presented by three-and-a-half billion persons, let alone 333 billion. More than two-thirds of the present human population is poor in the material sense and is malnourished. The affluent one-third is, with breathtaking rapidity, becoming even more affluent. Two separate and distinct societies have emerged in the world, and they are becoming increasingly distinct and separated. Numerically the largest is the culture of the poor, composed of some 2500 million persons. Numerically the smallest is the culture of the rich, composed of some 1000 million persons. On the surface, the rich countries would appear to have it made; in historical perspective, their average per capita incomes are enormous. Their technological competence is unprecedented. Yet, they have problems that might well prove insoluble.

The most serious problem confronting the rich countries today is nationalism. We fight among each other and arm ourselves in order to do so more effectively. The Cold War has become a way of life, as is reflected in military budgets. Today the governments of the United States and the Soviet Union spend more on their respective military establishments than they do on either education or health—indeed a scandalous situation but, even worse, an explosive one.

All of the rich countries are suffering from problems of growth. Although the rates of population proliferation in these areas are not large, per capita consumption is increasing rapidly. Today an average "population unit" in the United States is quite different from one in the primitive world. Originally, a unit of population was simply a human being whose needs could be met by "eating" 2500 calories and 60 grams of protein a day. Add to this some simple shelter, some clothing, and a small fire, and his needs were taken care of. A population unit today consists of a human being wrapped in tons of steel, copper, aluminum, lead, tin, zinc, and plastics. This new creature requires far more than food to keep it alive and functioning. Each day it gobbles up 60 pounds of coal or its equivalent, 3 pounds of raw steel, plus many pounds of other materials. Far from getting all of this food from his own depleted resources, he ranges abroad, much as the hunters of old, and obtains raw supplies in other parts of the world, more often than not in the poorer countries.

Industrial societies the world over are changing with unprecedented speed as the result of accelerated technological change, and they are becoming increasingly complex. All of them are encountering severe problems with their cities, which were designed within the framework of one technology and are falling apart at the seams within the framework of another.

The technological and social complexities of industrial society—composed as it is of vast interlocking networks of mines, factories, transportation systems, power grids, and communication networks, all operated by people—make it extremely vulnerable to disruption. Indeed, during the past year we have seen that the United States is far more vulnerable to labor strikes than North Vietnam is to air strikes. This vulnerability may eventually prove to be our undoing.

A concomitant of our affluence has been pollution. That which goes into a system must eventually come out; as our society has consumed more, it has excreted more. Given adequate supplies of energy and the necessary technology, such problems can be handled from a technical point of view. But it is by no means clear that we are about to solve these problems from a social or political point of view.

Although we know that theoretically we can derive our sustenance from the leanest of earth substances, such as seawater and rock, the fact remains that with respect to the raw materials needed for a highly industrialized society the research essential to the development of the necessary technology has hardly begun. Besides, it is less expensive for the rich countries to extract their sustenance from the poor ones.

As to the poor countries with their rapidly increasing populations, I fail to see how, in the long run, they can lift themselves up by their own bootstraps. In the absence of outside help commensurate with their needs, I suspect they will fail, and the world will become permanently divided into the rich and the poor—at least until such time as the rich, in their stupidity, blow themselves up.

One of the most difficult problems in the poor countries is that of extremely rapid population growth. If an economy grows only as fast as its population, the average well-being of the people does not improve—and indeed this situation prevails in many parts of the world. Equally important, rapid growth produces tremendous dislocations—physical, social, and economic. It is important to understand that the major population problem confronting the poor countries today is not so much the actual number of people as it is rapid growth rates. Clearly, if development is to take place, birth rates must be reduced.

Unfortunately, it is not clear just how birth rates can be brought down in these areas. Even with perfect contraceptives, there must be motivation upon the part of individuals, and in many areas this appears to be lacking. Some people say that economic development is necessary to produce the motivation, and they might be right. In any event, the solution will not be a simple one.

Although I am pessimistic about the future, I do not consider the situation to be by any means hopeless. I am convinced that our problems both here and abroad are soluble. But if they are ever solved, it will be because all of us reorient our attitudes away from those of our parents and more toward those of our children. I am convinced that young people today more often than not have a clearer picture of the world and its problems than do their elders. They are questioning our vast military expenditures and ask whether the Cold War is really necessary. They question the hot war in which we have become so deeply involved. They are questioning our concepts of nationalism, materialism, and laissez faire. It is just such questioning on the part of the young that gives me hope.

If this questioning persists, I foresee the emergence of a new human attitude in which people the world over work together to transform anarchy into law, to decrease dramatically military expenditures, to lower rates of population growth to zero, and to build an equitable world economy, so that all people can lead free and abundant lives in harmony with nature and with each other.

31 Declaration on the Human Environment

The United Nations Conference on the Human Environment

Having met at Stockholm from 5 to 16 June 1972, and having considered the need for a common outlook and for common principles to inspire and guide the peoples of the world in the preservation and enhancement of the human environment

The United Nations Conference on the Human Environment

Proclaims that

1. Man is both creature and molder of his environment, which gives him physical sustenance and affords him the opportunity for intellectual, moral, social and spiritual growth. In the long and tortuous evolution of the human race on this planet a stage has been reached when, through the rapid acceleration of science and technology, man has acquired the power to transform his environment in countless ways and on an unprecedented scale. Both aspects of man's environment, the natural and the man-made, are essential to his well-being and to the enjoyment of basic human rights—even the right to life itself.

2. The protection and improvement of the human environment is a major issue which affects the well-being of peoples and economic development throughout the world; it is the urgent desire of the peoples of the whole world and the duty of all Governments.

3. Man has constantly to sum up experience and go on discovering, inventing, creating and advancing. In our time, man's capability to transform his surroundings, if used wisely, can bring to all peoples the benefits of development and the opportunity to enhance the

quality of life. Wrongly or heedlessly applied, the same power can do incalculable harm to human beings and the human environment. We see around us growing evidence of man-made harm in many regions of the earth: dangerous levels of pollution in water, air, earth and living beings; major and undesirable disturbances to the ecological balance of the biosphere; destruction and depletion of irreplaceable resources; and gross deficiencies harmful to the physical, mental and social health of man, in the man-made environment, particularly in the living and working environment.

4. In the developing countries most of the environmental problems are caused by underdevelopment. Millions continue to live far below the minimum levels required for a decent human existence, deprived of adequate food and clothing, shelter and education, health and sanitation. Therefore, the developing countries must direct their efforts to development, bearing in mind their priorities and the need to safeguard and improve the environment. For the same purpose, the industrialized countries should make efforts to reduce the gap between themselves and the developing countries. In the industrialized countries, environmental problems are generally related to industrialization and technological development.

5. The natural growth of population continuously presents problems for the preservation of the environment, and adequate policies and measures should be adopted, as appropriate, to face these problems. Of all things in the world, people are the most precious. It is the people that propel social progress, create social wealth, develop science and technology and, through their hard work, continuously transform the human environment. Along with social progress and the advance of production, science and technology, the capability of man to improve the environment increases with each passing day.

6. A point has been reached in history when we must shape our actions throughout the world with a more prudent care for their environmental consequences. Through ignorance or indifference we can do massive and irreversible harm to the earthly environment on which our life and well-being depend. Conversely, through fuller knowledge and wiser action, we can achieve for ourselves and our posterity a better life in an environment more in keeping with human needs and hopes. There are broad vistas for the enhancement of environmental quality and the creation of a

good life. What is needed is an enthusiastic but calm state of mind and intense but orderly work. For the purpose of attaining freedom in the world of nature, man must use knowledge to build in collaboration with nature a better environment. To defend and improve the human environment for present and future generations has become an imperative goal for mankind—a goal to be pursued together with, and in harmony with, the established and fundamental goals of peace and of world-wide economic and social development.

7. To achieve this environmental goal will demand the acceptance of responsibility by citizens and communities and by enterprises and institutions at every level, all sharing equitably in common efforts. Individuals in all walks of life as well as organizations in many fields, by their values and the sum of their actions, will shape the world environment of the future. Local and national governments will bear the greatest burden for large-scale environmental policy and action within their jurisdictions. International co-operation is also needed in order to raise resources to support the developing countries in carrying out their responsibilities in this field. A growing class of environmental problems, because they are regional or global in extent or because they affect the common international realm, will require extensive co-operation among nations and action by international organizations in the common interest. The Conference calls upon Governments and peoples to exert common efforts for the preservation and improvement of the human environment, for the benefit of all the people and for their posterity.

Declaration of Principles

States the common conviction that

Principle 1. Man has the fundamental right to freedom, equality and adequate conditions of life, in an environment of a quality that permits a life of dignity and well-being, and he bears a solemn responsibility to protect and improve the environment for present and future generations. In this respect, policies promoting or perpetuating *apartheid*, racial segregation, discrimination, colonial and other forms of oppression and foreign domination stand condemned and must be eliminated.

Principle 2. The natural resources of the earth including the air, water, land, flora and fauna and especially representative samples of natural ecosystems must be safeguarded for the benefit of present and future generations through careful planning or management, as appropriate.

Principle 3. The capacity of the earth to produce vital renewable resources must be maintained and, wherever practicable, restored or improved.

Principle 4. Man has a special responsibility to safeguard and wisely manage the heritage of wildlife and its habitat which are now gravely imperiled by a combination of adverse factors. Nature conservation including wildlife must therefore receive importance in planning for economic development.

Principle 5. The non-renewable resources of the earth must be employed in such a way as to guard against the danger of their future exhaustion and to ensure that benefits from such employment are shared by all mankind.

Principle 6. The discharge of toxic substances or of other substances and the release of heat, in such quantities or concentrations as to exceed the capacity of the environment to render them harmless, must be halted in order to ensure that serious or irreversible damage is not inflicted upon ecosystems. The just struggle of the peoples of all countries against pollution should be supported.

Principle 7. States shall take all possible steps to prevent pollution of the seas by substances that are liable to create hazards to human health, to harm living resources and marine life, to damage amenities or to interfere with other legitimate uses of the sea.

Principle 8. Economic and social development is essential for ensuring a favorable living and working environment for man and for creating conditions on earth that are necessary for the improvement of the quality of life.

Principle 9. Environmental deficiencies generated by the conditions of underdevelopment and natural disasters pose grave problems and can best be remedied by accelerated development through the transfer of substantial quantities of financial and technological assistance as a supplement to the domestic effort of the developing countries and such timely assistance as may be required.

Principle 10. For the developing countries, stability of prices and adequate earnings for primary commodities and raw material are essential to environmental management since economic factors as well as ecological processes must be taken into account.

Principle 11. The environmental policies of all States should enhance and not adversely affect the present or future development potential of developing countries, nor should they hamper the attainment of better living conditions for all, and appropriate steps should be taken by States and international organizations with a view to reaching agreement on meeting the possible national and international economic consequences resulting from the application of environmental measures.

Principle 12. Resources should be made available to preserve and improve the environment, taking into account the circumstances and particular requirements of developing countries and any costs which may emanate from their incorporating environmental safeguards into their development planning and the need for making available to them, upon their request, additional international technical and financial assistance for this purpose.

Principle 13. In order to achieve a more rational management of resources and thus to improve the environment, States should adopt an integrated and co-ordinated approach to their development planning so as to ensure that development is compatible with the need to protect and improve the human environment for the benefit of their population.

Principle 14. Rational planning constitutes an essential tool for reconciling any conflict between the needs of development and the need to protect and improve the environment.

Principle 15. Planning must be applied to human settlements and urbanization with a view to avoiding adverse effects on the environment and obtaining maximum social, economic and environmental benefits for all. In this respect projects which are designed for colonialist and racist domination must be abandoned.

Principle 16. Demographic policies, which are without prejudice to basic human rights and which are deemed appropriate by Governments concerned, should be applied in those regions where the rate of population growth or excessive population concentrations are likely to have adverse effects on the environment or development, or where low population density may prevent improvement of the human environment and impede development.

Principle 17. Appropriate national institutions must be entrusted with the task of planning, managing or controlling the environmental resources of States with the view to enhancing environmental quality.

Principle 18. Science and technology, as part of their contribution to economic and social development, msut be applied to the identification, avoidance and control of environmental risks and the solution of environmental problems and for the common good of mankind.

Principle 19. Education in environmental matters, for the younger generation as well as adults, giving due consideration to the under-privileged, is essential in order to broaden the basis for an enlightened opinion and responsible conduct by individuals, enterprises and communities in protecting and improving the environment in its full human dimension. It is also essential that mass media of communications avoid contributing to the deterioration of the environment, but, on the contrary, disseminate information of an educational nature, on the need to protect and improve the environment in order to enable man to develop in every respect.

Principle 20. Scientific research and development in the context of environmental problems, both national and multinational, must be promoted in all countries, especially the developing countries. In this connection, the free flow of up-to-date scientific information and transfer of experience must be supported and assisted, to facilitate the solution of environmental problems; environmental technologies should be made available to developing countries on terms which would encourage their wide dissemination without constituting an economic burden on the developing countries.

Principle 21. States have, in accordance with the Charter of the United Nations and the principles of international law, the sovereign right to exploit their own resources pursuant to their own environmental policies, and the responsibility to ensure that activities within their jurisdiction or control do not cause damage to the environment of other States or of areas beyond the limits of national jurisdiction.

Principle 22. States shall co-operate to develop further the international law regarding liability and compensation for the victims of pollution and other environmental damage caused by activities within the jurisdiction or control of such States to areas beyond their jurisdiction.

Principle 23. Without prejudice to such criteria as may be agreed upon by the international community, or to standards which will have to be determined nationally, it will be essential in all cases to consider the systems of values prevailing in each country, and the extent of the applicability of standards which are valid for the most advanced countries but which may be inappropriate and of unwarranted social cost for the developing countries.

Principle 24. International matters concerning the protection and improvement of the environment should be handled in a co-operative spirit by all countries, big or small, on an equal footing. Co-operation through multilateral or bilateral arrangements or other appropriate means is essential to effectively control, prevent, reduce and eliminate adverse environmental effects resulting from activities conducted in all spheres, in such a way that due account is taken of the sovereignty and intersts of all States.

Principle 25. States shall ensure that international organizations play a co-ordinated, efficient and dynamic role for the protection and improvement of the environment.

Principle 26. Man and his environment must be spared the effects of nuclear weapons and all other means of mass destruction. States must strive to reach prompt agreement, in the relevant international organs, on the elimination and complete destruction of such weapons.

Additional Readings

American Academy of Arts and Sciences. 1967. Toward the year 2000: work in progress. *Daedalus* 96(3):639-1002.

Bahr, H. M., Chadwick, B. A., Thomas, D. L., eds. 1972. *Population, resources, and the future: Non-Malthusian perspectives.* Provo, Utah: Brigham Young University.

Boughey, A. S. 1971. *Man and the environment: an introduction to the human ecology and evolution.* New York: Macmillan.

Brown, H. 1954. *The challenge of man's future.* New York: Viking Press.

Brown, H., and Hutchings, E., Jr., eds. 1970. *Are our descendants doomed?* New York: Viking Press.

Brown, L. R. 1972. *World without borders.* New York: Random House.

Brown, L. R., and editors of *The Futurist.* 1972. An overview of world trends. *The Futurist* 6(6):225-32.

Caldwell, L. K. 1970. *Environment: a challenge to society*. Garden City, N. Y.: Natural History Press.
Clark, C. 1967. *Population growth and land use*. New York: St. Martin's Press.
Cole, H. S. D., Freeman, C., Jahoda, M., and Pavitt, K. L. R. 1973. *Models of doom: a critique of the limits of growth*. New York: Universe Books.
Darling, F. F., and Milton, J. P., eds. 1966. *Future environments of North America*. Garden City, N. Y.: Natural History Press.
Dunbar, M. J. 1971. *Environment and good sense*. Montreal: McGill-Queen's University Press.
Ehrlich, P. R., and Ehrlich, A. H. 1972. *Population resources, and environment—issues in human ecology*. 2nd ed. San Francisco: W. H. Freeman.
Farvar, M. T., and Milton, J. P., eds. 1972. *The careless technology: ecology and international development*. Garden City, N. Y.: Natural History Press.
Goldsmith, E., Allen, R., Allaby, M., Davoll, J., and Lawrence, S. 1972. *Blueprint for survival*. Boston: Houghton Mifflin.
Hardin, G. 1971. Nobody ever dies of overpopulation. *Science* 171:12.
Helfrich, H. W., Jr., ed. 1970. *The environmental crisis*. 2 vols. New Haven, Conn.: Yale University Press.
Jackson, W., ed. 1971. *Man and the environment*. Dubuque, Iowa: Wm. C. Brown.
Langer, W. L. 1972. Checks on population growth: 1750-1850. *Scientific American* 226(2):92-99.
McHale, J. 1969. *The future of the future*. New York: George Braziller.
McHarg, I. L. 1969. *Design with nature*. Garden City, N. Y.: Doubleday/Natural History Press.
Meadows, D. H., Meadows, D. L., Randers, J., and Behrens, W. W., III. 1972. *The limits of growth*. New York: Universe Books.
Moncrief, L. W. 1970. The cultural basis for our environmental crisis. *Science* 170:508-12.
Nichols, D. R., and Campbell, C. C., eds. 1971. *Environmental planning and geology*. U. S. Department of Housing and Urban Development and the U. S. Geological Survey. Washington, D. C.: U. S. Government Printing Office.
Odum, H. T. 1971. *Environment, power, and society*. New York: Wiley-Interscience.
Pitts, J. N., Jr., and Metcalf, R. L., eds. 1971. *Advances in environmental science and technology*, vol. 2. New York: Wiley-Interscience.
Roslansky, J. D., ed. 1967. *The control of the environment*. Amsterdam: North-Holland Publishing Co.
Sauer, R. C., ed. 1971. *Voyages: scenarios for a ship called earth*. New York: Zero Population Growth/Ballantine Books.
Scientific American. 1970. *The biosphere*. San Francisco: W. H. Freeman.
Seaborg, G. T. 1973. Science, technology, and development: a new world outlook. *Science* 181:13-19.

Smith, G-H., ed. 1971. *Conservation of natural resources*. 4th ed. New York: Wiley.

Strong, M. F., ed. 1973. *Who speaks for the earth?* New York: W. W. Norton.

Study of Critical Environmental Problems (SCEP). 1970. *Man's impact on the global environment*. Cambridge, Mass.: MIT Press.

Taylor, G. R. 1970. *The doomsday book*. Greenwich, Conn.: Fawcett Publications.

Toffler, A. 1970. *Future shock*. New York: Random House.

Troost, C. J., and Altman, H. 1972. *Environmental education: a sourcebook*. New York: Wiley.

Urban, G. R., ed. 1972. *Can we survive our future? A symposium*. New York: St. Martin's Press.

Wagar, J. A. 1970. Growth versus the quality of life. *Science* 168:1179-84.

White, L. T., Jr. 1967. The historical roots of our ecologic crisis. *Science* 155:1203-7.

Wittwer, S. W. 1969. Food supply: the fruits of research. *Technology Review* 71(5):18-25.

Appendix I

GEOLOGIC TIME CHART

Era	Period	Epoch	Duration (Millions of years)	Millions of years before present
CENOZOIC	Quaternary	Holocene / Pleistocene	2	
				2
	Tertiary	Pliocene / Miocene / Oligocene / Eocene / Paleocene	63	
				65
MESOZOIC	Cretaceous		70	
				135
	Jurassic		55	
				190
	Triassic		35	
				225
PALEOZOIC	Permian		55	
				280
	Pennsylvanian		40	
				320
	Mississippian		25	
				345
	Devonian		55	
				400
	Silurian		40	
				440
	Ordovician		60	
				500
	Cambrian		70	
				570
PRECAMBRIAN ERAS	No widely recognized period names		4130	
				4700

Appendix II

Periodic Table

The periodic table lists all the elements, arranged in order of increasing atomic numbers and grouped by similar physical and chemical characteristics into "periods." The table is based on the chemical law that the physical or chemical properties of the elements are periodic or regularly repeated functions of their atomic weights. Elements of similar electron configuration fall into one column; and two neighboring elements in a particular column tend to display similar chemical properties. The extreme left column contains elements that are highly reactive, and the extreme right column contains elements that are chemically nonreactive.

PERIODIC TABLE OF THE ELEMENTS

1 H																	2 He
3 Li	4 Be											5 B	6 C	7 N	8 O	9 F	10 Ne
11 Na	12 Mg											13 Al	14 Si	15 P	16 S	17 Cl	18 Ar
19 K	20 Ca	21 Sc	22 Ti	23 V	24 Cr	25 Mn	26 Fe	27 Co	28 Ni	29 Cu	30 Zn	31 Ga	32 Ge	33 As	34 Se	35 Br	36 Kr
37 Rb	38 Sr	39 Y	40 Zr	41 Nb	42 Mo	43 Tc	44 Ru	45 Rh	46 Pd	47 Ag	48 Cd	49 In	50 Sn	51 Sb	52 Te	53 I	54 Xe
55 Cs	56 Ba	57-71 La* Series	72 Hf	73 Ta	74 W	75 Re	76 Os	77 Ir	78 Pt	79 Au	80 Hg	81 Tl	82 Pb	83 Bi	84 Po	85 At	86 Rn
87 Fr	88 Ra	89-103 Act Series															

*Lathanide Series	57 La	58 Ce	59 Pr	60 Nd	61 Pm	62 Sm	63 Eu	64 Gd	65 Tb	66 Dy	67 Ho	68 Er	69 Tm	70 Yb	71 Lu
†Actinide Series	89 Ac	90 Th	91 Pa	92 U	93 Np	94 Pu	95 Am	96 Cm	97 Bk	98 Cf	99 Es	100 Fm	101 Md	102 No	(103) Lr

Transuranium elements are shown in shaded squares. The actinide series of elements as a group occupies a single square in the main figure. The lanthanide series of elements also occupies a single square in the larger chart.

Nuclide Designation (Subscripts and Superscripts)

In accordance with recommendations of the International Union of Pure and Applied Chemistry, the following designations are used for nuclides:

The *MASS NUMBER* of a nuclide is placed as a *superscript* to the left of the symbol for the chemical element of the nuclide, rather than to its right, as formerly; for example ^{14}N, rather than N^{14} for nitrogen-14.

The *ATOMIC NUMBER* is placed as a left *subscript*; for example, $^{14}_{6}C$ for carbon-14, or $^{235}_{92}U$ for uranium-235.

The state of IONIZATION is shown as a *right superscript*; for example, Ca^{++} or SO_4^{--}.

The number of *NEUTRONS* in the nucleus is shown as a *right subscript*; for example, $^{40}_{20}Ca_{20}$ for the isotope of calcium-40 containing 20 protons (its atomic number) (left subscript), and 20 neutrons (right subscript) in its nucleus.

Excited states are shown either as part of the *left superscript*, or sometimes the *right superscript*; for example: ^{110m}Ag or $^{110}Ag^m$ indicates an excited state of a silver-110 nucleus: He* indicates an excited state of a helium atom.

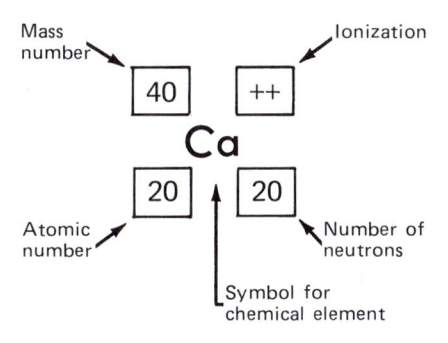

```
          Mass                    Ionization
          number
              ┌────┐   ┌────┐
              │ 40 │   │ ++ │
              └────┘   └────┘
                    Ca
              ┌────┐   ┌────┐
              │ 20 │   │ 20 │
              └────┘   └────┘
          Atomic                  Number of
          number                  neutrons
                   Symbol for
                   chemical element
```

ALPHABETICAL LIST OF ELEMENTS AND SYMBOLS

Element	Symbol	Atomic number	Atomic weight*	Element	Symbol	Atomic number	Atomic weight*
Actinium	Ac	89	227	Molybdenum	Mo	42	95.95
Aluminum	Al	13	26.98	Neodymium	Nd	60	144.26
Americium	Am	95	243	Neon	Ne	10	20.182
Antimony	Sb	51	121.75	Neptunium	Np	93	237
Argon	Ar	18	39.942	Nickel	Ni	28	58.71

*Atomic weight of the most abundant or best known isotope, or (in the case of radioactive isotopes) the isotope with the longest half-life, relative to atomic weight of Carbon-12 = 12.

Element	Symbol	Atomic number	Atomic weight*	Element	Symbol	Atomic number	Atomic weight*
Arsenic	As	33	74.91	Niobium			
Astatine	At	85	210	(Columbium)	Nb	41	92.91
Barium	Ba	56	137.35	Nitrogen	N	7	14.007
Berkelium	Bk	97	249	Nobelium	No	102	254
Beryllium	Be	4	9.013	Osmium	Os	76	190.2
Bismuth	Bi	83	208.99	Oxygen	O	8	15.999
Boron	B	5	10.82	Palladium	Pd	46	106.4
Bromine	Br	35	79.913	Phosphorus	P	15	30.973
Cadmium	Cd	48	112.40	Platinum	Pt	78	195.08
Calcium	Ca	20	40.08	Plutonium	Pu	94	242
Californium	Cf	98	251	Polonium	Po	84	210
Carbon	C	6	12.010	Potassium	K	19	39.098
Cerium	Ce	58	140.12	Praseodymium	Pr	59	140.91
Cesium	Cs	55	132.90	Promethium	Pm	61	147
Chlorine	Cl	17	35.455	Protactinium	Pa	91	231
Chromium	Cr	24	52.01	Radium	Ra	88	226
Cobalt	Co	27	58.94	Radon	Rn	86	222
Copper	Cu	29	63.54	Rhenium	Re	75	186.21
Curium	Cm	96	247	Rhodium	Rh	45	102.90
Dysprosium	Dy	66	162.50	Rubidium	Rb	37	85.48
Einsteinium	Es	99	254	Ruthenium	Ru	44	101.1
Erbium	Er	68	167.26	Samarium	Sm	62	150.34
Europium	Eu	63	152.0	Scandium	Sc	21	44.96
Fermium	Fm	100	253	Selenium	Se	34	78.96
Fluorine	F	9	19.00	Silicon	Si	14	28.09
Francium	Fr	87	223	Silver	Ag	47	107.875
Gadolinium	Gd	64	157.25	Sodium	Na	11	22.990
Gallium	Ga	31	69.72	Strontium	Sr	38	87.63
Germanium	Ge	32	72.60	Sulfur	S	16	32.064
Gold	Au	79	197.0	Tantalum	Ta	73	180.94
Hafnium	Hf	72	178.49	Technetium	Tc	43	99
Helium	He	2	4.003	Tellurium	Te	52	127.60
Holmium	Ho	67	164.93	Terbium	Tb	65	158.92
Hydrogen	H	1	1.0079	Thallium	Tl	81	204.38
Indium	In	49	114.81	Thorium	Th	90	232.04
Iodine	I	53	126.90	Thulium	Tm	69	168.93
Iridium	Ir	77	192.2	Tin	Sn	50	118.69
Iron	Fe	26	55.85	Titanium	Ti	22	47.90
Krypton	Kr	36	83.80	Tungsten			
Lanthanum	La	57	138.91	(Wolfram)	W	74	183.85
Lawrencium	Lr	103	257	Uranium	U	92	238.06
Lead	Pb	82	207.20	Vanadium	V	23	50.95
Lithium	Li	3	6.940	Xenon	Xe	54	131.29
Lutetium	Lu	71	174.98	Ytterbium	Yb	70	173.03
Magnesium	Mg	12	24.32	Yttrium	Y	39	88.92
Manganese	Mn	25	54.94	Zinc	Zn	30	65.38
Mendelevium	Md	101	256	Zirconium	Zr	40	91.22
Mercury	Hg	80	200.60				

APPENDIX II   337

Isotopes of Some of the Elements*

Element	Isotopes (Mass Numbers)
Hydrogen	1, 2, *3*
Helium	4, 3, *6*
Lithium	7, 6, *8, 9*
Carbon	12, 13, **14**, *11, 10, 15*
Nitrogen	14, 15, *13, 16, 17, 12*
Oxygen	16, 18, 17, *15, 14, 19*
Fluorine	19, *18, 17, 20, 21*
Sodium	23, *22, 24, 25, 21, 20*
Magnesium	24 26, 25, *28, 27, 23*
Aluminum	27, *26, 29, 28, 25, 24*
Sulfur	32, 34, 33, 36, *35, 37, 31*
Chlorine	35, 37, *36, 39, 38, 33, 34, 32*
Potassium	39, 41, **40**, *43, 42, 44, 38, 37*
Calcium	40, 44, 42, 48, 43, 46, *41, 45, 47, 49, 39*
Iron	56, 54, 57, 58, *55, 59, 52, 53*
Cobalt	59, *60, 57, 56, 58, 55, 61, 62, 54*
Nickel	58, 60, 62, 61, 64, *59, 63, 66, 57, 65, 56*
Copper	63, 65, *67, 64, 61, 60, 62, 58, 66, 68*
Zinc	64, 66, 68, 67, 70, *65, 72, 62, 71, 69, 63*
Bromine	79, 81, *77, 82, 76, 83, 75, 74, 84, 80, 78, 85, 87, 88*
Silver	107, 109, *105, 106, 111, 113, 112, 103, 104, 115, 108, 114, 110*
Tin	120, 118, 116, 119, 117, 124, 122, 112, 114, 115, *123, 113, 125, 121, 108, 127, 126, 111, 109*
Iodine	127, *129, 125, 126, 131, 124, 133, 123, 130, 135, 132, 121, 134, 128, 122, 137, 138, 139*
Barium	138, 137, 136, 135, 134, 130, 132, *133, 140, 131, 128, 129, 126, 141, 142, 143*
Platinum	195, 194, 196, 198, 192, **190**, *188, 191, 197, 189, 187, 199*
Gold	197, *195, 196, 199, 198, 194, 193, 192, 191, 200, 189, 201, 187, 203*
Mercury	202, 200, 199, 201, 198, 204, 196, *203, 197, 195, 192, 193, 191, 189, 205*
Lead	208, 206, 207, **204**, *202,* **210**, *203, 200,* **212**, *201, 209, 199,* **211, 214**, *198*
Bismuth	209, **210**, *207, 205, 206, 204, 203, 201, 202,* **212**, *213, 200, 199,* **214, 215**, *198,* **211**

*Stable isotopes in ordinary type. Naturally radioactive isotopes in **boldface**. Other radioisotopes in *italics*. Natural isotopes given in order of abundance. All other isotopes given in order of length of half-life.

Element	Isotopes (Mass Numbers)
Radon	**222**, *211, 210, 209, 221, 212, 208,* **220**, **219**, *218, 217, 216, 215*
Radium	**226, 228**, *225,* **223**, **224**, *227, 213, 222, 221, 220, 219*
Thorium†	*232, 223, 224, 225, 226,* **227**, **228**, *229,* **230**, **231**, *233,* **234**
Uranium	**238, 235, 234**, *236, 233, 232, 230, 237, 231, 240, 229, 239, 228, 227*
Neptunium	*237, 236, 235, 234, 239, 238, 240, 231, 233, 241, 232*
Plutonium	*244, 242, 239, 240, 238, 241, 236, 237, 246, 245, 234, 243, 232, 235*
Americium†	*243, 241, 242, 240, 239, 238, 245, 237, 244, 246*
Curium†	*248, 245, 246, 243, 244, 242, 247, 241, 240, 238*
Berkelium†	*247, 249, 245, 246, 248, 244, 243, 250*
Californium†	*251, 249, 250, 252, 248, 254, 253, 246, 247, 245, 244*
Einsteinium†	*254, 253, 245, 246, 248, 249, 250, 251, 252, 255, 256*
Fermium†	*257, 253, 252, 255, 248, 249, 250, 251, 254, 256*
Mendelevium	*256, 255*
Nobelium†	*254, 255, 256*
Lawrencium	*257*

†Not listed in order of length of half-life.

Appendix III

UNITS AND CONVERSIONS

Prefixes for International System of Units

Multiples and Submultiples	Prefixes	Symbols
1,000,000,000,000 = 10^{12}	tera	T
1,000,000,000 = 10^{9}	giga	G
1,000,000 = 10^{6}	mega	M
1,000 = 10^{3}	kilo	k
100 = 10^{2}	hecto	h
10 = 10	deka	da
0.1 = 10^{-1}	deci	d
0.01 = 10^{-2}	centi	c
0.001 = 10^{-3}	milli	m
0.000001 = 10^{-6}	micro	μ
0.000000001 = 10^{-9}	nano	n
0.000000000001 = 10^{-12}	pico	p

Units of Measure

Linear Measure

1 mile (mi)	=	5280 feet (ft)
1 chain (ch)	=	66 ft
1 rod (rd)	=	16.5 ft
1 fathom (fm)	=	6 ft
1 nautical mile	=	6076.115 ft
1 kilometer (km)	=	1000 meters (m)
1 km	=	10^{3} m
1 centimeter (cm)	=	0.01 m = 10^{-2} m
1 millimeter (mm)	=	0.001 m = 10^{-3} m
1 angstrom (Å)	=	0.0000000001 m = 10^{-10} m
1 micron (μ)	=	0.001 mm

Area Measure

1 square mile	=	640 acres
1 acre	=	43,560 square feet
1 acre	=	4840 square yards
1 acre	=	160 square rods
1 mile square	=	1 section
6 miles square	=	1 township = 36 square miles

1 square meter	=	10,000 square centimeters (cm)
100 square meters	=	1 are (a)
100 ares	=	1 hectare (ha)
100 hectares	=	1 square kilometer

Volume and Cubic Measure

1 quart	=	2 pints = 57.75 cubic inches
4 quarts	=	1 gallon = 231 cubic inches
1 cubic foot	=	1728 cubic inches
1 cubic yard	=	27 cubic feet
1 barrel (oil)	=	42 gallons
1 barrel (proof spirits)	=	40 gallons
1 cubic foot	=	7.48 gallons
1 cubic inch	=	0.554 fluid ounce
1 gallon (U.S.)	=	128 U.S. fluid ounces = 0.833 British gallon
1 liter	=	0.001 cubic meter = 1 cubic decimeter
1 liter	=	1000 milliliters
10 milliliters	=	100 milliliters
1 milliliter	=	approximately 1 cubic centimeter (cc)
1 cubic meter (m^3)	=	1,000,000 cubic centimeters

Weights and Masses

1 short ton	=	2000 pounds
1 long ton	=	2240 pounds
1 pound (avoirdupois)	=	7000 grains
1 ounce (avoirdupois)	=	437.5 grains
1 gram	=	15.432 grains
1000 grams	=	1 kilogram
1000 kilograms	=	1 metric ton

Force

1 dyne (d) = the force that will produce an acceleration of 1 centimeter / second^2 when applied to a 1-gram mass.

1 newton (nt) = the force that will produce an acceleration of 1 meter/second^2 when applied to a 1-kilogram mass.

1 nt = 100,000 d = 1 X 10^5 d

Energy and Power

1 erg = the work done by a force of 1 dyne when its points of application moves through a distance of 1 centimeter in the direction of the force.

1 erg = 9.48 X 10^{-11} British thermal unit (BTU)
1 erg = 7.367 X 10^{-8} foot-pounds
1 erg = 2.778 X 10^{-14} kilowatt-hours
1 kilowatt-hour = 3413 BTU = 3.6 X 10^{13} ergs = 860,421 calories (cal)

APPENDIX III

1 BTU = 2.930 × 10⁻⁴ kilowatt-hours = 1.0548 × 10¹⁰ ergs = 252 calories (cal)
1 watt* = 3.413 BTU/hour
1 watt = 1.341 × 10⁻³ horsepower
1 watt = 1 joule per second
1 watt = 14.34 calories per minute
1 joule* = 1 × 10⁷ ergs
1 joule = 1 newton-meter

*The watt and the joule are the internationally acceptable units for power and energy, respectively.

Heat

1 calorie (cal) = the amount of heat that will raise the temperature of 1 gram of water 1 degree Celsius with the water at 4 degrees Celsius.

1 calorie (gram) = 3.9685 × 10⁻³ BTU = 4.186 × 10⁷ ergs

Pressure

1 millibar (mb) = 1000 dynes per cm²
1 atmosphere (atm) = 76 cm mercury = 14.70 lb/in² = 1013 millibars (mb)

Additional Conversions

1 gallon of water = 8.3453 pounds of water
1 gallon per minute = 8.0208 cubic feet per minute
1 acre-foot = 1233.46 m³

Temperature

To change from Fahrenheit (F) to Celsius (C)

$$°C = \frac{(°F - 32°)}{1.8}$$

To change from Celsius (C) to Fahrenheit (F)

$$°F = (°C \times 1.8) + 32°$$

English-Metric Conversions

1 inch	= 25.4 millimeters
1 foot	= 0.3048 meter
1 yard	= 0.9144 meter
1 mile	= 1.609 kilometers
1 sq inch	= 6.4516 sq centimeters
1 sq foot	= 0.0929 sq meter
1 sq yard	= 0.836 sq meter
1 sq mile	= 259 hectares
1 acre	= 0.4047 hectare
1 cubic inch	= 16.39 cubic centimeters

1 cubic foot	=	0.0283 cubic meter
1 cubic yard	=	0.7646 cubic meter
1 quart (liq)	=	0.946 liter
1 gallon (U.S.)	=	0.003785 cubic meter
1 ounce (avdp)	=	28.35 grams
1 pound (avdp)	=	0.4536 kilogram
1 short ton	=	907.2 kilograms
1 horsepower	=	0.7457 kilowatt

Metric-English Conversions

1 millimeter	=	0.0394 inch
1 meter	=	3.281 feet
1 meter	=	1.094 yards
1 kilometer	=	0.6214 mile
1 sq centimeter	=	0.155 sq inch
1 sq meter	=	10.764 sq feet
1 sq meter	=	1.196 sq yards
1 hectare	=	2.471 acres
1 hectare	=	0.003861 sq mile
1 cu centimeter	=	0.061 cu inch
1 cu meter	=	35.3 cu feet
1 cu meter	=	1.308 cu yards
1 liter	=	1.057 quarts
1 cu meter	=	264.2 gallons (U.S.)
1 gram	=	0.0353 ounce (avdp)
1 kilogram	=	2.205 pounds (avdp)
1 metric ton	=	2205 pounds (avdp)
1 kilowatt	=	1.341 horsepower

Appendix IV

SURFACE WATER CRITERIA FOR PUBLIC WATER SUPPLIES

Constituent or characteristic	Permissible criteria	Desirable criteria
Physical:		
Color (color units)	75	<10
Odor	Narrative	Virtually absent
Temperature*	do	Narrative
Turbidity	do	Virtually absent
Microbiological:		
Coliform organisms	10,000/100 ml[1]	<100/100 ml[1]
Fecal coliforms	2000/100 ml[1]	<20/100 ml[1]
Inorganic chemicals:	(mg/l)	(mg/l)
Alkalinity	Narrative	Narrative
Ammonia	0.5 (as N)	<0.01
Arsenic*	0.05	Absent
Barium*	1.0	do
Boron*	1.0	do
Cadmium*	0.01	do
Chloride*	250	<25
Chromium,* hexavalent	0.05	Absent
Copper*	1.0	Virtually absent
Dissolved oxygen	≥4 (monthly mean) ≥3 (individual sample)	Near saturation
Fluoride*	Narrative	Narrative
Hardness*	do	do
Iron (filterable)	0.3	Virtually absent
Lead*	0.05	Absent
Manganese* (filterable)	0.05	do

*The defined treatment process has little effect on this constituent.

[1] Microbiological limits are monthly arithmetic averages based upon an adequate number of samples. Total coliform limit may be relaxed if fecal coliform concentration does not exceed the specified limit.

[2] As parathion in cholinesterase inhibition, it may be necessary to resort to even lower concentrations for some compounds or mixtures.

Note: The presence of the word "narrative" in the table indicates that the committee compiling the table could not arrive at a single numerical value which would be applicable throughout the country for all conditions.

Constituent or characteristic	Permissible criteria	Desirable criteria
Nitrates plus nitrites*	10 (as N)	Virtually absent
pH (range)	6.0–8.5	Narrative
Phosphorus*	Narrative	do
Selenium*	0.01	Absent
Silver*	0.05	do
Sulfate*	250	<50
Total dissolved solids* (filterable residue)	500	<200
Uranyl ion*	5	Absent
Zinc*	5	Virtually absent
Organic chemicals:		
Carbon chloroform extract* (CCE)	0.15	<0.04
Cyanide*	0.20	Absent
Methylene blue active substances*	0.5	Virtually absent
Oil and grease*	Virtually absent	Absent
Pesticides:		
Aldrin*	0.017	do
Chlordane*	0.003	do
DDT*	0.042	do
Dieldrin*	0.017	do
Endrin*	0.001	do
Heptachlor*	0.018	do
Heptachlor epoxide*	0.018	do
Lindane*	0.056	do
Methoxychlor*	0.035	do
Organic phosphates plus carbamates*	0.1^2	do
Toxaphene	0.005	do
Herbicides:		
2,4-D plus 2,4,5-T, plus 2,4,5-TP*	0.1	do
Phenols*	0.001	do
Radioactivity:	(pc/l)	(pc/l)
Gross beta*	1000	<100
Radium-226*	3	<1
Strontium-90*	10	<2

Glossary

ABYSSAL PLAINS. A flat region of the ocean floor at the base of a continental rise with a slope of less than 1:1000.
ACID SOILS. Soils having a pH below 7.0. Soils may be naturally acid by their origin, by leaching, or may become acid from decaying leaves or from soil additives.
AD VALOREM TAX. Tax in proportion to estimated value of goods.
AIR STABILITY. Upward and downward motions of parcels of air that are forced back to their original positions by the action of the surrounding environment indicate stable air. If the parcels of air are accelerated in their vertical movements, then the air is unstable and favors vertical motions. Stability criteria are directly related to the amount and distribution of water vapor in the air.
ALBEDO. The percentage of incident electromagnetic radiation reflected by a surface. It is the reflectivity of a body compared with that of a perfectly diffusing surface at the same distance from the sun, and normal to the incident radiation (remote sensing). Sometimes used to mean the flux of the reflected radiation, e.g., the earth albedo is 0.64 calories per square centimeter.
ALKALINE EARTH METALS. These elements include Be, Mg, Ca, Sr, and Ba. The geochemistries of the alkaline earth elements are dominated by their ease of oxidation. With the exception of Mg, and to a lesser extent Ca, these elements are concentrated in the continental crust.
ALKALINITY. The number of milliequivalents of hydrogen ion that is neutralized by one liter of seawater at $20°C$ (oceanography).
ALPHA-PARTICLE. (1) Nucleus of a helium atom. (2) Positively charged particle with two protons and two neutrons emitted from an atomic nucleus during radioactive decay.
ALTIMETER. An instrument for determining height above ground or sea level based on the decrease in atmospheric pressure with increase in altitude.
ANDESITE. Fine-grained igneous rock that is the extrusive equivalent of diorite. Phenocrysts, if present, are mainly the andesine variety of plagioclase and one or more of the mafic minerals.
ANHYDRITE. A mineral, $CaSO_4$, anhydrous calcium sulfate, common in evaporite beds.
ANION. A negative ion.
ANOMALY. A deviation from uniformity or regularity in geophysical quantities; a difference between observed and computed value.
ANORTHOSITE. A group of essentially monomineralic plutonic igneous rocks composed mainly of plagioclase feldspar which is usually

labradorite. Anorthosites occur as large nonstratiform plutonic bodies, as stratiform intrusions, and have been identified in lunar rock samples.

ANTICYCLONIC GYRE. A great, closed, circular motion of water in an ocean basin centered on an atmospheric high-pressure system. Flow of water generated by earth's rotation, prevailing winds, and convective flow of warm surface water poleward. The water turns clockwise in the Northern Hemisphere and counterclockwise in the Southern Hemisphere.

AQUIFER. Permeable rock strata below the surface through which ground water moves, which is generally capable of producing water for a well.

ARTESIAN AQUIFER. A confined aquifer which is bounded above and below by impermeable beds.

ASTHENOSPHERE. Layer of the earth below the crust, which is weak and in which isostatic adjustments take place. Magmas may be generated within this layer and seismic waves are strongly attenuated. It is equivalent to the upper mantle.

ATMOSPHERIC PRESSURE. The force per unit area in any part of the earth's atmosphere. Normal pressure at sea level is variously defined as 76.0 cm or 29.92 inches of mercury, 1033.3 cm or 33.9 ft of water, 3 grams or 1,013,250.0 dynes per cm^2, 14.66 pounds per square inch, or 1.01325 bars.

ATOMIC ABSORPTION SPECTRUM. The absorption spectrum seen when the unexcited atoms of a vaporized sample selectively absorb certain wavelengths of light passed through the sample.

AUTHIGENIC. Formed or generated in place; often refers to a mineral (such as quartz or feldspar) formed after deposition of the original sediment.

BENCH MARK. A permanent marker which designates a point of known elevation.

BENTHONIC. Pertaining to those forms of marine life that are bottom dwelling or the ocean bottom itself.

BIOCOENOSE. Group of organisms that live closely together and form a natural ecologic unit.

BIOHERM. A moundlike or domelike mass of rock built up by the remains of sedentary organisms, such as corals, stromatoporoids, and algae.

BIOSTROME. A distinctly bedded, blanketlike mass of rock built by and composed mainly of the remains of sedentary organisms.

BRECCIA. A general term for a rock made up of coarse angular fragments as distinguished from conglomerate which is composed of rounded rock fragments. There are sedimentary as well as volcanic and other types of breccia.

CALCAREOUS. Containing calcium carbonate. When applied to a rock name it implies that a considerable percentage (up to 50 percent) of the rock is calcium carbonate.

CALCIUM CARBONATE COMPENSATION DEPTH. The level in the ocean below which the rate of calcium carbonate solution exceeds the rate of its deposition.

CALDERA. A large, basin-shaped volcanic depression, more or less circular, the diameter of which is many times greater than that of the included vent or vents, no matter what the steepness of the walls or form of the floor.

CALICHE. (1) A hard soil layer cemented by calcium carbonate and found in arid and semiarid regions of southwestern U.S. and Mexico. (2) Nitrate deposits of the Atacama Desert of Chile and Peru which are cemented with soluble salts of sodium.

CARBON-14 DATING. A method of determining age by measuring the concentration of carbon-14 remaining in an organic material. Useful in the range of 500 to 40,000 years, although it may be extended to 70,000 years by enrichment techniques (radiocarbon dating).

CHLORINITY. The chloride content of seawater, measured by mass, or grams per kilogram of seawater, and including all the halides.

CHLORITE. A group of platy, greenish minerals of general formula: $(Mg,Fe^{+2},Fe^{+3})_6 AlSi_3 O_{10}(OH)_8$. Chlorite is characterized by prominent ferrous iron and by the absence of calcium and alkalies; chromium and manganese may also be present. Chlorites are associated with and resemble the micas (the tabular crystals of chlorite cleave into small thin flakes or scales that are flexible, but not elastic as those of mica), and are widely distributed, esp. in low-grade metamorphic rocks, or as alteration products of ferromagnesian minerals in igneous rocks.

COCCOLITH. A general term applied to various structural elements or plates with many different shapes and averaging 3 microns in diameter. They are made of aragonite or calcite and constitute the outer skeletal remains of a coccolithophore.

COCCOLITHOPHORE. Any of numerous, mostly marine planktonic flagellate organisms that produce coccoliths; variously classified as algae and protozoans (coccolithophorid).

COLLOID. A particle-size range of less than 0.00024 mm, i.e., smaller than clay size. Also any fine-grained material in suspension, or any such material that can be easily suspended.

CONNATE. Originating at the same time as the adjacent material. Usually refers to water and volatiles entrapped in sediments at the time of deposition.

CONNATE WATER. Water entrapped in the interstices of a sedimentary rock at the time of its deposition.

CONTINENTAL DRIFT. General term for theory of displacement of large plates of continental (sialic) crust, moving freely across a substratum of oceanic (simatic) crust (continental displacement).

CONVECTION. General term for the movement of subcrustal or mantle material as a result of heat variations. Also applies to similar movements within the earth's atmosphere and oceans.

CONVERGENCE. An increase in volume of an air mass; greater inflow than outflow of air.

COSMIC RAYS. Atomic nuclei of very high energy from outer space that bombard the earth's atmosphere where most of them are absorbed. Secondary cosmic rays with less energy reach the earth's surface and are part of the natural background radiation.

CRATON. A relatively large part of the earth's crust which has attained stability, and which has been little deformed for a prolonged period.

CREEP. (1) Slow deformation (strain) of solid rock resulting from a small, constant stress acting over a long period of time. (2) Slow downslope movement of regolith or soil.

CRUST. The outermost layer or shell of the earth, generally above the Mohorovicic discontinuity.

CRUSTAL PLATE. One of the six major blocks into which the lithosphere is divided, according to the scheme of global tectonics. Plates are about 100 km thick.

CRYSTALLIZATION AGES. Age expressed in years calculated from the quantitive determination of the radioactive elements and their decay products since the minerals in a rock crystallized.

DEW-POINT. The temperature to which air must be cooled, at constant pressure and constant water-vapor content, in order for saturation to occur; the temperature at which the saturation pressure is the same as the existing vapor pressure.

DIATOM OOZE. Sediment, found on the bottom of the sea, consisting of clay minerals and the siliceous skeletal remains of microscopic single-celled plants known as diatoms.

DIPOLE FIELD. A mathematically simple magnetic field, having an axis of symmetry, with magnetic field lines pointing outward along one half of the axis (positive pole) and inward along the negative half of the axis (negative pole).

DISSOLVED OXYGEN. The amount of dissolved oxygen, in parts per million by weight, present in water, generally expressed in mg/l. A critical factor for fish and other aquatic life, and for self-purification of a surface-water body after inflow of oxygen-consuming pollutants.

DOPPLER EFFECT. Change in the observed frequency of electromagnetic or other waves caused by relative motion between the source and the observer.

DYNAMO THEORY. The earth's main magnetic field is thought to be sustained by self-exciting dynamo action in the fluid core. The conducting liquid is supposed to flow in such a pattern that the electric current induced by its motion through the magnetic field sustains that field.

EARTHQUAKE. A local trembling, shaking, undulating, or sudden shock of the surface of the earth, sometimes accompanied by fissuring or by permanent change of level.

ECCENTRICITY. The condition, degree, amount, or instance of deviation from a center. The distance of a center of figure of a body from an axis about which it turns.

ECLIPTIC. The great circle cut out of the celestial sphere by a plane containing the orbit of the earth. The plane of the ecliptic is inclined to the plane of the equator by an angle of approximately 23.5°.

ECLOGITE FACIES. Rocks formed by regional metamorphism at extremely high pressures (10,000 bars) and temperatures of 600° to 700°C. The high density mineral association includes omphacite (pyroxene) and garnet plus rutile, kyanite, enstatite, and diamond.

EJECTA. Glass, shock-metamorphosed rock fragments, and other material thrown out of an explosion or impact crater during formation. Such material may be distributed around a crater in distinctive patterns forming "ejecta rays" or "ejecta loops."

ELECTRICAL CONDUCTIVITY. A measure of the ease with which a conduction current can be caused to flow through a material under the influence of an applied electric field. It is measured in mhos per meter and is the reciprocal of resistivity.

ENVIRONMENT. (1) The aggregate of external conditions that influence the life of an individual or population. (2) The aggregate of all the surrounding conditions, influences, or forces affecting a locus of sedimentation.

EOLIAN. Pertaining to the wind. Includes deposits whose constituents were transported by wind, landforms eroded by wind, and sedimentary structures made by the wind.

EPICENTER. The point of the earth's surface directly above an earthquake focus.

EPIDEMIOLOGY. That branch of medical science which is concerned with the study of disease as it appears in natural surroundings, and as it affects a community of people rather than a single individual.

EQUINOXES. Equinoxes occur at the two points in the earth's orbit where the ecliptic intersects the celestial equator. They are on or about March 21 and September 22, at which times the sun appears vertically above the equator.

EQUIVALENT (CHEMICAL). The weight in grams of a substance which combines with or displaces one gram of hydrogen, obtained by dividing the formula weight by the valence.

ESTUARINE. Pertaining to, formed, or living in an estuary; esp. said of deposits and of the sedimentary or biological environment of an estuary.

EUGEOSYNCLINE. A geosyncline in which volcanism is associated with clastic sedimentation; the volcanic part of an orthogeosyncline, located away from the craton.

EXCHANGE CAPACITY. Ability of a substance to have ion exchange. It is measured by the quantity of exchangeable ions in a given unit of the material.

FAULT. A fracture or fracture zone along which the opposite sides have been relatively displaced.

FAULT BLOCK. A crustal unit formed by block faulting; it is bounded by faults, either completely or in part. It behaves as a unit during block faulting and tectonic activity. An example is the Sierra Nevada of western United States.

FIELD REVERSAL. A change in the earth's magnetic field between normal polarity and reversed polarity (geomagnetic reversal).

FLUORESCENCE SPECTRUM. The spectrum produced when a fluorescent material is induced to emit radiation of one kind when irradiated with radiation of another kind as in X-ray fluorescence spectroscopy.

FLUX. (1) A stream of flowing water; a flood or an outflow. (2) The number of radioactive particles in a given volume times their mean velocity.

FLUX-GATE MAGNETOMETER. A magnetometer that measures the component of the magnetic field along the axis of its sensor.

FOCAL DEPTH. Distance from the focus of an earthquake to the epicenter.

FOOD CHAIN. The passage of energy and materials from producers through a progressive sequence of plant-eating and meat-eating consumers.

FORAMINIFER. Any protozoan belonging to the order Foraminiferida, characterized by a shell of agglutinated particles or secreted calcite; commonly found in marine or brackish environments.

FORMATION. (1) A lithologically distinctive product of essentially continuous sedimentation selected from a local succession of strata as a convenient unit for mapping, description, and reference. (2) Something naturally formed, commonly differing conspicuously from adjacent objects or material, or being noteworthy for some other reason.

GAL. An acceleration of one centimeter per second. A milligal is 0.001 gal.

GAMMA. Unit of magnetic field strength. It is equal to 10^{-5} oersted.

GAMMA-RAY LOG. A radioactivity log obtained by recording the naturally emitted gamma rays emitted from the rocks traversed in a borehole and plotting the data as a function of depth.

GAMMA-RAY SPECTROMETER. An instrument for measuring the energy distribution, or spectrum, of gamma rays, whether from natural or artificial sources. It is used as an airborne remote-sensing technique for potassium, thorium, and uranium.

GAS CHROMATOGRAPHY. A process for separating gases or vapors from one another by passing them over a solid (gas-solid chromatography) or liquid (gas-liquid chromatography) phase. The gases are repeatedly absorbed and released at differential rates resulting in separation of their components.

GAUSS. The cgs (centimeter gram second) unit for magnetic induction or magnetic flux density. The field one centimeter from a straight wire carrying 5 amps is one gauss.

GEOMAGNETIC REVERSALS. A change of the earth's magnetic field between normal polarity and reversed polarity.

GEOPHYSICAL LOGS. A log obtained by lowering an instrument into a borehole and recording continuously on a meter at the surface some physical property of the rock material being logged.

GEOSTROPHIC CURRENT. A wind or ocean current in which the horizontal pressure force is exactly balanced by the equal but opposite Coriolis force.

GEOSYNCLINE. A mobile elongate or basinlike downwarping of the crust of the earth which is subsiding as sedimentary and volcanic rocks accumulate to thicknesses of thousands of meters. A geosyncline may be part of a tectonic cycle in which orogeny follows.

GEOTECTONICS. A branch of geology dealing with the gross features of the upper part of the earth's crust including the assembling of structural or deformational features, a study of their mutual relations, their origin and their historical evolution.

GEOTHERMAL ENERGY. Useful energy that can be extracted from naturally occurring steam and hot water found in the earth's volcanic and young orogenic zones, whose surface manifestations include hot springs, fumaroles, and geysers. All such areas under development are in areas of late Cenozoic volcanic activity.

GLOBAL TECTONICS. Tectonics on a global scale, such as tectonic processes related to very large-scale movement of material within the earth.

GLOBIGERINA OOZE. A calcaereous ooze whose skeletal remains are foraminiferal tests, predominantly of the genus *Globigerina*.

HALF-LIFE. The time necessary for a radioactive substance to lose half its radioactivity during decay.

HEAT FLOW. The product of the thermal conductivity of a substance and the thermal gradient in the direction of the flow of heat.

HUMUS. The generally dark, more or less stable part of the organic matter of the soil which is so well decomposed that the original sources cannot be identified (sometimes used incorrectly for the total organic matter of the soil, including relatively undecomposed material).

HYDRODYNAMIC. The branch of science that deals with the cause and effect of regional subsurface fluid migration. The branch of hydraulics that relates to flow of liquids through pipes and openings.

HYPERFUSIBLE. Any substance capable of lowering the melting ranges in end-stage magmatic fluids.

INFRARED. Pertaining to the portion of the electromagnetic spectrum with wavelengths just beyond the red end of the spectrum, in the range from 0.7 to about 1.0 micrometer.

INORGANIC. Pertaining or relating to matter that is not organic. Involving neither organic life nor the products of organic life.

INVERSION. A reversal of the gradient of a meteorological element, e.g., an increase rather than a decrease of temperature with height.

ION EXCHANGE. The reversible replacement of certain ions by others, without loss of crystal structure.

ISLAND ARCS. A chain of islands, e.g., the Aleutians, rising from the deep sea floor and near to the continents, a primary arc expressed as a curved belt of islands. Its curve is generally convex toward the open ocean.

ISOBATH. A line on a map that connects points of equal water depth.

ISOTOPE. Alternate form of an element. Elements having the same number of protons but different numbers of neutrons in their nuclei. Isotopes have the same atomic number but differing atomic weights. Though the isotopes of an element have basically the same chemical properties, due to the similarity in electron configuration, they have slightly different physical properties by which they can be separated.

ISOTOPIC. Pertaining or relating to an isotope.

KEY BEDS. A well-defined, easily identifiable stratum or body of strata that has sufficiently distinctive characteristics (such as lithology or fossil content) to facilitate correlation in field mapping or subsurface work.

LAGOON. A marsh, shallow pond, or lake, especially one into which the sea flows.

LAURASIA. Hypothetical continent in the Northern Hemisphere which supposedly separated during Mesozoic time to form the present northern continents.

LEACH. To wash or to drain by percolation. To dissolve minerals by percolating solutions.

LEFT LATERAL SEPARATION. Movement of a lateral fault along which, in plan view, the side opposite the observer appears to have moved to the left.

LEVEE. An embankment beside a river to prevent overflow.

LIBRATION. The small, angular change in the face that a body, e.g., the moon, presents toward the earth. Only a tiny part is due to dynamic rotational motion (physical libration). In the case of the moon, it is due primarily to the fact that although the moon is in synchronous rotation, its rotation is uniform while its rate of revolution varies due to orbital eccentricity producing longitudinal geometric libration; and also the fact that its rotational axis is not exactly perpendicular to the plane of its orbit.

LITHOLOGY. The character of a rock described in terms of its structure, color, mineral composition, grain-size, and arrangement of its component parts.

LITHOSPHERE. The solid portion of the earth, as compared with the atmosphere and the hydrosphere; the crust of the earth, as compared with the mantle and the core.

LITTORAL. (1) Pertaining to the benthic ocean environment or depth zone between high water and low water; also pertaining to the organisms of that environment. (2) Pertaining to the depth zone between the shore and a depth of about 200 m. In this meaning, the term includes the neritic zone.

LOG. A detailed record of the progress made in drilling a well. It includes rock types, fossils, fluids, and structures encountered and is determined by using geophysical devices lowered into the borehole and examination of cuttings and cores.

LOGISTIC CURVE. A growth curve used to describe functions which continually increase, gradually at first, more rapidly in the middle growth period, and slowly again, reaching a maximum at the end of the growth.

LOGNORMAL DISTRIBUTION. A frequency distribution whose logarithm follows a normal distribution.

LONG-WAVE TERRESTRIAL RADIATION. Infrared radiation emitted from the earth's surface, including the oceanic surface.

LOST CIRCULATION. The condition during a drilling operation when the drilling mud escapes into the porous or cavernous sidewalls of the borehole and does not return to the surface.

LOW-VELOCITY ZONE. Zone in the upper mantle of the earth, variously defined as from 60 to 250 km in depth, in which velocities are about 6 percent lower than in the outermost mantle.

MAGNETOMETER. An instrument that measures the earth's magnetic field and its changes, or the magnetic field of a particular rock (from which its magnetization is deduced).

MAGNETOPAUSE. The outer limit of the earth's atmosphere which is located many earth radii from the earth. It is also the outer limit of the magnetosphere which is the region where the magnetic field rather than collisions controls the motions of charged particles.

MAGNETOSPHERE. The region around the earth to which the earth's magnetic field is confined, due to interaction between the solar wind and the geomagnetic field. On the sunlit side, the magnetosphere is approximately hemispherical, with a radius of about ten earth radii under quiet conditions; it may be compressed to about six earth radii by magnetic storms. On the side opposite the sunlit side, the magnetosphere extends in a "tail" of several hundred earth radii.

MANTLE. (1) The layer of the earth between the crust and the core. (2) Mantle rock is the soil or other unconsolidated rock material which is more commonly referred to as overburden or regolith.

MARE. (1) One of the several dark, low-lying, level, relatively smooth, plainslike areas of considerable extent on the surface of the moon, having fewer large craters than the highlands, and composed of mafic or ultramafic volcanic rocks, e.g., Mare Imbrium. (2) A dark area, on the surface of Mars, whose origin is not definitely known.

MASCONS. A large-scale, high-density, lunar mass concentration below a ringed mare.

MASSIFS. A massive topographic and structural feature in an orogenic belt, commonly formed of rocks more resistant than those of its surroundings. These rocks may be protruding bodies of basement rocks, consolidated during earlier orogenies, or younger plutonic bodies.

MASS SPECTROMETER. An instrument for producing and measuring, usually by electrical means, a mass spectrum. It is especially useful for determining molecular weights and relative abundances of isotopes within a compound.

MEMBER. A rock-stratigraphic unit of subordinate rank, comprising some specially developed part of a varied formation (such as a subdivision of only local extent, or a lithologically unified subdivision distinguished from adjacent parts of the formation by color, hardness, composition, or similar features), and not defined by specified shape or extent.

MERCATOR PROJECTION. An equatorial, cylindrical, conformal map projection in which the equator is represented by a straight line true to scale; the meridians by parallel straight lines, equally spaced, and the parallels by straight lines perpendicular to the meridians and parallel with the equator.

MERCURY. A heavy, silver- to tin-white hexagonal mineral. The native metallic element Hg. It is the only metal that is liquid at ordinary temperatures. Native mercury is found as minute fluid globules disseminated through cinnabar or deposited from the waters of certain hot springs, but it is an unimportant source of the metal.

MESOSPHERE. The lower mantle; it is not involved in the earth's tectonic processes.

METALLIFEROUS DEPOSIT. A mineral deposit from which a metal or metals can be extracted by metallurgical processes.

METASOMATIC. Pertaining to the process of metasomatism and to its results. The term is especially used in connection with the origin of ore deposits.

METASOMATISM. Gases contained within a rock body or introduced from external sources. The process of nearly simultaneous capillary solution and deposition by which a new mineral of partly or wholly different chemical composition may grow in the body of an old mineral or mineral aggregate.

METEORIC. (1) Pertaining to, dependent on, derived from, or belonging to the earth's atmosphere; e.g., "meteoric erosion" caused by rain, wind, or other atmospheric forces. (2) Relating to or composed of meteors or meteoroids.

METEORIC WATER. Water of recent atmospheric origin.

GLOSSARY 355

METEOROIDS. One of the countless solid objects moving in interplanetary space, distinguished from asteroids and planets by their smaller size but considerably larger than an atom or molecule. A meteoroid that has fallen through the earth's atmosphere without being completely vaporized is known as a meteorite.

MICROWAVE. Region of the electromagnetic spectrum in the millimeter and centimeter wavelengths. Passive sensing systems operating at these wavelengths are called microwave systems; active systems are called radar.

MID-OCEAN RIFT. The deep, central cleft in the crest of the mid-oceanic ridge, about 25-50 km in width, with a mountainous rather than a flat floor.

MILLIEQUIVALENT. 1/1000 of an equivalent.

MILLIGAL. See gal.

MINERAL FUELS. Coal and petroleum, including natural gas.

MIOGEOSYNCLINE. A geosyncline in which volcanism is not associated with sedimentation; the nonvolcanic aspect of an orthogeosyncline, located near the craton.

MODIFIED MERCALLI SCALE. One of the earthquake intensity scales, having twelve divisions ranging from I (not felt by people) to XII (damage nearly total). It is a revision of the Mercalli scale made by Wood and Newmann in 1931.

MONSOON. A seasonal type of wind system in which its direction changes with the seasons, e.g., over the Arabian Sea where the winds are from the northeast for six months and then from the southeast for the next six months.

MONTMORILLONITE. Clay minerals which have a theoretical composition of essentially $Al_4 Si_8 O_{20}(OH)_4 \cdot nH_2 O$.

NEBULA. An interstellar cloud of gas or dust.

NEUTRON ACTIVATION ANALYSIS. A method for identifying and measuring chemical elements using neutron bombardment to make the sample radioactive followed by determination of the radiations that are characteristic of the type and quantity of atoms present.

NEUTRON LOG. A radioactivity log that measures the intensity of radiation (neutrons or gamma rays) artificially produced when rocks around a borehole are bombarded by neutrons from a synthetic source. It indicates the pressure of fluids and when used with the gamma-ray log porous and nonporous formations can be differentiated.

NOBLE (RARE, INERT) GASES. These gases constitute Group O of the periodic table and are Ar, He, Kr, Ne, Rn, and Xe. The noble gases are products of radioactive decay and are colorless, odorless, and tasteless. They are characterized by closed shells or subshells of electrons.

NUCLIDE. A general term applicable to all atomic forms of the elements. Often erroneously used as a synonym for "isotope." Whereas isotopes are the various forms of a single element (hence are a family of nuclides) and all have the same atomic number and number of protons, nuclides comprise all the isotopic forms of all the elements.

OFFSET. In a fault, the horizontal component of displacement, measured parallel to the strike of the fault.

OXIDATION. The chemical process of oxidation is a change in the availability of the outer orbital electrons of the element, specifically, loss of electrons from an ion constitutes oxidation.

OXYGEN 18 CONTENT. See oxygen isotope fractionation.

OXYGEN ISOTOPE FRACTIONATION. Temperature-dependent isotopic fractionation which changes the oxygen-18/oxygen-16 isotope ratio in the carbonate shells of marine organisms. The ratio is used as an indication of water temperature at the time of deposition of the shell.

OZONE. A faintly blue, irritating gas (O_3) occurring in minute quantities near the earth's surface and in large quantities in the stratosphere as a product of the action of ultraviolet light of short wavelengths on ordinary oxygen. Ozone, concentrated in a layer about 15 miles above the earth, shields the earth from UV radiation in the range of 2400-3000 Å by absorption.

PALEOMAGNETISM. The study of natural remanent magnetization in order to determine the intensity and direction of the earth's magnetic field in the geologic past.

PANTHALASSA. The hypothetical proto-ocean surrounding Pangaea. Supposed by some geologists to have combined all the oceans or areas of oceanic crust of the earth at an early time in the geologic past, across which the present continents were gradually displaced to their present positions from the original proto-continent.

PELAGIC. (1) Pertaining to the water of the ocean as an environment. (2) Said of marine organisms whose environment is the open ocean, rather than the bottom or shore areas. (3) Pertaining to the deeper part of a lake, characterized by deposits of mud or ooze and by the absence of aquatic vegetation.

PENEPLAIN. A low, nearly featureless, and almost plain land surface of considerable area which presumably has been reduced by the processes of long-continued subaerial erosion.

PENEPLANATION. The act or process of formation and development of a peneplain.

PERIHELION. Point in the orbit of any member of the solar system when the object is closest to the sun.

PERMEABILITY. The property or capacity of a porous rock, sediment, or soil for transmitting a fluid without impairment of the structure of the medium; it is a measure of the relative ease of fluid flow under unequal pressure. The customary unit of measurement is the millidarcy.

PERTURBATIONS. The influences which the attractions of the other members of the solar system have on the motion of the object under consideration are known as perturbations.

PETROLEUM. Gaseous, liquid, or solid substances, occurring naturally and consisting chiefly of chemical compounds of carbon and hydrogen.

pH. A measure of acidity or alkalinity. It is the negative logarithm of the hydrogen ion activity; pH7 indicates an H^+ concentration (activity) of 10^{-7} mole/litre.

PHANEROZOIC. That part of geologic time for which, in the corresponding rocks, the evidence of life is abundant, esp. of higher forms. Comprised of the Paleozoic, Mesozoic, and Cenozoic eras.

PHENOCRYSTS. A term for a relatively large, conspicuous crystal in a porphyritic rock.

PHOTOCHEMICAL REACTIONS. Reactions in which light supplies the energy necessary for the activation of the reacting molecules.

PHOTON. The carrier of a quantum of electromagnetic energy. Photons have an effective momentum but no mass or electrical charge.

PHOTOSYNTHESIS. The chemical process during which green plants convert carbon dioxide to organic food substances.

PHOTOVOLTAIC. Pertaining to a class of radiation detector that functions on a voltage change.

PHYTOPLANKTON. The plant forms of plankton, e.g., diatoms.

PIEZOMETRIC SURFACE (POTENTIOMETRIC SURFACE). An imaginary surface representing the static head of ground water and defined by the level to which water will rise in a well. The water table is a particular potentiometric surface.

PLANKTONIC. Pelagic organisms which float are said to be planktonic.

PLASMA. An electrically neutral gaseous mixture of positive and negative ions. High-temperature plasmas are used in controlled fusion experiments.

PLATE TECTONICS. Global tectonics based on an earth model characterized by a small number of large, broad, thick plates (blocks composed of areas of both continental and oceanic crust and mantle) each of which "floats" on some viscous underlayer in the mantle and moves more or less independently of the others and grinds against them like ice floes in a river, with much of the dynamic activity concentrated along the periphery of the plates which are propelled by sea-floor spreading. The continents form a part of the plates and move with them, like logs frozen in the ice floes.

PLUTON. An igneous intrusion.

POLONIUM. A radioactive element which has an atomic number of 84. The most common isotope has an atomic weight of 210 and is in the uranium radioactive series.

POROSITY. The proportion, usually stated as a percentage of the total volume of a rock material or regolith that consists of pore space or voids.

PRESSURE HEAD. Hydrostatic pressure expressed as the height of a column of water that the pressure can support, expressed with reference to a specific level such as land surface.

PRIMORDIAL. Original or existing at the beginning.

RADIOACTIVE EQUILIBRIUM. The condition of equilibrium in which the rate of decay of the parent isotope is exactly matched by the rate of decay of every intermediate daughter isotope. When equilibrium has been established, the concentrations of intermediate daughters remain virtually constant. Specimens which contain the natural radioactive elements and which have been in radioactive equilibrium for a very long time are said to be in secular equilibrium.

RADIOACTIVE WASTES. Equipment and materials (from nuclear operations) which are radioactive and for which there is no further use. Wastes are generally classified as high-level (having radioactivity concentrations of hundreds to thousands of curies per gallon or cubic foot), low-level (on the order of 1 microcurie per gallon or cubic foot), or intermediate (between these ranges).

RADIOACTIVITY. The spontaneous decay of the atoms of certain isotopes into new isotopes, which may be stable or undergo further decay until a stable isotope is finally created; radioactive decay. Radioactivity is accompanied by the emission of alpha particles, beta particles, and gamma rays and by the generation of heat.

RADIOCHEMISTRY. The chemical study of irradiated and naturally occurring radioactive materials and their behavior. It includes their use in tracer studies and other chemical problems.

RADIOGENIC AGE. See radiometric age.

RADIOLARIAN. Any actinopod belonging to the subclass Radiolaria, characterized mainly by a siliceous skeleton and a marine pelagic environment. Their stratigraphic range is Cambrian to present. In some classifications the radiolarians are grouped with the rhizopods.

RADIOMETER. A radiation-measuring instrument having substantially equal response to a band of wavelengths, usually either in the infrared or visible.

RADIOMETRIC AGE. An age expressed in years and calculated from the quantitative determination of radioactive elements and their decay products.

RADIONUCLIDES. A radioactive nuclide. The term radioisotope is loosely used synonymously.

RADON. A radioactive element, one of the heaviest gases known. Its atomic number is 86, and its atomic weight is 222. It is a daughter of radium in the uranium radioactive series. (Symbol Rn.)

REDUCTION. The chemical process of reduction is a change in the availability of the outer orbital electrons of the element, specifically, gain of electrons by an ion constitutes reduction.

REGOLITH. The layer or mantle of loose, incoherent rock material, of whatever origin, that nearly everywhere forms the surface of the land and rests on the bedrock.

REMANENT MAGNETIZATION. That component of a rock's magnetization that has a direction fixed relative to the rock and which is independent of moderate, applied magnetic fields such as the earth's magnetic field.

REVERSAL TIME SCALE. A time scale based upon the record of geomagnetic reversals.

REVETMENTS. A facing of stone, concrete, or other material, built to protect an embankment (as of a stream or lake) or a shore structure from wave erosion.

RICHTER SCALE. The range of numerical values of earthquake magnitude, devised in 1935 by the seismologist C. F. Richter. Very small earthquakes, or microearthquakes, can have negative magnitude values. In theory there is no upper limit to the magnitude of an earthquake. However, the strength of earth materials produces an actual upper limit of slightly less than 9.

RILLES. One of several relatively long (up to several hundred kilometers), narrow (1-2 km), trenchlike or cracklike valleys commonly occurring on the moon's surface. Rilles may be extremely irregular with meandering courses ("sinuous rilles"), or they may be relatively straight depressions ("normal rilles"); they have relatively steep walls and usually flat bottoms. Rilles are essentially youthful features and apparently represent fracture systems originating in brittle material.

SALINITY. The total quantity of dissolved salts in sea water, measured by weight in parts per thousand, with the following qualifications: all the carbonate has been converted to oxide, all the bromide and iodide to chloride, and all the organic matter has been completely oxidized. Salinity is usually computed from some other factor, such as chlorinity.

SAMPLE LOGS. A log showing the rocks penetrated in drilling a borehole or well, compiled by a geologist from information obtained through microscopic study of drilling samples (core and cuttings) recovered at the surface, and plotted on a log strip subdivided into units of depth. It shows the sequence and characteristics of the strata penetrated in drilling and consists of colors and/or symbols with a written, abbreviated description of the lithology.

SANITARY LANDFILL. A disposal area for solid wastes where the wastes are compacted and covered by a layer of impermeable material, such as clay, daily.

SAVANNA. An open, grassy, essentially treeless plain, esp. as developed in tropical or subtropical regions. Usually there is a distinct wet and dry season; what trees and shrubs are found there are drought-resistant.

SCARP. A cliff, escarpment, or steep slope of some extent formed by a fault or a cliff of steep slope along the margin of a plateau, mesa, or terrace.

SEA OF TETHYS. A sea which existed for long periods of geologic time between the northern and southern continents of the Eastern Hemisphere.

SEA-FLOOR SPREADING. A hypothesis that the oceanic crust is increasing by convective upwelling of magma along the midoceanic ridges or world rift system, and a moving away of the new material at a rate of from one to ten centimeters per year. This movement provides the source of power in the hypothesis of plate tectonics. This hypothesis supports the continental displacement hypothesis.

SEDIMENTARY ROCK. Rock formed by the accumulation of sediment in water (aqueous deposits) or from air (eolian deposits). A characteristic feature of sedimentary deposits is a layered structure known as stratification or bedding.

SEISMIC. Pertaining to an earthquake or earth vibration, including those that are artificially induced.

SEISMOGRAPHS. An instrument for recording earthquake or seismic waves. The record made by a seismograph is called a seismogram.

SEISMOMETERS. An instrument that detects earth motions. It is the detector part of the seismograph system, and does not by itself contain a recording unit.

SHOCK METAMORPHISM. The totality of observed permanent physical, chemical, mineralogic, and morphologic changes produced in rocks and minerals by the passage of transient high-pressure shock waves acting over short time intervals ranging from a few microseconds to a fraction of a minute. The only known natural mechanism for producing shock-metamorphic effects is the hypervelocity impact of large meteorites, but the term also includes identical effects produced by shock waves generated in small-scale laboratory experiments and in nuclear and chemical explosions.

SILICEOUS. Said of a rock containing abundant silica, esp. free silica rather than as silicates.

SLICKENSIDES. A polished and smoothly striated surface that results from friction along a fault plane. A surface bearing slickensides is said to be "slickensided."

SOIL. (1) The unconsolidated material above the bedrock that forms as a result of weathering by organic and inorganic processes. (2) In pedology, the weathered material that will support rooted plants. (3) In engineering geology, soil is equivalent to regolith.

SOIL HORIZON. A soil layer distinguished from other soil layers by physical properties such as color, structure, and texture, or by chemical composition.

SOLAR WIND. The motion of interplanetary plasma or ionized particles away from the sun and towards the earth, near which it interacts with the earth's magnetic field.

SOLSTICE. The two points in the earth's orbit which correspond to the maximum deviations of the ecliptic from the celestial equator. When these points are occupied, the sun is farthest north and south of the equator — namely, June 21 (summer solstice) and December 22 (winter solstice).

SPALLATION. Ejection of atomic particles from a nucleus following collision of an atom and a high-energy particle (e.g., a cosmic ray) which results in the formation of a different isotope that is not a fission product.

SPECTROMETER. An instrument for producing, observing, and measuring wavelengths and indices of refraction of rays in a spectrum.

SPECTRUM. An array of visible light arranged according to its constituent wavelengths (colors) by being passed through a prism or diffraction grating.

STEPPE. An extensive, treeless grassland area in southeastern Europe and Asia developing in the semiarid mid-latitudes of that region. They are generally considered drier than the prairie which develops in the subhumid mid-latitudes of the U.S.

STRATOSPHERE. (1) The second lowest layer of the atmosphere; characterized by more or less isothermal conditions and a highly stable stratification. The stratosphere extends from about 7 to 20 miles above the earth's surface and contains little moisture or dust but most of the ozone. (2) In oceanography, the waters of the ocean below the thermocline.

STRIKE-SLIP FAULT. A fault in which the movement is parallel to the strike of the fault.

SUBDUCTION. The process of one crustal block descending beneath another, by folding or faulting or both. The concept was originally used by Alpine geologists.

SUBDUCTION ZONES. An elongate region along which a crustal block descends relative to another crustal block, e.g., the descent of the Pacific plate beneath the Andean plate along the Andean trench.

SUBMARINE FANS. A terrigenous, cone- or fan-shaped deposit located seaward of large rivers and submarine canyons.

SUBSIDENCE. A local mass movement that involves principally the gradual downward settling or sinking of the solid earth's surface with little or no horizontal motion and that does not occur along a free surface (not the result of a landslide or failure of a slope). The movement is not restricted in rate, magnitude, or area involved. Subsidence may be due to natural geologic processes such as solution, erosion, oxidation, thawing, lateral flow, or compaction of subsurface materials; earthquakes, slow crustal warping, and volcanism (withdrawal of fluid lava beneath a solid crust); or man's activity such as removal of subsurface solids, liquids, or gases and wetting of some types of moisture-deficient loose or porous deposits.

SUPERHEATED. The addition of more heat than necessary to complete a given phase change.

TALUS. The heap of coarse rock waste at the foot of a cliff or a sheet of waste covering a slope below a cliff.

TECTONICS. Branch of geology dealing with the broad architecture of the earth's crust including the origin, relationship, and history of regional structural and deformational features.

TEKTITES. A small (usually walnut-sized), rounded, pitted, jet black to olive-greenish or yellowish body of silicate glass of nonvolcanic origin, found usually in groups in several widely separated areas of the earth's surface and bearing no relation to the associated geologic formations. Most tektites have a uniformly high silica (68-82%) and very low water contents (average 0.005%); their composition is unlike that of obsidian and more like that of shale. They have various shapes (teardrop, dumbbell, canoe) strongly suggesting modelling by aerodynamic forces, and they average a few grams in weight (largest weighs 3.2 kg). Tektites are believed to be of extraterrestrial origin (e.g., moon splash, formed as gravity-escaping ejecta following large lunar impacts), or alternatively the product of large hypervelocity meteorite impacts on terrestrial rocks.

TEXTURE. The physical appearance of a rock, as shown by size, shape, and arrangement of the particles in the rock.

THANATOCOENOSE. A group of organisms brought together after death.

THERMAL CONDUCTIVITY. A measure of the ability of a material to conduct heat. Typical values of thermal conductivity for rocks range from 3 to 15 millicalories/cm-sec-$°C$. Rocks with abundant quartz have high thermal conductivities. Poorly consolidated sediments have lower thermal conductivities.

THERMOHALINE CIRCULATION. The gain or loss of heat by the surface layer of the ocean and changes in salinity, due to evaporation or precipitation, produce density changes which can give rise to vertical movements of water in the ocean.

THORIUM. A naturally radioactive element with atomic number 90 and, as found in nature, an atomic weight of approximately 232. The fertile thorium 232 isotope is abundant and can be transmuted to fissionable uranium 233.

TIME-STRATIGRAPHIC BOUNDARIES. Boundaries of rock units which are synchronous, i.e., based on geologic time.

TRACE ELEMENTS. Elements present in minor amounts in the earth's crust or that occur in minor quantities in plant or animal tissue. Also known as minor elements and accessory elements.

TRAJECTORIES. Paths of a seismic wave; a line representing the locus of points determined by experiment or computation.

TRANSCURRENT FAULT. A large strike-slip fault in which the fault surface is steeply inclined.

TRANSFORM FAULTING. A strike-slip fault characteristic of midoceanic ridges and along which the ridges are offset. Analysis of transform faults is based on the concept of sea-floor spreading.

TRIPLE JUNCTION. Area on the earth's surface where three lithospheric plates meet. Triple junctions may be formed where three ridges or three subduction zones join.

TROPOSPHERE. In oceanography, the waters of the ocean above the thermocline. In atmospheric sciences, the lowest main layer of the atmosphere; characterized by a steep lapse rate and a low degree of hydrostatic stability, with frequent overturnings. The average thickness is 7 miles.

TSUNAMIS. A gravitational sea wave produced by any large scale, short-duration disturbance of the ocean floor, principally by a shallow submarine earthquake, but also by submarine earth movement, subsidence, or volcanic eruption, characterized by great speed of propagation (up to 950 km/hr), long wavelength (up to 200 km), long period (varying from 5 min to a few hours, generally 10-60 min), and low observable amplitude on the open sea although it may pile up to great heights (30 m or more) and cause considerable damage on entering shallow water along an exposed coast, often thousands of kilometers from the source. Mistakenly used as a synonym for tidal wave.

TURBIDITE. A sediment or rock deposited from, or inferred to have been deposited from, a turbidity current. It is characterized by graded bedding, moderate sorting, and well-developed primary structures.

TURBIDITY FLOW. A tonguelike flow of dense, muddy water moving down a slope; the flow of a turbidity current.

ULTRAVIOLET. Radiation beyond the visible spectrum at its violet end; having a wavelength shorter than those of visible light and longer than those of X rays.

UPLIFT. A structurally high area in the crust, produced by positive movements that raise or upthrust the rocks, as in a dome or arch.

UPWELLING. The rising of cold, heavy subsurface water toward the surface, esp. along the western coasts of continents (as along the coast of southern California); the displaced surface water is transported away from the coast by the action of winds parallel to it or by diverging currents. Upwelling may also occur in the open ocean where cyclonic circulation is relatively permanent, or where southern trade winds cross the equator.

URANIUM. A radioactive element with the atomic number 92 and, as found in natural ores, an average atomic weight of 238. The two principle isotopes are uranium 235 (0.7 percent of natural uranium), which is fissionable, and uranium 238 (99.3 percent of natural uranium) which is fertile.

X-RAY SPECTROMETER. An instrument for producing, recording, and analyzing an X-ray spectrum by reflecting X rays from a given sample, measuring the angle of diffraction, and thence determining the wavelengths of the X rays.

Credits and Acknowledgments

Cover

Apollo 17 view of the earth; photo courtesy of NASA.

Title Page

Apollo 8 view of the earth over the moon's horizon; photo courtesy of NASA.

Part I

Page 1—Glacier; photo by John Mercer. p. 2—The American factory, one cause of air pollution; photo by Preston Somers.

Page 7—Reprinted from *The Science Teacher*, vol. 38, no. 9, December 1971, pp. 12-16, with permission of the author and the National Science Teachers Association./ p. 20—Extracted from *The Environmental Future*, ed. by Nicolas Polunin, pp. 11-19, 1972, by permission of the author and Macmillan Press Limited (London and Basingstoke) and Barnes & Noble (New York)./ p. 28—Reprinted from *Science*, vol. 178, 13 October 1972, pp. 190-191, with permission of the author and publisher. Copyright 1972 by the American Association for the Advancement of Science./ p. 34—Reprinted from *Man's Impact on the Global Environment,* Study of Critical Environmental Problems, pp. 10-19, by permission of the MIT Press, Cambridge, Massachusetts. Copyright 1970 by the Massachusetts Institute of Technology./ p. 44—Reprinted from *Environmental Quality*, Second Annual Report on Environmental Quality, August 1971, pp. 191-196, with permission of the Council on Environmental Quality.

Part II

Page 51—Irish moss plant, an ocean food source; photo courtesy of Environment Canada, Graphic Services. p. 52—Dead fish in the polluted waters of Lake Michigan, near Chicago; photo courtesy of the Environmental Protection Agency, John Sterling.

Page 56—From *The Encyclopedia of Oceanography* by Rhodes W. Fairbridge, pp. 187-191. © 1966 by Litton Educational Publishing, Inc. Reprinted by permission of Van Nostrand Reinhold Company and the author./ p. 68—From *The Encyclopedia of Oceanography* by Rhodes W. Fairbridge, pp. 469-474. © 1966 by Litton Educational Publishing, Inc. Reprinted by permission of Van Nostrand Reinhold Company./ p. 77.—Reprinted from *American Scientist,* vol. 59, no. 4, July-August 1971, pp. 420-424, with permission of the author and *American Scientist.*

Part III

Page 89—Gemini XI photo of the Near East; photo courtesy of NASA. p. 90—The Cerro Negro volcano erupting; photo courtesy of the U.S. Geological Survey.

Page 93—Reprinted with permission of the author and publisher from *The Journal of Geological Education*, vol. 17, no. 1, February 1969, pp. 6-16. This paper was prepared under the sponsorship of the Council on Education in the Geological Sciences, a project of the American Geological Institute. The author wishes to thank H. H. Hess, D. P. McKenzie, and W. J. Morgan for valuable discussions, and Susan Vine for preparing the diagrams. This work was supported in part by the U.S. Office of Naval Research—contract no. NOO 014-67A-0151-0005AA./ p. 112—Reprinted from *Journal of College Science Teaching*, vol. 2, no. 1, October 1972, pp. 16-19, with permission of the author and the National Science Teachers Association.

Part IV

Page 125—Jupiter and two of its moons; photo from Ryan Photographic Service, Inc., Glenn Dale, Maryland./ p. 126—Simulation of a UFO over a Minnesota park; photo by Dennis Tasa.

Page 130—Extracted from *American Scientist,* vol. 60, no. 2, March-April 1972, pp. 162-174, with permission of the author and *American Scientist.*/p. 150—Reprinted from *The Science Teacher*, vol. 40, no. 3, pp. 23-26, with permission of the author and the National Science Teachers Association./p. 160—Reprinted from *Science*, vol. 179, 2 February 1973, pp. 463-465, with permission of the author and publisher. Copyright 1973 by the American Association for the Advancement of Science./p. 167—Extracted from *Science*, vol. 175, 21

CREDITS AND ACKNOWLEDGMENTS

January 1972, pp. 294-305, with permission of senior author and publisher. Copyright 1972 by the American Association for the Advancement of Science.

Part V

Page 173—Bear Lake and Hallet's Peak, Colorado; photo courtesy of Union Pacific Railroad./p. 174—Destruction from taconite tailings; photo courtesy of the Minneapolis Tribune.

Page 178—Reprinted from *Environmental Geology Notes*, no. 42, February 1971, Illinois State Geological Survey, Urbana, Illinois. Originally prepared for "Voice of America," Earth Science Series, Dr. Charles F. Park, Jr., Coordinator./p. 189—Extracted from "Hydrology for Urban Land Planning—A Guidebook on the Hydrologic Effects of Urban Land Use," *U. S. Geological Survey Circular* 554, 18 pp., 1968./p. 193— From U. S. Department of Interior "news release" of November 24, 1968./p. 199—Reprinted from *California Geology*, vol. 24, no. 8, August 1971, pp. 148-149. Originally from U. S. Geological Survey./p. 203—Extracted from *Special Distribution Publication* 60, Kansas Geological Survey, 20 pp./p. 209—Reprinted from *Geotimes* September 1965, pp. 14-15, with permission of the authors and the publisher. Copyright © 1965 by the American Geological Institute./p. 213— Reprinted from *Environment*, vol. 14, no. 1, 1972, pp. 33-39, by permission of the Committee for Environmental Information. Copyright © 1972 by the Committee for Environmental Information./p. 227—Reprinted from *Ground Water*, vol. 10, no. 5, 1972, pp. 47-49, with permission of the author and the publisher. Copyright © 1972 by the National Water Well Association.

Part VI

Page 237—Geothermal power plant in Sonoma, California; photo courtesy of Pacific Gas and Electric Company. p. 238—An open pit mine in Tucson, Arizona; photo courtesy of the Environmental Protection Agency, Cornelius Keyes.

Page 246—Extracted from *American Scientist*, vol. 60, no. 1, January-February 1972, pp. 32-40, with permission of the author and *American Scientist*./p. 266—Reprinted from *Science*, vol. 178, no. 4059, October 1972, p. 355, with permission of the publisher. Copyright 1972 by the American Association for the Advancement of Science./p. 268—Reprinted from *The Science Teacher*, vol. 39, no. 3, March 1972, pp. 40-43, with permission of the author and the National Science Teachers Association./p. 278—Reprinted from *The Science Teacher*, vol. 39, no. 3, March 1972, pp. 36-39, with permission of the author and the National Science Teachers Association./p. 287—From *Where There Is Life* by Paul B. Sears, pp. 142-152. Copyright © 1962, 1966, 1970 by Paul B. Sears. Reprinted by permission of the publisher, Dell Publishing Co., Inc. Permission also granted by George Allen & Unwin Ltd., London./p. 296—Reprinted from *The Science Teacher*, vol. 39, no. 2, February 1972, pp. 33-36, with permission of the author and the National Science Teachers Association.

Part VII

Page 307—Apollo 8 photo of the earth (Western hemisphere); photo courtesy of NASA. p. 308—Aerial view of Phoenix, Arizona; photo courtesy of the Bureau of Reclamation, E. E. Hertzog.

Page 312—Reprinted from *Environmental Quality*, Second Annual Report of the Council on Environmental Quality, August 1971, pp. 234-235, with permission of the Council on Environmental Quality./p. 314—Copyright © 1971 by Saturday Review, Inc. First appeared in *Saturday Review*, June 26, 1971. Used with permission.

Glossary

Definitions of selected geologic terms extracted by permission from *Glossary of Geology* copyright © 1972, American Geological Institute.

Man's Finite Earth

Man's Finite

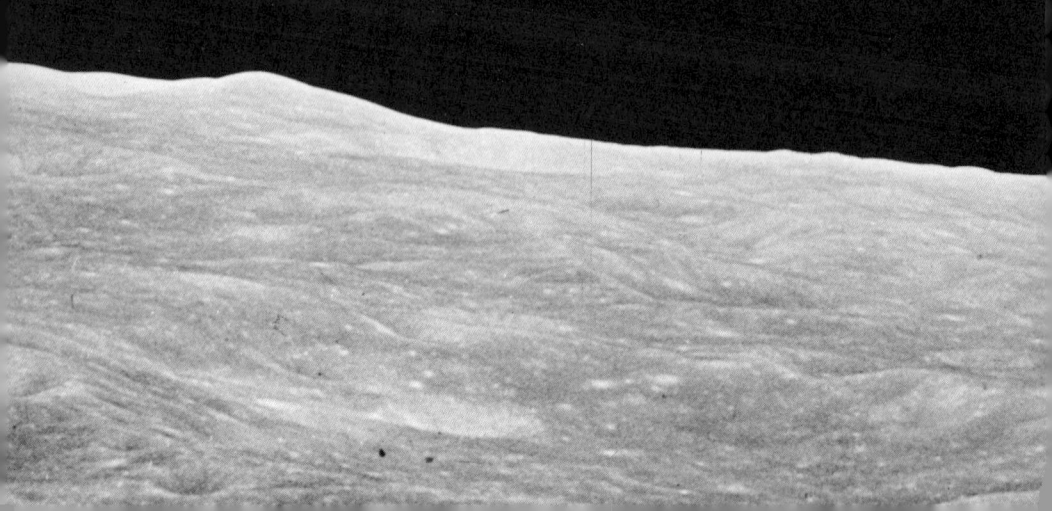